Creating the Twentieth Century

Creating the Twentieth Century

Technical Innovations of 1867–1914
and Their Lasting Impact

Vaclav Smil

UNIVERSITY PRESS

2005

OXFORD
UNIVERSITY PRESS

Oxford University Press, Inc., publishes works that further
Oxford University's objective of excellence
in research, scholarship, and education.

Oxford   New York
Auckland   Cape Town   Dar es Salaam   Hong Kong   Karachi
Kuala Lumpur   Madrid   Melbourne   Mexico City   Nairobi
New Delhi   Shanghai   Taipei   Toronto

With offices in
Argentina   Austria   Brazil   Chile   Czech Republic   France   Greece
Guatemala   Hungary   Italy   Japan   Poland   Portugal   Singapore
South Korea   Switzerland   Thailand   Turkey   Ukraine   Vietnam

Published by Oxford University Press, Inc.
198 Madison Avenue, New York, New York 10016
www.oup.com

Oxford is a registered trademark of Oxford University Press

Library of Congress Cataloging-in-Publication Data

Smil, Vaclav.
Creating the twentieth century : technical innovations of 1867–1914 and their lasting
impact / Vaclav Smil.
p. cm.
Includes bibliographical references and index.
ISBN-13: 978-0-19-516874-7
ISBN: 0-19-516874-7
1. Technological innovations—History—19th century. 2. Technological
innovations—History—20th century. I. Title.

T173.8.S615 2004
609'.034—dc22   2004054757

Preface

This book has been on my mind for more than three decades. My first musings about the technically exceptional nature of the two pre-WWI generations go back to the late 1960s, before Eva and I escaped from a newly invaded province of the Soviet Empire to Pennsylvanian ridge and valley countryside. I worked on some of its topics (for other publications) during the late 1980s and throughout the 1990s, and I finally began to write it in 2002. In the words that my favorite composer used in dedicating his quartets, it is the result of *lunga, e laboriosa fattica*—and yet I wish that the task could continue. I have a selfish and an objective reason for this: further immersion in the world of pre-WWI innovations would bring more revelations, surprises, and confirmations, and I would also like more space as there are many topics that I addressed only cursorily, many reflections and considerations that I had to leave out. At the same time, I have always followed Faraday's great dictum— work, finish, publish—and so here is my incomplete and imperfect story of one of the greatest adventures in history, my homage to the creators of a new world.

My great intellectual debt to hundreds of historians, engineers, and economists without whose penetrative work this volume could not have been written is obvious, and I must thank Douglas Fast for completing the unusually challenging job of preparing more than 120 images that are an integral part of this book. And I offer no apologies for what some may see as too many numbers: without quantification, there is no real appreciation of the era's fundamental and speedy achievements and of the true magnitude of its accomplishments. Metric system and scientific units and prefixes (listed and defined below) are used throughout. Finally, what not to expect. This is neither a world history of the two pre-WWI generations seen through a prism of technical innovations nor an economic history of the period written with an engineering slant.

The book is not intended to be either an extended argument for technical determinism in human affairs or an uncritical exaltation of an era. I am quite content to leave its genre undefined: it is simply an attempt to tell a story of amazing changes, of the greatest discontinuity in history, and to do so from a multitude of perspectives in order to bring out the uniqueness of the period and to be reminded of the lasting debt we owe to those who invented the fundamentals of the modern world. Or, to paraphrase Braudel's (1950) remarks offered in a different context, I do not seek a philosophy of this great discontinuity but rather its multiple illumination.

Contents

Units and Abbreviations

Units

A	ampere (unit of current)
C	degree of Celsius (unit of temperature)
g	gram (unit of mass)
h	hour
hp	horsepower (traditional unit of power = 745.7 W)
Hz	hertz (unit of frequency)
J	joule (unit of energy)
K	degree of Kelvin (unit of temperature)
lm	lumen (unit of luminosity)
m	meter
Pa	pascal (unit of pressure)
s	second
t	metric ton (= 1,000 kg)
V	volt (unit of voltage)
W	watt (unit of power)

Prefixes

n	nano	10^{-9}
μ	micro	10^{-6}
m	milli	10^{-3}
c	centi	10^{-2}
h	hecto	10^{2}
k	kilo	10^{3}
M	mega	10^{6}
G	giga	10^{9}
T	tera	10^{12}
P	peta	10^{15}
E	exa	10^{18}

Creating the Twentieth Century

☙ I

The Great Inheritance

> "You must follow me carefully. I shall have to controvert
> one or two ideas that are almost universally accepted . . ."
>
> "Is not that a rather large thing to expect us to begin
> upon?" said Filby, an argumentative person with red hair.
>
> "I do not mean to ask you to accept anything without
> reasonable ground for it. You will soon admit as much as I
> need from you . . ."
>
> H. G. Wells, *The Time Machine* (1894)

Imagine an exceedingly sapient and durable civilization that began scanning a bubble of space, say 100 light years in diameter, for signs of intelligent life about half a billion years ago. Its principal surveillance techniques look for any emissions of organized electromagnetic radiation as opposed to the radio frequencies emitted from stars or light that originates from natural combustion of carbon compounds (wildfires) or from lightning. For half a billion years its probes that roam the interstellar space have nothing to report from nine planets

FRONTISPIECE 1. Technical advances that began to unfold during the two pre-WWI generations and that created the civilization of the 20th century resulted in the first truly global human impacts. Some of these are detectable from space, and nighttime images of Earth (here the Americas in the year 2000) are perhaps the most dramatic way to show this unprecedented change. Before 1880 the entire continents were as dark as the heart of Amazon remains today. This image is based on NASA's composite available at http://antwrp.gsfc.nasa.gov/apod/ap001127.html.

that revolve around an unremarkable star located three-fifths of the way from the center of an ordinary-looking spiral galaxy that moves inexorably on a collision course with one of its neighbors. And then, suddenly, parts of that star system's third planet begin to light up, and shortly afterward they begin to transmit coherent signals as two kinds of radiation emanating from Earth's surface provide evidence of intelligent life.

A closer approach of the probes would reveal an organized pattern of night-time radiation in the visible (400–700 nm) and near-infrared (700–1,500 nm) part of the electromagnetic spectrum produced by electric lights whose density is highest in the planet's most affluent and heavily populated regions (see the frontispiece to this chapter). And from scores of light years away one can detect a growing multitude of narrow-band, pulsed, modulated signals in frequencies ranging from less than 30 kHz (very low radio band) to more than 1 GHz (radar bands). These signals have their origins in fundamental scientific and technical advances of the 1880s and 1890s, that is, in the invention of durable incandescent electric lights, in the introduction of commercial generation and transmission of electricity, in production and detection of Hertzian waves, and in the first tentative wireless broadcasts. And all of these capabilities became considerably developed and commercialized before WWI.

That was the time when the modern world was created, when the greatest technical discontinuity in history took place. This conclusion defies the common perception of the 20th century as the period of unprecedented technical advances that originated in systematic scientific research and whose aggressive deployment and commercialization brought profound economic, social, and environmental transformations on scales ranging from local to global. The last two decades of the 20th century witnessed an enormous expansion of increasingly more affordable and more powerful computing and of instantaneous access to the globe-spanning World Wide Web, and they have been singled out as a particularly remarkable break with the past. This is as expected according to those who maintain that the evolution of our technical abilities is an inherently accelerating process (Turney 1999; Kurzweil 1990).

This common impression of accelerating technical innovation seems to be borne out by a number of exponential trends, perhaps most notably by the fact that the number of transistors per microchip has been doubling approximately every 18 months since 1972. This trend, known as Moore's law, was predicted in 1965 by Intel's cofounder, and it continues despite repeated forecasts of its imminent demise (Intel 2003; Moore 1965). In 1972 Intel's 8008 chip had 2,500 transistors; a decade later there were more than 100,000 components on a single memory microchip; by 1989 the total surpassed 1 million, and by the year 2000 the Pentium 4 processor had 42 million transistors (figure 1.1).

Given these realities, it is not surprising that many people believe that this acceleration has already amounted to a technical revolution that had ushered in the age of the New Economy dependent on machines capable of extending,

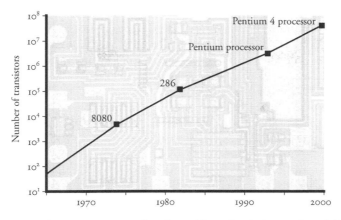

FIGURE 1.1. Gordon E. Moore, a co-founder and later the chairman of Intel Corporation, predicted in 1965 that the number of transistors per integrated circuit would double every year. Later he revised the rate to every 18 months, and the actual doubling time between 1971 and 2001 was almost exactly two years. This rate is not based on any laws of physics, and it can continue for years, but not for decades. Plotted from data in Moore (1965) and Intel (2003).

multiplying, and leveraging our mental abilities (Banks 2001; Litan and Rivlin 2001; Donovan 1997; Kurzweil 1990). As a result, it is only a matter of time, perhaps as early as 2040, before the thinking machines cut loose, develop hyperintelligence, and engineer our demise (Moravec 1999; Kurzweil 1999). Younger readers of this book will have a chance to verify the fate of this charming forecast, but my concern is with the past—and here the verdict is clear: those commonly held perceptions of accelerating innovation are ahistorical, myopic perspectives proffered by the zealots of electronic faith, by the true believers in artificial intelligence, e-life forms, and spiritual machines.

One of the best quantitative confirmations of this judgment concludes unequivocally that the New Economy has not measured up to the truly great inventions of the past (Gordon 2000). Indubitably, the 20th century was exceedingly rich in innovations and microelectronics has expanded immensely our capacities for problem analysis and information transfer. And yet many techniques whose everyday use keeps defining and shaping the modern civilization had not undergone any fundamental change during the course of the 20th century. Their qualitative gains (higher efficiency, increased reliability, greater convenience of their use, lower specific pollution rates) took place without any change of basic, long-established concepts.

In this book I demonstrate that the fundamental means to realize nearly all of the 20th-century accomplishments were put in place even before the century began, mostly during the three closing decades of the 19th century and in the years preceding WWI. That period ranks as history's most remarkable discontinuity not only because of the extensive sweep of its innovations but also because of the rapidity of fundamental advances that were achieved during that time. This combination makes it a unique event—but one with kindred precedents as the process of incremental gains that has dominated the course of technical change throughout the human history is recurrently, but infrequently and unpredictably, interrupted by concentrated spurts of astonishing creativity. These two processes can be seen as just the most recent, historical, demonstrations of a dichotomy present in the grand pattern of the biospheric evolution, namely, the contrast between phyletic gradualism and punctuated equilibrium (Simpson 1983; Eldredge and Gould 1972).

Gradualism is a process marked by even and slow transformations that take place over all, or a very large part, of the ancestral species range, with new organisms arising as modified descendants of original populations. In contrast, the other phylogenetic process entails rapid development of new organisms that appear by the splitting of lineages in a very small subpopulation of the ancestral form and in a restricted, often peripheral, part of the area. Existence of these two different patterns is well documented in the fossil record, and it belies the claims of those who maintain that technical progress is "the continuation of evolution by other means, and is itself and evolutionary process. So it, too, speeds up" (Kurzweil 1999:32).

The idea of accelerating evolution implies the existence of a grand evolutionary trend, as well as a tacit assumption of a purpose and a goal. None of these conclusions can be justified by studying the evolutionary record. Moreover, there has been nothing inevitable about the course the biosphere's evolution during the past 4 billion years (Smil 2002). From a paleontologist's perspective, evolution has been a

> staggeringly improbable series of events, sensible enough in retrospect and subject to rigorous explanation, but utterly unpredictable and quite unrepeatable . . . Wind back the tape of life . . . let it play again from an identical starting point, and the chance becomes vanishingly small that anything like human intelligence would grace the replay. (Gould 1989: 14)

And there is yet another critical consideration. Although during the past 50,000 years our species has undoubtedly experienced a rapid mental development (seen as a key example of accelerating evolution that culminated in the cognitive powers of *Homo sapiens*), the fundamental precondition of our existence has not changed. Our survival still depends irrevocably on incessant functioning of biospheric services that range from decomposition of wastes to pollination of plants. Bacteria, archaea, and fungi do the first set of those tasks, and insects the second. None of these organisms has shown any accelerated evolution, and recently decoded genomes show that bacteria and archaea closely resemble their ancestors that lived billions of years ago (Smil 2002).

Technical analogies of evolutionary gradualism can be illustrated by numerous examples documenting slow improvements in the efficiency and power of waterwheels, in greater maneuverability of sails, or in higher productivity and lower use of fuel in metal smelting (Smil 1994). Basalla (1988) argued that this continuous and cumulative process of technical change contrasts with scientific advances that are characterized by discrete events or, as Kuhn (1962) concluded, by abrupt shifts in fundamental paradigms. But the evolution of technical capacities does not consist merely of these gradual changes, as long periods of incremental improvements are punctuated by spells of relatively rapid advances. During these spells, the innovating societies improve their technical abilities and expand their productive capabilities far beyond the prevailing norm, and do so in historically very short periods of time.

Students of long waves would claim that such accelerations come at surprisingly regular intervals. Mensch (1979) is the leading proponent of the idea that major innovations will appear in clusters during the depression phases of long economic waves, and his decadal analysis for the years 1740–1960 shows distinct peaks for the 1760s, 1830s and 1840s, 1880s, and 1930s. Marchetti (1986) used these findings to support his conclusion about 50-year pulsation in human affairs. But Mensch's conclusion that capital will be risked on new innovations only when the profits of used-up techniques are unbearably low was criticized by Clark, Freeman, and Soete (1981) as contradicting all economic theories as well as the evidence of case histories of innovation. I do not see these statistical studies based on long lists of items as particularly useful. Although the intended aim may be to include only major advances, extensive lists will inevitably mix marginal and fundamental items.

The ballpoint pen will have obviously different socioeconomic impacts than the continuous casting of steel, and the world economy would miss diesel engines a great deal more than the absence of watertight cellophane; all of these items are from Mensch's undifferentiated list, which, by the way, does *not* include the synthesis of ammonia from its elements—as I will demonstrate in this book, this was perhaps the most far-reaching of all modern technical innovations. That is why my approach to singling out the periods of truly epochal innovations with enormous long-lasting impacts—be they artistic or technical— is guided primarily by qualitative considerations rather than by considering statistically outstanding concatenations of undifferentiated advances.

Artistic examples of these saltations are generally better known than the technical leaps. Periclean Athens (460–429 B.C.E.) and Florence of the first decade of the 16th century are perhaps the two most outstanding examples of such advances. Just to think that in 1505 an idler on the Piazza della Signoria could have in a matter of days bumped into Leonardo da Vinci, Raphael, Michelangelo, and Botticelli; such an efflorescence of creative talent may be *never* again repeated in any city. My second, modern choice, of an extraordinary concatenation of artistic talents would be the fin de siècle Paris—where Émile Zola's latest installment of his expansive and gripping Rougon-Macquart cycle could be read just before seeing Claude Monet's canvases that transmuted the landscape around Giverny into images of shimmering light and later the same day hearing Claude Debussy's intriguing *L'Après-midi d'un Faune*.

In technical terms there are two saltation periods of human history that stand apart as the times of the two most astounding, broad, and rapid innovation spurts. The first one, purely oriental, took place during the Han dynasty China (207 B.C.E.–9 C.E.); the second one, entirely occidental in both its genesis and its nearly instant flourish, unfolded in Europe and North America during the two generations preceding WWI. In both instances, those widespread and truly revolutionary innovations not only changed the course of the innovating societies but also were eventually translated into profound global impacts. The concatenation of Han advances laid strong technical foundations for the development of the world's most persistent empire—which was, until the 18th century, also the world's richest economy—and for higher agricultural and manufacturing productivities far beyond its borders (Needham et al. 1965, 1971; Temple 1986).

The dynasty's most important innovations were in devising new mechanical devices, tools, and machines and advancing the art of metallurgy. The most remarkable new artifacts included the breast-band harness for horses and prototypes of efficient collar harnesses, wooden moldboard ploughs with curved shares made from nonbrittle iron, multitube seed drills, cranks, rotary winnowing fans, wheelbarrows, and percussion drills (figure 1.2). Metallurgical innovations included the use of coal in ironmaking, production of liquid iron, decarburization of iron to make steel, and casting of iron into interchangeable molds. But this innovative period was spread over two centuries, and some of its products were not adopted by the rest of the Old World for centuries, or even for more than a millennium after their initial introduction in China (Smil 1994).

The Unprecedented Saltation

In contrast, *the impact of the late 19th and the early 20th century advances was almost instantaneous,* as their commercial adoption and widespread diffusion were very rapid. Analogy with logic gates in modern computers captures the importance of these events. Logic gates are fundamental building blocks of digital circuits that receive binary inputs (0 or 1) and produce outputs only if a specified combination of inputs takes place. A great deal of potentially very useful scientific input that could be used to open some remarkable innovation gates was accumulating during the first half of the 19th century. But it was

FIGURE 1.2. Sichuanese salt well made with a percussion drill, one of the great inventions of the Han dynasty in China. The same technique was used to drill the first U.S. oil well in Pennsylvania in 1859. Reproduced from a Qing addition to Song's (1673) survey of China's techniques.

only after the mid-1860s when so many input parameters began to come together that a flood of new instructions surged through Western society and our civilization ended up with a very different program to guide its future.

The most apposite evolutionary analogy of this great technical discontinuity is the Cambrian eruption of highly organized and highly diversified terrestrial life. This great evolutionary saltation began about 533 million years ago and produced—within a geologically short spell of just 5–10 million years, or less than 0.3% of the evolutionary span—virtually all of the animal lineages that are known today (McMenamin and McMenamin 1990; Bowring et al. 1993). Many pre-WWI innovations were patented, commercialized, and ready to be diffused in just a matter of months (telephone, lightbulbs) or a few years (gasoline-fueled cars, synthesis of ammonia) after their conceptualization or experimental demonstration. And as they were built on fundamental scientific principles, it is not only their basic operating modes that have remained intact but also many specific features of their pioneering designs are still very much recognizable among their most modern upgrades.

The era's second key attribute is *the extraordinary concatenation of a large number of scientific and technical advances*. The first category of these scientific advances embraces those fundamentally new insights that made it possible to introduce entirely new industries, processes, and products. Certainly the most famous example of this kind is a fundamental extension of the first law of thermodynamics that was formulated by Albert Einstein as a follow-up of his famous relativity paper: "An inertial mass is equivalent with an energy content μc^2" (Einstein 1907). By 1943 this insight was converted into the first sustained fission reaction; 1945 saw the explosions of the first three fission bombs (Alamogordo, Hiroshima, Nagasaki), and by 1956 the first commercial fission reactor began generating electricity (Smil 2003).

But during the 20th century everyday lives of hundreds of millions of people were much more affected by Heinrich Hertz's discovery of electromagnetic waves much longer than light but much shorter than sound. This adventure started in 1886 with detecting the spark-generated waves just across a lecture room in Karlsruhe. Soon the reach progressed to Marconi's Morse signals on land, between ships, and across the Atlantic, then to Fessenden's pioneering radio broadcasts and, after WWI, to television—and to much more. By the year 2000 the Hertzian waves made possible such wonders as finding one's place on the planet with global positioning systems or alerting the owners of colorful Nokia phones to the arrival of new messages by chirping operatic tunes—be it in Hong Kong's packed subways or on mountain peaks in the Alps.

And then there were new scientific insights that did not launch new products or entirely new industries but whose broad theoretical reach has been helpful in understanding a variety of everyday challenges and has been used to construct better devices and more efficient machines. An outstanding example was the realization that a ratio calculated by multiplying characteristic distance and velocity of a moving fluid by its density and dividing that product by the fluid's viscosity yields a dimensionless number whose magnitude provides fundamental information about the nature of the flow. Osborne Reynolds (1842–1912), a priest in the Anglican Church and the first professor of engineer in Manchester, found this relationship in 1883 after experiments with water flowing through glass tubes (Rott 1990).

Low Reynolds numbers correspond to smooth laminar flow that is desirable in all pipes as well as along the surfaces of ships or airplanes. Turbulence sets in with higher Reynolds numbers, and completely turbulent flows are responsible for cavitation of ship propellers (chapter 2 tells how Charles Parsons solved this very challenge), vibration of structures, erosion of materials, and relentless noise. A great deal of the roar you hear when sitting in the aft section of a jet airplane does not come from the powerful gas turbines but from the air's turbulent boundary layer that keeps pounding the plane's aluminum alloy skin. Preventing cavitation, vibration, erosion, and noise and operating with optimized Reynolds numbers brings many great rewards.

The period's fundamental technical advances include, above all, large-scale electricity generation and transmission and the inventions of new prime movers and energy converters. Internal combustion engines and electric motors have eventually become the world's most common mechanical prime movers. Transformers and rectifiers ensure the most efficient use of electricity in energizing many specialized assemblies and machines whose sizes range from microscopic (components of integrated circuits) to gargantuan (enormous excavators and construction cranes), from stationary designs (in manufacturing, commerce, and households) to the world's fastest trains. Yet another group of technical advances includes production processes whose commercialization put on the market new, or greatly improved, products or procedures whose use in turn boosted other technical capabilities and transformed economic productivities and private consumption alike.

By far the most important new materials were inexpensive high-quality steels and aluminum produced electrolytically by Hall-Héroult process. Henry Ford's moving assembly line and the liquefaction of air are excellent examples of new processes whose adoption changed the nature of industrial production. The last process is also a perfect illustration of how ubiquitous and indispensable are the synergies of these inventions. Less than two decades after its discovery, the liquefaction of air found one of its most massive (and entirely unanticipated) uses as the supplier of nitrogen for the Haber-Bosch synthesis of ammonia. High crop yields and deep greens of suburban lawns are thus linked directly to air liquefaction through nitrogen fertilizers. And, most remarkably, in many cases the ingredients necessary for completely new systems fell almost magically in place just as they were needed. The most notable concatenation brought together incandescing filaments, efficient dynamos and transformers, powerful steam turbines, versatile polyphase motors, and reliable cables and wires for long-distance transmission to launch the electric era during a mere dozen or so years.

The third remarkable attribute of the pre-WWI era is *the rate with which all kinds of innovations were promptly improved after their introduction*—made more efficient, more convenient to use, less expensive, and hence available on truly mass scales. For example, as I detail in chapter 2, efficiency of incandescent lights rose more than six-fold between 1882 and 1912 while their durability was extended from a few hundred to more than 1,000 hours. There were similarly impressive early gains in the efficiency of steam turbines and electricity consumption in aluminum electrolysis.

The fourth notable characteristic of the great pre-WWI technical discontinuity is *the imagination and boldness of new proposals.* There is no better testimony to the remarkable pioneering spirit of the era than the fact that so many of its inventors were eager to bring to life practical applications of devices and processes that seemed utterly impractical, even impossible, to so many of their contemporaries. Three notable examples illustrate these attitudes of widely shared disbelief. On March 29, 1879, just nine months before Thomas Edison demonstrated the world's first electrical lighting system, *American Reg-*

ister concluded that "it is doubtful if electricity will ever be used where economy is an object" (cited in Ffrench 1934:586). The same year, the Select Committee on Lighting by Electricity of the British House of Commons heard an expert testimony that there is not "the slightest chance" that electricity could be "competing, in general way, with gas," and *The Engineer* wrote on November 9, 1877, that "electricity for domestic illumination would never, in our view, prove as handy as gas. An electric light would always require to keep in order a degree of skilled attention which few individuals would possess" (quoted in Beauchamp 1997:136). Henry Ford reminisced that the Edison Company objected to his experiments with internal combustion and that its executives offered to hire him "only on the condition that I would give up my gas engine and devote myself to something really useful" (Ford 1922:34). And three years before the Wright brothers took off above the dunes at Kitty Hawk in North Carolina on December 17, 1903 (figure 1.3), Rear Admiral George W. Melville (1901) concluded that "outside of the proven impossible, there probably could be found no better example of the speculative tendency carrying man to the verge of chimerical than in his attempts to imitate the birds" (p. 825).

Finally, there is *the epoch-making nature of these technical advances,* the proximate reason for writing this book: most of them are still with us not just as

FIGURE 1.3. Orville Wright is piloting while Wilbur Wright is running alongside as their machine lifts off briefly above the sands of the Kitty Hawk, North Carolina, at 10:35 A.M. on December 17, 1903. Library of Congress image (LC-W86-35) is reproduced from the Wrights' glass negative.

inconsequential survivors or marginal accoutrements from a bygone age but as the very foundations of modern civilization. Such a profound and abrupt discontinuity with such lasting consequences has no equivalent in history. The closest analogy in the more recent human prehistory was obviously the emergence of the first settled agricultural societies nearly 10,000 years ago. But the commonly used term of Agricultural Revolution is a misnomer for that gradual process during which foraging continued to coexist first with incipient and then with slowly intensifying cultivation (Smil 1994). In contrast, the pre-WWI innovations tumbled in at a frenzied pace.

When seen from the vantage point of the early 21st century, there is no doubt that the two generations between the late 1860s and the beginning of WWI remain the greatest technical watershed in human history. Moreover, as stressed at the outset, this was the first advance in nearly 4.5 billion years of the planet's evolution that led to the generation of cosmically detectable signals of intelligent life on Earth: a new civilization was born, one based on synergy of scientific advances, technical innovation, aggressive commercialization, and intensifying, and increasingly efficient, conversions of energy.

The Knowledge Economy

Technical advances of the antiquity, Middle Ages, and the early modern era had no scientific foundation. They had to be based on observations, insights, and experiments, but they were not guided by any coherent set of accumulated understanding that could at least begin to explain why some devices and processes work while others fail. They involved an indiscriminate pursuit of ideas that opened both promising paths of gradual improvements, be it of waterwheels or sails, as well as cul-de-sacs of *lapis philosophorum* or *perpetuum mobile*. Even the innovations of the early decades of the Industrial Revolution conformed to this pattern. Writing at the very beginning of the 19th century, Joseph Black noted that "chemistry is not yet a science. We are very far from the knowledge of first principles. We should avoid everything that has the pretensions of a full system" (Black 1803:547). Mokyr's (1999) apt characterization is that the first Industrial Revolution created a chemical industry without chemistry, an iron industry without metallurgy, and power machinery without thermodynamics.

In contrast, most of the technical advances that appeared during the two pre-WWI generations had their basis in increasingly sophisticated scientific understanding, and for the first time in history, their success was shaped by close links and rapid feedbacks between research and commercialization. Naturally, other innovations that emerged during that period had still owed little to science as they resulted from random experimenting or serendipity. This is not surprising as we have to keep in mind that the new process of scientifically based technical developments was unfolding along the underlying trend of traditionally incremental improvements.

The first foundations of new knowledge economy appeared during the 17th century, their construction accelerated during the 18th century, and the process matured in many ways before 1870. Its genesis, progress, grand features, and many fascinating details are best presented by Mokyr (2002). Here I will illustrate these changes by a few examples describing the evolution of prime movers that I have studied in some detail (Smil 1991, 1994). Prime movers are those energy converters whose capacities and efficiencies determine the productive abilities of societies as well as their tempo of life; they are also critical for energizing chemical syntheses whose accomplishments help to form our surroundings as well as to expand the opportunities for feeding and healing ourselves.

At the beginning of the 18th century, both the dominant and the largest prime movers were the same ones as in the late antiquity. Muscles continued to be the most common prime movers: humans could sustain work at rates of just between 50 and 90 W, while draft mammals could deliver no more than 300 W for small cattle, 400–600 W for smaller horses, and up to 800 W for heavy animals. Capacity of European waterwheels, the most powerful prime movers of the early modern era, averaged less than 4 kW. This means that it took until 1700 to boost the peak prime mover ratings roughly 40-fold (figure 1.4), and there was no body of knowledge to understand the conversion of food and feed into mechanical energy and to gauge the efficiency of this transformation.

By 1800 there was still no appreciation of thermal cycles, no coherent concept of energy, no science of thermodynamics, no understanding of metabolism. Antoine Lavoisier's (1743–1794) suggestion of the equivalence between heat output of animals and men and their feed and food intake was only the first step on a long road of subsequent studies of heterotrophic metabolism. But there was important practical progress as James Watt (1736–1819) converted Newcomen's steam engine from a machine of limited usefulness and very low efficiency to a much more practical device capable of about 20 kW that began revolutionizing many tasks in coal mining, metallurgy, and manufacturing (Thurston 1878; Dalby 1920). Watt also invented a miniature recording steam gauge, and this indicator made it possible to study the phases of engine cycles.

By 1870 thermodynamics was becoming a mature science whose accomplishments helped to build better prime movers and to design better industrial processes. This transformation started during the 1820s with Sadi Carnot's (1796–1832) formulation of essential principles that he intended to be applicable to all imaginable heat engines (Carnot 1824). This was a bold claim but one that was fully justified: thermodynamic studies soon confirmed that no heat engine can be more efficient than a reversible one working between two fixed temperature limits and that the highest theoretical efficiency of this Carnot cycle cannot surpass 65.32%. Another important insight published during

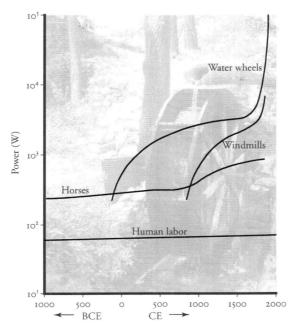

FIGURE 1.4. Maximum power of prime movers, 1000 B.C.E. to 1700 C.E. Waterwheels became the most powerful inanimate prime movers of the pre-industrial era and lost this primacy to steam engines after 1730.

the 1820s was Georg Simon Ohm's (1789–1854) explanation of electricity con-
duction in circuits and the formulation of what later became the eponymous
law relating current, potential, and resistance (Ohm 1827). Correct understanding of this relationship was a key to devising a commercially affordable system
of electric lighting as it minimized the mass of expensive conductors.

Understanding of energy conversions progressed rapidly during the 1840s
and 1850s with the brilliant deductions of Robert Mayer (1814–1878) and James
Prescott Joule (1818–1889). When Mayer worked as a physician on a Dutch
ship bound for Java, he noted that sailors had a brighter venous blood in the
tropics. He correctly explained this fact by pointing out that less energy, and
hence less oxygen, was needed for basal metabolism in warmer climates. Mayer
saw muscles as heat engines energized by oxidation of blood, and this led to
the establishment of mechanical equivalent of heat and hence to a formulation
of the law of conservation of energy (Mayer 1851) later known as the first law
of thermodynamics. A more accurate quantification of the equivalence of work
and heat came independently from Joule's work: in his very first attempt, he
was able to come within less than 1% of the actual value (Joule 1850).

Soon afterward, William Thomson (Lord Kelvin, 1824–1907) described a universal tendency to the dissipation of mechanical energy (Thomson 1852). Rudolf Clausius (1822–1888) formalized this insight by concluding that the energy content of the universe is fixed and that its conversions result in an inevitable loss of heat to lower energy areas: energy seeks uniform distribution, and the entropy of the universe tends to maximum (Clausius 1867). This second law of thermodynamics—the universal tendency toward disorder and heat death—became perhaps the most influential, as well as a much misunderstood, cosmic generalization. And although its formulation did not end those futile attempts to build *perpetuum mobile* machines it made it clear why that quest will never succeed.

There is perhaps no better illustration of the link between the new theoretical understanding and astonishing practical results than Charles Parsons's invention and commercialization of the steam turbine, the most powerful commonly used prime mover of the 20th (and certainly at least of the first half of the 21st) century (figure 1.5 for details, see chapter 2). Parsons, whose father was an astronomer and a former president of the Royal Society, received mathematical training at the Trinity College in Dublin and at Cambridge, joined an engineering firm as a junior partner, and proceeded to build the first model steam turbine because thermodynamics told him it could be done. He prefaced his Rede lecture describing his great invention by noting "that the work was initially commenced because calculation showed that, from the known data, a successful steam turbine ought to be capable of construction. The practical development of this engine was thus commenced chiefly on the basis of the data of physicists . . ." (Parsons 1911:1).

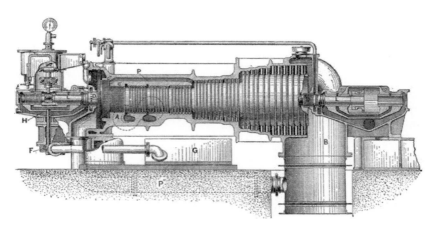

FIGURE 1.5. Longitudinal cross section through the casing of 1-MW-capacity Parsons's steam turbine designed in 1899. Reproduced from Ewing (1911).

Pre-1870 gains in chemical understanding were, comparatively, even greater as the science started from a lower base. Brilliant chemists of the late 18th century—Antoine Lavoisier, Wilhelm Scheele, Joseph Priestley—began to systematize the fragmentary understanding of elements and compounds, but there was no unifying framework for their efforts. Early 19th-century physics at least had a solid grasp of mechanics, but large parts of chemistry still had the feel of alchemy. Then came a stream of revolutionary chemical concepts. First, in 1810, John Dalton put atomic theory on a quantitative basis, and in 1828 Friedrich Wöthler (1800–1882) prepared the first organic compound when he synthesized urea. Beginning in the 1820s, Justus von Liebig (1803–1873; figure 1.6) established standard practices of organic analysis and attributed the generation of CO_2 and water to food oxidation, thus providing a fundamentally correct view of heterotrophic metabolism (Liebig 1840).

After 1860, Friedrich August Kekulé (1829–1896) and his successors made sense of the atomic structure of organic compounds. In 1869 Dimitrii Mendeleev (1834–1907) published his magisterial survey of chemistry and placed all known, as well as yet unknown, elements, in their places in his periodical table. This achievement remains one of the fundamental pillars of the modern understanding of the universe (Mendeleev 1891). These theoretical advances were accompanied by the emergence of chemical engineering, initially led by the British alkali industry, and basic research in organic synthetic chemistry brought impressively rapid development of the coal-tar industry. Its first success was the synthesis of alizarin, a natural dye derived traditionally from the root of madder plant (*Rubia tinctorum*).

Badische Anillin- & Soda-Fabrik (BASF) synthesized the dye by using the method invented by Heinrich Caro (1834–1910), the company's leading researcher; his method was nearly identical to the process proposed in England by William Henry Perkin (1838–1907). Synthetic alizarin has been used ever since to dye wool and also to stain microscopic specimens. German and British pat-

FIGURE 1.6. What Justus von Liebig (1803–1873) helped to do so admirably for chemistry, other early 19th-century scientists did in their disciplines: made them the foundations of a new knowledge economy. Photo from author's collection.

ents for the process were filed less than 24 hours apart, on June 25 and 26, 1869 (Brock 1992). This was not as close a contest as the patenting of telephone by Alexander Graham Bell and Elisha Gray, which they did just a few hours apart. These two instances illustrate the intensity and competitiveness of the era's quest for innovation.

Remarkable interdisciplinary synergies combining new scientific understanding, systematic experiments, and aggressive commercialization can be illustrated by developments as diverse as the birth of the electric era and the synthesis of ammonia from its elements. Edison's lightbulb was not a product (as some caricatures of Edison's accomplishments would have it) of intuitive tinkering by an untutored inventor. Incandescent electric lights could not have been designed and produced without combining deep familiarity with the state-of-the-art research in the field, mathematical and physical insights, a punishing research program supported by generous funding by industrialists, a determined sales pitch to potential users, rapid commercialization of patentable techniques, and continuous adoption of the latest research advances.

Fritz Haber's (1868–1934) discovery of ammonia synthesis was the culmination of years of research effort based on decades of previously unsuccessful experiments, including those done by some of the most famous chemists of that time (among them two future Nobel Prize winners, Wilhelm Ostwald and Walter Nernst). Haber's success required familiarity with the newly invented process of air liquefaction, willingness to push the boundaries of high-pressure synthesis, and determination, through collaboration with BASF, not to research the process as just another laboratory curiosity but to make it the basis of commercial production (Stoltzenberg 1994; Smil 2001).

And Carl Bosch (1874–1940), who led BASF's development of ammonia synthesis, made his critical decision as a metallurgist and not as a chemist. When, at the crucial management meeting in March 1909, the head of BASF's laboratories heard that the proposed process will require pressures of at least 100 atmospheres, he was horrified. But Bosch remained confident: "I believe it can go. I know exactly the capacities of steel industry. It should be risked" (Holdermann 1954:69). And it was: Bosch's confidence challenged the German steelmakers to produce reaction vessels of unprecedented size able to operate at previously unheard of pressures—but less than five years later these devices were in operation at the world's first ammonia plant.

The Age of Synergy

At this point I must anticipate, and confront, those skeptics and critics who would insist on asking—despite all of the arguments and examples given so far in support of the Age of Synergy—how justified is my conclusion to single out this era, how accurate is my timing of it? As De Vries (1994:249) noted, history is full of "elaborate, ideal constructions that give structure and coherence to our historical narratives and define the significant research questions." And while all of these historiographical landmarks—be it the Renaissance or the Industrial Revolution—are based on events that indubitably took place, these concepts tend to acquire life of their own, and their interpretations tend to shift with new insights. And so the concept of the Renaissance, the era seen as the opening chapter of modern history, came to be considered by some historians a little more than an administrative convenience, "a kind of blanket under which we huddle together" (Bouwsma 1979:3).

Even more germane for any attempt to delimit the Age of Synergy is the fact that some historians have questioned the very concept of its necessary precursor, the Industrial Revolution. Its dominant interpretation as an era of broad economic and social change (Mokyr 1990; Landes 1969; Ashton 1948) has been challenged by views that see it as a much more restricted, localized phenomenon that brought significant technical changes only to a few industries (cotton, ironmaking) and left the rest of the economy in premodern stagnation until the 1850s: Watt's steam engines notwithstanding, by the middle of the 19th century the British economy was still largely traditional (Crafts and Harley 1992).

Or as Musson (1978:141) put it, "the typical British worker in the mid-nineteenth century was not a machine-operator in a factory but still a traditional craftsman or labourer or domestic servant." At the extreme, Cameron (1982) argued that the change was so small relative to the entire economy that the very title of the Industrial Revolution is a misnomer, and Fores (1981) went even further by labeling the entire notion of a British industrial revolution a myth. Temin (1997) favors a compromise, seeing the Industrial Revolution's technical progress spread widely but unevenly. I am confident that the concept of the Age of Synergy is not a mental construct vulnerable to a devastating criticism but an almost inevitable identification of a remarkable reality.

The fact that the era's epoch-making contributions were not always recognized as such by the contemporary opinion is not at all surprising. Given the unprecedented nature of the period's advances, many commentators simply did not have the requisite scientific and technical understanding to appreciate the reach and the transforming impact of new developments. And so during the early 1880s most people thought that electricity will merely substitute faint light- bulbs for similarly weak gas lights: after all, the first electric lights were explicitly designed to match the luminosity of gas jets and gave off only about 200 lumens, or an equivalent of 16 candles (compared to at least 1,200 lm for today's 100 W incandescent bulb). And even one of the era's eminent innovators and a pioneer of electric industry did not think that gas illumination is doomed.

William Siemens (see figure 4.4) reaffirmed in a public lecture on November 15, 1882 (i.e., after the first two Edison plants began producing electricity), his long-standing conviction that gas lighting "is susceptible of great improvement, and is likely to hold its own for the ordinary lighting up of our streets and dwellings" (Siemens 1882:69). A decade later, during the 1890s, people thought that gasoline-fueled motor cars were just horseless carriages whose greatest benefit may be to rid the cities of objectionable manure. And many people also hoped that cars will ease the congestion caused by slow-moving, and often uncontrollable, horse-drawn vehicles (figure 1.7).

And given the fact that such fundamental technical shifts as the widespread adoption of new prime movers have invariably long lead times—for example, by 1925 England still derived 90% of its primary power from steam engines (Hiltpold 1934)—it is not surprising that even H. G. Wells, the era's most famous futurist, maintained in his first nonfictional prediction that if the 19th century needs a symbol, then that symbol will be almost inevitably a railway steam engine (Wells 1902a). But in retrospect it is obvious that by 1902 steam engines were a symbol of a rapidly receding past. Their future demise was already irrevocably decided as it was only a matter of time before the three powerful, versatile, and more energy-efficient prime movers that were invented during the 1880s—steam turbines, gasoline-powered internal combustion engines, and electric motors—would completely displace wasteful steam engines.

I also readily concede that, as is so often the case with historical periodizations, other, and always somewhat arbitrary, bracketings of the era that created the 20th century are possible and readily defensible. This is inevitable

FIGURE 1.7. A noontime traffic scene at the London Bridge as portrayed in *The Illustrated London News*, November 16, 1872.

given the fact that technical advances have always some antecedents and that many claimed firsts are not very meaningful. Patenting dates also make dubious markers as there have been many cases when years, even more than a decade, elapsed between the filing and the eventual grant. And too many published accounts do not specify the actual meanings of such claims as "was invented" or "was introduced." Consequently, different sources will often list different dates for their undefined milestones, and entire chains of these events may be then interpreted to last only a few months or many years, or even decades.

Petroski (1993) offers an excellent example of a dating conundrum by tracing the patenting and the eventual adoption of zipper, a minor but ubiquitous artifact that is now produced at a rate of hundreds of millions of units every year. U.S. Patent 504,038 for a clasp fastener was granted to Whitcomb L. Judson of Chicago in August 1893, putting the invention of this now universally used device squarely within the Age of Synergy. But an examiner overlooked a very similar idea was patented in 1851 by Elias Howe, Jr., the inventor of sewing machine. And it took about 20 years to change Judson's idea of the slide fastener to its mature form when Gideon Sundback patented his "new and improved" version in 1913, and about 30 years before that design became commercially successful. In this case there is a span of some seven decades between the first, impractical invention and widespread acceptance of a perfected design. Moreover, there is no doubt that patents are an imperfect measure of invention as some important innovations were not patented and as many organizational and managerial advances are not patentable.

Turning once again to the Industrial Revolution, we see that its British, or more generally its Western, phase has been dated as liberally as 1660–1918 and as restrictively as 1783–1802. The first span was favored by Lilley (1966), who divided the era into the early (1660–1815) and the mature (1815–1918) period; the second one was the time of England's economic take-off as defined, with a specious but suspect accuracy, by Rostow (1971). Other dates can be chosen, most of them variants on the 1760–1840 span that was originally suggested by Toynbee (1884). Ashton (1948) opted for 1760–1830; Beales (1928) preferred 1750–1850, but he also argued for no terminal date. Leaving aside the appropriateness of the term "revolution" for what was an inherently gradual process, it is obvious that because technical and economic take-offs began at different times in different places there can be no indisputable dating even if the determination would be limited to European civilization and its overseas outposts.

This is also true about the period that some historians have labeled as the second Industrial Revolution. This clustering of innovations was recently dated to between 1870 and 1914 by Mokyr (1999) and to 1860–1900 by Gordon (2000). But in Musson's (1978) definition the second Industrial Revolution was most evident between 1914 and 1939, and King (1930) was certain that, at least in the United States, it started in 1922. And in 1956, Leo Brandt and Carlo Schmid described to the German Social Democratic Party Congress the principal features, and consequences, of what they perceived was the just un-folding second Industrial Revolution that was bringing the reliance on nuclear energy (Brandt and Schmid 1956). Three decades later, Donovan (1997) used the term to describe the recent advances brought by the use of Internet during the 1990s, and e-Manufacturing Networks (2001) called John T. Parsons—who patented the first numerical machine control tool—"father of the Second Industrial Revolution."

Moreover, some high-tech *enamorati* are now reserving the term only for the ascent of future nanomachines that not only will cruise through our veins but also will eventually be able to build complex self-replicating molecular structures: according to them the second Industrial Revolution is just about to take place. Consequently, I am against using this constantly morphing and ever-advancing term. In contrast, constraining the singular Age of Synergy can be done with greater confidence. My choice of the two generations between the late 1860s and the beginning of WWI is bracketed at the beginning by the introduction of first practical designs of dynamos and open-hearth steel-making furnaces (1866–1867), the first patenting of sulfite pulping process (1867), introduction of dynamite (1866–1868), and the definite formulation of the second law of thermodynamics (1867). Also in 1867 or 1868 the United States, the indisputable overall leader of the great pre-WWI technical saltation, became the world's largest economy by surpassing the British gross domestic product (Maddison 1995).

Late in 1866 and in the early part of 1867, several engineers concluded independently that powerful dynamos can be built without permanent magnets by using electromagnets, and this idea was publicly presented for the first time in January 1867. These new dynamos, after additional improvements during the 1870s, were essential in order to launch the electric era during the 1880s. The open-hearth furnace was another fundamental innovation that made its appearance after 1867: in 1866 William Siemens and Emile Martin agreed to share the patent rights for its improved design; the first units became opera-tional soon afterward, and the furnaces eventually became the dominant pro-ducers of the steel during most of the 20th century. Tilghman's chemical wood pulping process (patented in 1867) opened the way for mass production of inexpensive paper. And by 1867 Alfred Nobel was ready to produce his dy-namite, a new, powerful explosive that proved to be another epoch-making innovation, in both the destructive and constructive sense.

The formulation of the second law of thermodynamics in 1867 was such an epoch-making achievement that more than two decades later *Nature* edi-torialized (quite correctly as we can see in retrospect) that the theory "seems to have outrun not only our present experimental powers, but almost any conceivable extension which they may hereafter undergo" (Anonymous 1889a: 2). The year 1867, the beginning of my preferred time span of the Age of Synergy, also saw the design of the first practical typewriter and the publication of Marx's *Das Kapital*. Both of these accomplishments—an ingenious machine that greatly facilitated information transfer, and a muddled but extraordinarily influential piece of ideological writing—had an enormous impact on the 20th century that was suffused with typewritten information and that experienced the prolonged rise and the sudden fall of Marx-inspired Communism.

Inventors of the new technical age and the entrepreneurs who translated the flood of ideas into new industries and thus formed new economic and social realities (sometimes they were the same individuals) came from all regions of Europe and North America, but in national terms there can be no doubt about the leading role played by the United States. Surpassing the United Kingdom to become the world's largest economy during the late 1860s was only the beginning. Just 20 years later Mark Twain's introduction of a Yankee stranger (preparatory to his imaginary appearance at King Arthur's court) sounded as a fair description of the country's technical prowess rather than as an immodest boast of a man who

> learned to make everything; guns, revolvers, cannons, boilers, engines, all sorts of labor-saving machinery. Why, I could make anything a body wanted—anything in the world, it didn't make any difference what; and if there wasn't any quick new-fangled way to make a thing, I could invent one—and do it as easy as rolling off a log. (Twain 1889:20)

The most prominent innovations that marked the end of the era just before WWI were the successful commercialization of Haber-Bosch ammonia synthesis, introduction of the first continuously moving assembly line at the Ford Company, and Irving Langmuir's patenting of coiled tungsten filament that is still found in every incandescent lightbulb (figure 1.8). The first accomplishment was perhaps the most important technical innovation of the modern era: without it the world could not support more than about 3.5 billion people (Smil 2001). The second invention revolutionized mass production far beyond car assembly. The third one keeps brightening our dark hours. Last pre-WWI years also brought a number of fundamental scientific insights whose elaboration led to key innovations of the 20th century. Robert Goddard's concept of multiple-stage rocket (patented in 1914) led eventually to communication and Earth observation satellites and to the exploration of space. Niels Bohr's model of the atom was the conceptual harbinger of the nuclear era (Bohr 1913).

I readily concede that there are good arguments for both extending and trimming my timing of the Age of Synergy. Extending the 1867–1914 span just marginally at either end would make it identical to the Age of Energy (1865–1915), which Howard Mumford Jones defined for the United States simply as the half century that followed the end of the Civil War (Jones 1971). He used the term energy not in its strict scientific sense but in order to describe the period of an extraordinary change, expansion, and mobility. Trimming the span to 1880–

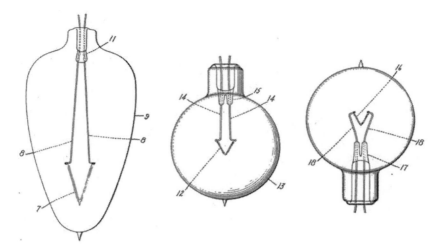

FIGURE 1.8. Three illustrations that accompanied Irving Langmuir's U.S. Patent 1,180,159 (1913) for an incandescent electric lamp show different kinds of coiled filaments that resulted in higher conversion of electricity to light. These drawings and the patent specification are available at the U.S. Patent and Trademark Office.

1910 would shorten the era to the three exceptionally inventive decades that were marked by an unprecedented concatenation of innovation, incipient mass production, and shifting consumption patterns. Other choices are possible, even if less defensible; for example, opting for the three decades between 1875 and 1905 would cover the time between the introduction of Otto's four-stroke internal combustion engine and the publication of Einstein's first relativity paper (Einstein 1905) that some interpretations could see as the start of a new era.

But one choice I am deliberately ignoring is the dating of well-known long-wave economic cycles whose idea was originally formulated by Kondratiev (1935), elaborated by Schumpeter (1939), and embraced by many economists after WWII (Freeman 1983; van Duijn 1983; Vasko, Ayres, and Fontvieille 1990). Datings of these waves vary. The second Kondratiev cycle commenced between 1844 and 1851 and ended between 1890 and 1896, which means that my preferred dating of the Age of Synergy (1867–1914) would be cut into two segments. Perez (2002) distinguished three stages of economic growth associated with technical advances during that period: Age of Steam and Railways (the second technical revolution) that lasted between 1829 and 1874, the Age of Steel, Electricity, and Heavy Engineering (third revolution, 1875–1907), and the Age of Oil, the Automobile, and Mass Production (fourth revolution, 1908–1971).

I ignore these periodizations not because I am questioning the validity and utility of the long-wave concept. I am actually an enthusiastic proponent of studying the cyclicity of human affairs, but the historical evidence forces me to conclude that the Age of Synergy was a profound technical singularity, a distinct discontinuity and not just another (second, or second and third) installment in a series of more-or-less regular repetitions, a period that belongs to two or three debatably defined economic cycles. My concern is not with the timing of successive sequences of prosperity, recession, depression, and recovery but with the introduction and rapid improvement and commercialization of fundamental innovations that defined a new era and influenced economic and social development for generations to come.

Even a rudimentary list of such epoch-defining artifacts must include telephones, sound recordings, lightbulbs, practical typewriters, chemical pulp, and reinforced concrete for the pre-1880 years. The astonishing 1880s brought reliable incandescent electric lights, electricity-generating plants, electric motors and trains, transformers, steam turbines, gramophone, popular photography, practical gasoline-fueled four stroke internal combustion engines, motorcycles, cars, aluminum production, crude oil tankers, air-filled rubber tires, first steel-skeleton skyscrapers, and prestressed concrete. The 1890s saw diesel engines, x-rays, movies, liquefaction of air, wireless telegraph, the discovery of radioactivity, and the synthesis of aspirin.

And the period between 1900 and 1914 witnessed mass-produced cars; the first airplanes, tractors, radio broadcasts, and vacuum diodes and triodes as well as tungsten lightbulbs, neon lights, common use of halftones in printing, stainless steel, hydrogenation of fats, and air conditioning. Removing these items and processes from our society would not only deprive it of a very large part of its anthropogenic environment we now take for granted but also would render virtually all 20th-century inventions useless as their production and/or functioning depend on an uninterrupted supply of electricity and the use of many high-performance materials, above all steel and aluminum.

Although my dating (1867–1914) of the Age of Synergy differs from Gordon's (2000) timing of the second Industrial Revolution (1860–1900), we share the conclusion that the development and diffusion of technical and organizational advances introduced during those years created a fundamental transformation in the Western economy as it ushered the world into the golden age of productivity growth that lasted until 1973 when OPEC's sudden quintupling of oil price introduced years of high inflation and low growth (figure 1.9). Neither the pre-1860 advances nor the recent diffusion and enthusiastic embrace of computers and the Internet are comparable with the epoch-making sweep and with the lasting impacts of that unique span of innovation that dominated the two pre-WWI generations.

An excellent confirmation of this conclusion comes from a list of the 20th century's greatest engineering achievements that the U.S. National Academy of Engineering (NAE) released to mark the end of the second millennium

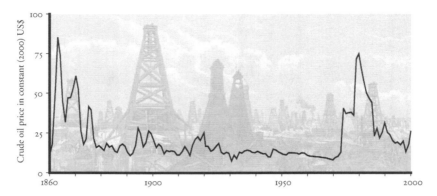

FIGURE 1.9. A century-long declining trend of crude oil prices (expressed in constant monies) ended with OPEC's first round of sudden price increases in 1973–1974. Based on a graph in BP (2003).

(NAE 2000). No fewer than 16 of the 20 listed categories of engineering achievements had not just their genesis but often also a considerable period of rapid pioneering development during the two pre-WWI generations. These achievements are headed by electrification, automobiles, and airplanes, include telephone and air conditioning and refrigeration (numbers 9 and 10), and conclude with petrochemical techniques and high-performance materials (numbers 17 and 20).

Another list, commissioned by the Lemelson-MIT Prize Program (an institution that recognizes and rewards innovation) and assembled by polling 1,000 American adults in November 1995, was topped by four inventions that were introduced and considerably developed before the WWI. With 34% of all responses, the automobile ranked as the most important invention of modern times, followed by the lightbulb (28%), telephone (17%), and aspirin, which was tied, with 6%, for the fourth place with the personal computer, the highest-ranking of all post-WWI innovations (Lemelson-MIT Program 1996). Finally, in van Duijn's (1983) list of 38 major innovations that shaped the course of six key 20th-century growth sectors (steel, telecommunication, cars, aircraft, illumination, and photography), 23 (60%) were invented between 1867 and 1914.

A few revealing comparisons are perhaps the best way to impress an uninitiated mind by the scope and impacts of the period's remarkable innovations. During the first generation of the 19th century, everyday life of most people was not significantly different from that of the early 18th century. Even a well-off New England farmer ploughed his fields with a heavy wooden plough pulled by slow oxen. Poorly sprung carriages and fully rigged sailing ships were the fastest means of transport, and information—leaving aside the limited network of optical telegraphs built in France and a few neighboring countries (Holzmann and Pehrson 1994)—traveled only as fast as people or animals did.

Illumination came in meager increments of tallow candles or smoky oil lamps. Recycling of organic wastes and planting of legumes were the only manageable sources of nitrogen in agriculture, and average staple grain yields were only marginally above the late medieval levels. One of the key inescapable social consequences of these realities was the fact that even a very large agricultural labor force was unable to feed adequately the slowly growing populations. Despite the fact that more than four fifths of the work force was required to produce food through taxing labor, large shares of rural population did not have even enough bread, meat was an occasional luxury for most people, and recurrent spells of serious food shortages were common in Europe while frequent famines were affecting Asia (Kiple and Ornelas 2000; Smil 1994).

One hundred years later, a single small tractor provided traction equal to that of a dozen of large horses, and the Haber-Bosch synthesis of ammonia made it possible to supply optimum amounts of the principal macronutrient to crops. Highest crop yields rose by roughly 50%, and average food supply in all industrializing countries was well in excess of even liberally defined nutritional requirements. The age of steam locomotives was about to give way to more efficient and more powerful diesel engines and electric motors, shipbuilders were installing new powerful steam turbines, and the luminosity of Edison's original lightbulb was greatly surpassed by new lights with incandescing tungsten wires.

I must close this brief recounting of fundamental pre-WWI advances by emphasizing that I have no interest in forcing the concept of the Age of Synergy or exaggerating the era's importance (although that would be hard to do). I am not promoting any simplistic, deterministic, techniques-driven interpretation of modern history. I would never argue that adoption of particular technical advances determines the fortunes of individual societies or that it eventually dictates convergence of their behavior. Identical machines and processes deployed as a part of a democratic, fairly transparent society that is governed by enforceable laws and engaged in promoting private enterprise will have very different effects compared to those they could bring when operating as a part of a dictatorship where autarky and the lack of accountability foster mismanagement.

Nor am I claiming that nearly all of the 20th-century technical accomplishments were either direct derivatives (albeit improved and mass-produced in unprecedented quantities) or at least indirect descendants of innovations introduced during the two pre-WWI generations. But it is indisputable that these enduring innovations were indispensable for creating the 20th century, and it is also possible to forecast with a high degree of confidence that many of them will not lose this importance for generations to come. And so it is in order to understand better where we have come from and where we are heading that I will portray the genesis of those enduring innovations and appraise their legacy. But before I begin this account of the pre-WWI advances and of their defining importance for the 20th century, I will take one more brief, comparative look at the era's accomplishments and their enduring quality.

The Distance Traveled

An effective way to appreciate the distance traveled between 1867 and 1914 is to contrast the state of our understanding of the world in the early 19th and the early 20th century with the realities at the very beginning of the 21st century. This could be done perhaps most impressively by contrasting the degree of comprehension that a competent scientist of one period would have when, just for the sake of this thought experiment, we would transport him 100 years into the future. If Antoine Lavoisier (figure 1.10), one of the founders of modern chemistry, were not guillotined during the French Revolution's Age of Terror in 1794, he would have been 70 years old in 1813. A few men alive at that time could have equaled his comprehensive understanding of natural sciences and technical advances—but for him the world of the year 1913 would hold countless inexplicable wonders.

Steam turbines in large power plants, high-voltage transmission lines, electric lights and motors, oil drilling rigs, refineries, internal combustion engines, cars, radio broadcasts, x-rays, high explosives, high-performance steels, synthetic organic compounds and fertilizers, aluminum, and airplanes—all of these and scores of other machine, materials, objects, and processes would stun and puzzle him, and most of them would be utterly incomprehensible to his lay contemporaries. In contrast, were one of the accomplished innovators of the early 20th century—Edison or Fessenden, Haber or Parsons—be transported from its first decade to

FIGURE 1.10. Antoine Lavoisier (1743–1794) —the greatest chemist of the 18th century and an accomplished polymath—would view most of today's technical achievements with utter incomprehension. Portrait from E. F. Smith Collection, Rare Book & Manuscript Library, University of Pennsylvania, Philadelphia.

2005, he would have deep understanding of most of them (and actually created some of them) and at least a highly competent familiarity with the rest of the items listed in the preceding paragraph. Moreover, he would not need a great deal of explanation to understand many devices and processes that he never saw at work but whose operation is so clearly derived from the foundations laid down before WWI.

This legacy of the pre-WWI era is definitely most obvious as far as energy sources and prime movers are concerned. As already stressed, no two physical factors are of greater importance in setting the pace and determining the ambience of a society than its energy sources and its prime movers. Global fossil fuel era began sometime during the 1890s when coal, increasing volumes of crude oil, and a small amount of natural gas began supplying more than half of the world's total primary energy needs (Smil 1994). By the late 1920s biomass energies (wood and crop residues) provided no more than 35% of the world's fuels, and by the year 2000 their share was about 10% of global energy use. The two prime movers that dominate today's installed power capacity—internal combustion engines and steam turbines—were also invented and rapidly improved before 1900. And an entirely new system for the generation, transmission and use of electricity—by far the most versatile form of energy—was created in less than 20 years after Edison's construction of first installations in London and New York in 1882.

The only new primary energy source that has made a substantial commercial difference during the 20th century was nuclear fission (in 2000 the world derived from it about 16% of electricity), but due to its arrested development it now contributes globally less heat than does the burning of wood and crop residues (Smil 2003). The only 20th-century prime mover that entered everyday use after 1914 was gas turbine. Its development took off during the 1930s, and it led both to stationary machines for electricity generation and to the era of jet-powered flight. Naturally, the stories of nuclear fission and gas turbines will feature prominently in this book's companion volume.

In order to appreciate the lasting legacy of the epoch-making innovations introduced between 1867 and 1914, in this book I concentrate on four classes of fundamental advances: formation, diffusion, and standardization of electric systems (chapter 2); invention and adoption of internal combustion engines (chapter 3); unprecedented pace of introducing new, high-performance materials and new chemical syntheses (chapter 4); and the birth of a new information age (chapter 5). These topical chapters are followed first by introducing some additional perspectives on that eventful era and then by a brief restatement of its lasting legacies (chapter 6). After closing with a look at two major trends that had governed the 20th century—creation of high-energy societies and mechanized mass production that is aimed at rising the standards of living—I end with a look at contemporary perceptions of the pre-WWI achievements (chapter 7).

And although this is a book about inventions, improvements, and applications of techniques and about the power of applied scientific understanding, I wish I could have followed concurrently those fascinating artistic developments that took place during the two pre-WWI generations. Not only were they remarkable by any historic standards, but, much as their technical counterparts, they had set in place many tastes, preferences, and sensibilities that could be felt during the 20th century. Moreover, the world of technique left many brilliant imprints on the world of arts. One of my great favorites is *Au Bonheur des Dames*, Zola's (1883) masterful portrait of the new world of spreading affluence and frenzied mass consumption set in the late 1860s that was closely modeled on his thorough studies of large Parisian department stores of the early 1880s (figure 1.11). And the restlessly kinetic pre-WWI futurist paintings by Giacomo Balla, Umberto Boccioni, or Marcel Duchamp evoke the speed and the dynamism of a new age driven by electricity and motors as no technical specification can.

What legacies of those two incomparable generations we will take with us into the 21st century? Closing decades of the 20th century had brought some unprecedented new technical capabilities, but by the year 2000 the New Economy was in retreat. I will not speculate what the post-1999 realities mean for the long term. But, averse as I am to engage in any long-range forecasting, I am on very solid ground when I keep pointing out repeated failures of enthusiastic forecasts that have been predicting the demise of some well-established techniques and imminent adoption of new technical and managerial approaches (Smil 2000a).

Imminent demise of internal combustion engine is perhaps the best example in the first category. Edison believed that engines will never really make it and that the future belongs to electric cars—and he spent almost the entire first decade of the 20th century in a stubborn quest to develop a battery whose energy density would rival that of gasoline. A century later we still do not have an electricity storage device that would even approach such a density—but the promise of electric cars is still with us as forecasts that promise how these vehicles will capture a significant share of the automobile market within a decade are being rescheduled for another decade once the original forecast fizzles. A no less notorious example of forecasting excess includes all those monotonously advancing predictions of electricity from nuclear fusion that see the success always 30–50 years ahead.

Most recent examples of unrealistic expectations include the forecasts of rapid diffusion of fuel cells that will usher a new age of hydrogen economy—in a clear disregard of the past experience, which shows that long lead times are necessary in order to accomplish such transitions (Smil 2003)—and the promise of paperless offices saving trees everywhere: in reality, the consumption of paper has been rising along with the diffusion of electronic word processing and data management. These and other similarly conspicuous forecasting failure only strengthen my conviction that many epoch-making innovations that were intro-

FIGURE I.II. Grandiose, elegantly domed, and richly decorated central hall of Grands Magasins du Printemps (owned by Jules Jaluzot & Cie.) on Boulevard Haussman in Paris, the city's premiere luxury department store. Its three wings and eight floors of diverse merchandise reflected the rise of affluent middle class during the two pre-WWI generations. Reproduced from *The Illustrated London News*, October 18, 1884.

duced before WWI and that had served us so well, and often with no fundamental modifications, during the 20th century have much greater staying power than may be commonly believed. We will continue to rely on them for much of the 21st century, and this realization makes it even more desirable, as well as more rewarding, to understand the genesis of these innovations.

SCIENTIFIC AMERICAN, New York.

∾ 2

The Age of Electricity

> There is a powerful agent, obedient, rapid, easy, which con-
> forms to every use, and reigns supreme on board my vessel.
> Everything is done by means of it. It lights it, warms it, and
> is the soul of my mechanical apparatus. This agent is elec-
> tricity.
>
> Captain Nemo to Pierre Aronnax in Jules Verne's
> *Twenty Thousand Leagues Under the Sea* (1870)

I do not know of any better description of electricity's importance in modern
society than taking this quotation from Jules Verne's famous science fiction
novel and substituting "in modern civilization" for "on board my vessel." In
1870, when Verne set down his fictional account of Nemo's global adventures,
various electric phenomena had been under an increasingly intensive study for
more than a century (Figuier 1888; Fleming 1911; MacLaren 1943; Dunsheath
1962). Little progress followed the pioneering 17th-century investigations by
Robert Boyle (1627–1691) and Otto von Guericke (1602–1686). But Pieter van
Musschenbroek's (1692–1761) invention of the Leyden jar (a condenser of static
electricity) and Benjamin Franklin's (1706–1790) bold and thoughtful experi-

FRONTISPIECE 2. Cross section of Edison's New York station (thermal capacity, 93 MW)
completed in 1902. Boilers are on the right; steam engine and dynamo hall on the left.
Four steel-plate stacks were 60 m tall. Reproduced from the cover page of *Scientific
American*, September 6, 1902.

ments in Boston (beginning in 1749) and later in Paris revived the interest in the properties of that mysterious force.

Electricity ceased to be a mere curiosity and became a subject of increasingly systematic research with the work by Luigi Galvani (1737–1798, of twitching frog legs fame, during the 1790s), Alessandro Volta (1745–1827; his pioneering paper that described the construction of the first battery was published in 1800), Hans Christian Ørsted (1777–1851, uncovered the magnetic effect of electric currents in 1819), and André Marie Ampère (1775–1836), who contributed the concept of a complete circuit and quantified the magnetic effects of electric current. Their research opened up many new experimental possibilities, and already in 1820 Michael Faraday (1791–1867), using Ørsted's discovery, built a primitive electric motor, and before 1830 Joseph Henry's (1797–1878) experimental electromagnets became powerful enough to lift briefly loads of as much as 1 t.

But it was Michael Faraday's discovery of the induction of electric current in a moving magnetic field in 1831 that eventually led to the large-scale conversion of mechanical energy into electricity (Faraday 1839). Its revelation is easily stated: the magnitude of the electromotive force that is induced in a circuit is proportional to the rate of change of flux. This discovery of how to generate alternating current (AC) was the logic gate that opened up the route toward practical use of electricity that was not dependent on bulky, massive, low-energy-density batteries. Eventually, this route had led to three classes of machines whose incessant, highly reliable, and remarkably efficient work makes it possible to permeate the modern world with inexpensive electricity: turbogenerators, transformers, and electric motors.

Shortly after Faraday's fundamental discoveries came the invention of telegraph that by the 1860s evolved into a well-established, globe-spanning way of wired communication. But by 1870 there were no commercially viable electric lights and, as the puzzled Aronnax noted when told about electric propulsion of the *Nautilus* by Nemo, until that time electricity's "dynamic force has remained under restraint, and has only been able to produce a small amount of power"—because they were no reliable means of large-scale electricity generation and because electric motors were limited to small, battery-operated units.

Experiments with electricity and the expanding telegraphy were energized by batteries, commonly known as Voltaic bimetallic piles whose low energy density ($<$ 10 Wh/kg) was suited only for applications that required limited power. Then, after millennia of dependence on just three basic sources of energy—combustion of fuels (biomass or fossil), animate metabolism (human and animal muscles), and conversion of indirect solar flows (water and wind)—everything changed in the course of a single decade. During the 1880s the combined ingenuity of inventors, support of investors, and commercially viable designs of enterprising engineers coalesced into a new energy system without whose smooth functioning there would be no modern civilization.

Practical carbon-filament electric lights, soon supplanted by incandescing finely drawn wires, illuminated nights. Parsons's invention of the steam turbine created the world's most powerful prime mover that had made the bulky and inefficient steam engines obsolete and allowed inexpensive generation of electricity on large scale. Transformers made it possible to transmit electricity over increasingly longer distances. Efficient induction motors, patented first in 1888 by Nikola Tesla, converted the flow of electrons into mechanical energy, and innovative electrochemistry began producing new materials at affordable prices. Economic and social transformations brought by electricity were so profound because no other kind of energy affords such convenience and flexibility, such an instant and effortless access to consumers: a flip of a switch, or even just a preprogrammed order.

To any homemaker or any laborer of the pre-electrical era, electricity was a miracle. In households it had eventually eliminated a large number of daily chores that ranged from tiresome—drawing and hauling water, washing and wringing clothes by hand, and ironing them with heavy wedges of hot metal— to light but relentlessly repetitive tasks (e.g., trimming wicks and filling oil lamps). Electricity on farms did away with primitive threshing of grains, hand-milking of cows, laborious preparation of feed by manual chopping or grinding, and pitchforking of hay into lofts. And, of course, in irrigated agricultures electric pumps eliminated slow and laborious water-raising techniques powered by people and animals. In workshops and factories, electricity replaced poorly lit premises first by incandescent and later by fluorescent lights while convenient, efficient, and precisely adjustable electric motors did away with dangerous transmission belts driven by steam engines. On railways, electricity supplanted inefficient, polluting steam engines.

No other form of energy can equal electricity's flexibility: it can be converted to light, heat, motion, and chemical potential, and hence it has been used extensively in every principal energy-consuming sector except commercial flying. Until 1998 the last sentence would have said simply "with the exception of flying"—but in that year AeroVironment's unmanned *Pathfinder* aircraft rose to 24 km above the sea level. In August 2001, a bigger *Helios*, a thin, long curved, and narrow flying wing (its span of just over 74 m is longer than that of a Boeing 747) driven by 14 propellers powered by 1 kW of bifacial solar cells, became the world's highest flying plane as it soared to almost 29 km (figure 2.1; AeroVironment 2004).

Besides this all-encompassing versatility, electricity is also a perfectly clean, as well as a silent, source of energy at the point of consumption, and its delivery can be easily and precisely adjusted in order to provide desirable speed and accurate control for flexible industrial production (Nye 1990; Schurr et al. 1990). Electricity can be converted without virtually any losses to useful heat, and large electric motors can transform more than 90% of it into mechanical energy; it can also generate temperatures higher than combustion of any fossil

FIGURE 2.1. Helios prototype flying wing, powered by photovoltaic cells, during its record-setting flight above the Hawaiian Islands on July 14, 2001. NASA photo ED 01-0209-6.

fuel, and once a requisite wiring is in place any new converters can be just plugged in. Perhaps the best proof of electricity's importance comes from simply asking what we would not have without it. The answer is just about everything in the modern world.

We use electricity to power our lights, a universe of electronic devices (from cell phones to supercomputers), a panoply of converters ranging from hand-held hair dryers to the world's fastest trains, and almost every life saver (modern synthesis and production of pharmaceuticals is unthinkable without electricity, vaccines need refrigeration, hearts are checked by electrocardiograms and during operations are bypassed by electric pumps), and most of our food is produced, processed, distributed, and cooked with the help of electric machines and devices. This chapter's goal is thus simple: to describe the genesis and evolution of electric systems that took place between the 1870s, the last pre-electric decade in history, and the first decade of the 20th century. During this span of less than two generations, we made an enormous progress as we put in place the foundation of a new energy system whose performance is now far ahead of anything we had before WWI but whose basic features remained remarkably constant throughout the 20th century.

Fiat Lux: Inventing Electric Lights

Thomas Edison was accustomed to keeping a brutal work pace and demanded others to follow, and his search for a durable filament that would produce more than an ephemeral glow was particularly frustrating (Josephson 1959; Israel 1998). But the often repeated dramatic story of continuous and suspenseful "death-watch" that took 40 hours of waiting for the first successful filament to stop incandescing is just a legend, derived from later reminiscences of Edison's assistant (Jehl 1937). Edison's laboratory records indicate that the lamp that made history by passing the 10-hour mark—the one with a very fine carbonized cotton filament, a piece of six-cord thread fastened to platinum wires—was attached to an 18-cell battery at 1:30 A.M. on October 22, 1879, and that it still worked by 3 P.M. and continued to do so for another hour after increased power supply overheated and cracked the bulb (Friedel and Israel 1986).

A filament incandescing for nearly 15 hours represented a major advance in the quest for electric light, but as far as materials and design procedures were concerned, it was a step clearly retracing previous research of other inventors. Edison's principal, and lasting, contribution to the development of electric light was in changing the basic operating conditions, not in discovering new components. Even then, the initial success had to be followed by numerous improvements and modifications, in Edison's own laboratory and by others, before reliable and reasonably long-lasting lightbulbs were ready for mass marketing. What is so remarkable is that so many basic material and operational components that were finessed during the 1880s are still very recognizably with us, and that many subsequent innovations of incandescent lighting have greatly improved its performance without changing their operating principles.

Early Electric Lights

The quest to use electricity for lighting began decades before Edison was born. The possibility to do so was first realized at the very beginning of the 19th century. In 1801 Humphry Davy (1778–1829) was the first scientist to describe the electric arc that arises as soon as two carbon electrodes are slightly separated. In 1808 Davy publicly demonstrated the phenomenon on a large scale at the Royal Institution in London by using 2,000 Voltaic cells to produce a 10-cm-long arc between the electrodes of willow charcoal (Davy 1840). First trials of arcs preceded by decades the introduction of large electricity generators. The world's premiere of public lighting took place in December 1844 when Joseph Deleuil and Léon Foucault briefly lit Place de la Concorde by a powerful arc (Figuier 1888).

The first dynamo-driven outdoor arc lights were installed during the late

1870s: starting in 1877, P. N. Yablochkov's (1847–1894) much admired electric candles were used to illuminate downtown streets and other public places in Paris (Grand Magasins du Louvre, Avenue de l'Opéra) and London (Thames Embankment, Holborn Viaduct, the British Museum's reading room). By the mid-1880s arc lamps were fairly common sights in many Western cities, but they were massive and complicated devices that required skilled installation and frequent maintenance. Because a continuous arc wears away the electrodes, a mechanism was needed to move the rods in order to maintain a steady arc, and also to rejoin them and separate them once the current was switched off and on.

Many kinds of self-regulating mechanisms were invented to operate arc lamps, and later designs did not need any complex regulators because they used several pairs of parallel upright carbons (separated by plaster), each able to operate for 90–120 minutes and switched on, manually or automatically, by bridging the two tips. But that still left considerable costs of the carbons and the labor for recarboning the devices. A typical 10-A arc lamp using, respectively, carbons of 15 and 9 mm for its positive and negative electrodes and operating every night from dusk to dawn would have consumed about 180 m of carbon electrodes a year (Garcke 1911a). Placing these 500-W lamps 50 m apart for basic street illumination would have required annual replacement of 3.6 km of carbons for every kilometer of road, a logistically cumbersome and costly proposition precluding the adoption of arc lamps for extensive lighting of public places.

Glass-enclosed arcs extended the lamp's useful life to as much as 200 hours, but they did not lower the cost of recarboning enough to make arcs a better choice than advanced gas lighting, and they also reduced the maximum luminosity, from as much as 18,000 to no more than 2,500 lumens (lm). As for the electricity supply, even a typical 10-A arc consuming 500–600 W had to be supplied by a line of just 50–60 V, a voltage too low for efficient large-scale electricity distribution. Obviously impractical for most indoor, and for all household, uses and uneconomical for the outdoor applications, arc lights could not become the sources of universal illumination. Less powerful and much more practical incandescent lights were the obvious choice to fill this enormous niche—and it was also Humphry Davy who demonstrated their possibility when he placed a 45-cm-long piece of platinum wire (diameter of 0.8 mm) in a circuit between the bars of copper and induced first red heat and then brilliant white light (Davy 1840).

A tedious account of inventors and their failed or promptly superseded attempts would be needed to review all of the activities that took place between 1820, when William de La Rue experimented with a platinum coil, and January 27, 1880, when Edison was granted his basic patent (U.S. Patent 223,898) for carbon-filament electric lamp (Edison 1880a; figure 2.2). More than a score of inventors in the United States, the United Kingdom, France, Germany, and Russia sought patents for the "subdivided" electric light for nearly four decades before Edison began his experiments (Pope 1894; Howell and Schroeder 1927; Bright 1949; Friedel and Israel 1986; Bowers 1998). In his February 1913 lecture at the New York Electrical Society, William J. Hammer, one of Edison's laboratory assistants at Menlo Park during the period of lightbulb experiments and later the electrician of the first Edison lamp factory, pleaded "that we should keep these names green in our memory because their work was work of importance, even though it did not directly result in the establishment of the commercial incandescent electric lamp" (Hammer 1913:2).

But more than a century later none of these names is known to the general public, and even the historians of lighting mention some of them primarily because the men were involved in complex litigations with Edison's company during the 1880s (Pope 1894; Dyer and Martin 1929; Covington 2002).

FIGURE 2.2. Drawing attached to Edison's fundamental U.S. Patent 223,898 for the incandescent light shows a tightly coiled filament, but carbonized materials used by Edison in 1879–1880 would have made that impossible, and the first lamps had simple loop filaments. The coiled metallic filament was patented only in 1913 by Irving Langmuir (see figure 1.8). Reproduced from Edison (1880a).

Their names are forgotten because many of their inventions entered the cul-de-sacs of subdivided lighting. They sought materials with very high melting points because the proportion of light to heat radiated by a filament rises with the increasing temperature, and the most efficient light will be produced by the highest filament temperatures. Carbon, with the highest melting point of all elements (3,650°C), is an excellent incandescing material, and they used it as solid rods or charcoal tubes but could not produce sustained light. Platinum has a relatively low melting point of (1,772°C), which can be raised by alloying. In addition, unsuccessful early lamp designs had imperfect vacuum, or their imperfectly sealed glass bulbs were filled with nitrogen.

Edison's R&D Dash

Edison (figure 2.3) began his lighting experiments—in August 1878, after a memorable trip to the Rockies and California, when he rode most of the way from Omaha to Sacramento on a cushioned cowcatcher of a locomotive (Josephson 1959)—by retracing many unproductive steps that preceded him. Much of his early effort was devoted to alloying platinum with iridium and testing various coatings, including oxides of titanium and zirconium, that further improved the filament's incandescence. Edison's first patent application for the incandescent light was made on October 5, 1878 (U.S. Patent 214,636, granted on April 22, 1879): it had a platinum-iridium alloy filament.

Ten days later, on October 15, 1878—after the initial capital was raised from investors by Edison's long-time friend, patent attorney Grosvenor P. Lowrey—an agreement was reached to set up Edison Electric Light Co. with the objective "to own, manufacture, operate and license the use of various apparatus in producing light, heat or power by electricity" (Josephson 1959:189). In December 1878 Edison boasted to a *New York Sun* reporter that "I am all right on my lamp. I don't care anything more about it" (quoted in Friedel and Israel 1986:42), but in reality, there was little progress during the winter months of 1878–1879 as improved versions of alloyed filaments were tested and as entirely new, but unpromising, designs (including luminous zircon and carbon rod resting on platinum) were tried.

At that time, Edison was only one of many inventors (and a late starter) racing to develop practical incandescent light: Moses Farmer, William Sawyer (financed by Albon Man), and Hiram Maxim (1840–1916, much more famous for his 1885 invention of machine gun) were his main competitors in the United States, and St. George Lane-Fox and Joseph Wilson Swan in the United Kingdom. What eventually set Edison apart was not primarily his legendary perseverance in pursuing a technical solution, or the financing he received

FIGURE 2.3. Thomas A. Edison in 1880 at the time of his work on incandescent lights and electric systems. Library of Congress photograph by Emil P. Spahn (LC-USZ62-98067).

from some of the richest men of his time, but the combination of a winning conceptualization and a fairly rapid realization of an entire practical commercial system of electric lighting (Friedel and Israel 1986; Israel 1998).

In that sense, he deservedly ranks ahead of even Joseph Wilson Swan (1828–1914), an English chemist and physicist who remains credited in the United Kingdom as the inventor of lightbulb. Swan rejected platinum as the best incandescing substance already during the 1850s and concentrated instead on producing suitable carbon filaments. By 1860 he made a lamp that contained the key ingredient of Edison's early promising models of 1879: Swan's carbon filament was made by packing pieces of paper or cardboard with charcoal powder and heating them in a crucible. The carbonized paper strip was then mounted in an (imperfectly) evacuated glass vessel and connected to a battery. Swan eventually abandoned these short-lived lights that could reach only red glow, but resumed his experiments during the late 1870s when better vacuum pumps became available.

A working lamp—demonstrated for the first time at a meeting of the Newcastle-on-Tyne Chemical Society on December 18, 1878—had platinum lead wires and a carbon filament, the same components as Edison's first longer lasting lamp revealed 10 months later (Bowers 1998). But Swan's lights operated, as did those of all other inventors before 1879, with filaments whose resistance was very low, ranging from less than 1 Ω to no more than 4–5 Ω. Their goal was stable, long-lasting incandescence, but any mass deployment of such lights would have required very low voltages and hence impracticably high currents, resulting in a ruinously large mass of transmission wires. Moreover, pre-Edisonian lamps were connected in series and supplied with a constant current from a dynamo, making it impossible to switch on the lights individually and forcing a shutdown of the entire system because of a single interruption.

Edison's key insight was that any commercially viable lighting system must minimize electricity consumption and hence must use high-resistance filaments with lights connected in parallel across a constant-voltage system. This concept, so contrary to the prevailing wisdom of the time, was at first questioned even by Francis R. Upton (1852–1921), the only highly trained scientist (mathematics and physics, at Princeton and Berlin University) in Edison's laboratory, who was hired specifically to formulate and to buttress the inventor's ideas in rigorous scientific terms. Forty years later Upton recalled how eminent electricians of that time maintained the subdivision of the electric current was commercially impossible, and how muddled was their understanding of the very concept (Jehl 1937). A few simple calculations illustrate the difference between the two approaches.

An incandescent lamp of 100 W whose filament had resistance of just 2 Ω and operated at 36 V (common pre-Edisonian ratings) would have required the current of 18 A (V/Ω). In contrast, voltage of 110 V and a lightbulb whose filament had resistance of 140 Ω—these being parameters chosen by Edison and Upton in order to reduce the cost of copper conduits and to make filaments stable (Martin 1922)—needed just 0.79 A for a 100-W light. This means that for the same transmission distance a system composed of the low-resistance lamps would have required 70 times the mass of identical conducting wires to distribute electricity than would the high-resistance arrangement. The former choice was obviously a nonstarter from the economic point of view. As Edison noted in his specification for the first patent concerning high-resistance lamps (U.S. Patent 227,229 submitted on April 12, 1879), "By the use of such high-resistant lamps I am enabled to place a great number in multiple arc without bringing the total resilience of all the lamps to such a low point as to require a large main conductor; but, on the contrary, I am enabled to use a main conductor of very moderate dimensions" (Edison 1880b:1).

This was a winning concept, but the patented lamp itself was only an improved, but impractical, version of his older platinum lights. The lamp had an alloyed platinum wire wound into a bobbin mounted inside a small vacuum tube (his technicians succeeded in creating a vacuum of nearly one millionth of the atmospheric pressure) that was placed inside a glass cover. A flexible metallic aneroid chamber underneath the vacuum tube expanded when the bobbin got very hot and interrupted briefly the electrical circuit and prevented the filament from overheating without affecting the uniformity of the light. This was an ingenious design, but the complex arrangement of connections, wires, and magnets and the placement of the vacuum bulb within another glass container were not obviously a basis for a sturdy and practical lamp.

Edison's Success and Competitors

Once Edison abandoned the idea of a thermostatic regulator, he turned not only to achieving a nearly perfect, and long-lasting, vacuum inside the lamp and to searching for filaments other than platinum, but also to designing a new telephone and building his first dynamo, the largest electricity generator of that time. He returned to his search for a better filament only in August 1879. Many different substances, including bones and sugar syrup, were subsequently carbonized in a small furnace. Edison's first success with carbonized cotton sewing thread in October 1879 was followed by intense activity to improve the design before its first public presentation. A filament made of carbonized cardboard (Bristol board) proved to be a much better choice than cotton thread, and new lamps lasted easily more than 100 hours.

But before the demonstration, scheduled for December 31, 1879, took place, the *New York Herald* scooped the event by publishing a full-page article in its December 21 issue. Authored by Marshall Fox, who spent two weeks in the

Menlo Park laboratory preparing the piece, the lengthy article was informative, accurate, and well written (it is reproduced in Jehl 1937). Fox conveyed well not only the incredulity of the event (electric light being produced from "a tiny strip of paper that a breath would blow away") and the glory of its final product ("the result is a bright, beautiful light, like the mellow sunset of an Italian autumn") but also the basic scientific preconditions and challenges of Edison's quest. Not surprisingly, it was later judged by Edison as "the most accurate story of the time concerning the invention" (Jehl 1937:381).

The first public demonstration of incandescent lighting took place, as planned, at Menlo Park on December 31, 1879. Edison used 100 cardboard filament lamps—long-stemmed and topped by an onion-shaped dome, each consuming 100 W and producing about 200 lm—to light the nearby streets, the laboratory buildings, and the railway station. In so many ways this was not an end but merely a frantic beginning of the quest for incandescent light. Other inventors rapidly switched to high-resistance designs, and new patents were filed at a dizzying rate. In 1880 Edison alone filed nearly 30 patents dealing with electric lamps and their components (TAEP 2002). Swan's 1880 British patents, based on the key components of his previous designs, incorporated high resistance and the parallel arrangement, and after they were upheld in British courts Edison decided not to challenge them. Instead, a joint Edison & Swan United Electric Light Co. was set up in 1883 to manufacture Swan's lamps in the United Kingdom.

In the United States, Edison had to embark on many-sided defense of his patents while also defending himself against numerous claims of patent infringement. Between 1880 and 1896 more than $2 million was spent in prosecuting more than 100 lawsuits of the latter kind (Dyer and Martin 1929). Edison's greatest loss came on October 8, 1883, when the U.S. Patent Office decided that his lightbulb patents rest on the previous work of William Sawyer. The U.S. Circuit Court of Appeals made its final decision in favor of the Edison lamp patent only on October 4, 1892, more than 12 years after the filing of the patent itself, leaving the inventor to complain that he had never enjoyed any benefits from his lamp patents.

But protracted litigations did not stop the search for better filaments. Edison decided to ransack the world for the most suitable materials, and his laboratory tested more than 6,000 specimens of any available plant fibers, ranging from rare woods to common grasses. Edison finally settled on a Japanese bamboo (Madake variety), whose core had a perfect cellular structure and yielded a strong and highly resistant carbonized filament. These bamboo filaments of the early 1880s lasted about 600 hours, a great improvement on 150 hours obtained with the cardboard lamps of December 1879. But cellulose filaments, introduced by Swan in 1881, proved to be the most popular carbon choice, and their improved versions became the industry standard by the early 1890s.

Different filament configurations were designed to direct more light downward for the lamps mounted with their base up, while other designs for standard mountings tried to diffuse the light more evenly in all directions. As for the bulbs, they were all initially free-blown from 2.5-cm tubing, first in short and later in longer pear shapes, and their glass widened gently after it emerged from the base collar. In contrast, mold-blown bulbs, first introduced by the General Electric Co. during the early 1890s, had a short but almost perpendicular step just above the collar, and only then they widened into a rounded shape. Early bulbs came in a profusion of shapes, ranging from spheres to ellipsoids and from elongated teardrops to lozenges, with straight or curving sides, with or without necks, all of them with a short tip as the glass was nipped off while mouth-blowing the bulb, and Edison himself designed many of these shapes (figure 2.4). The first electric decade also saw the first colored, ground glass, opal and etched lamps, and ornamental lamps of various shapes.

The earliest bases were wooden or, between 1881 and 1899, plaster of Paris. Porcelain was introduced in 1900, and glass insulation, standard in today's bulbs, was introduced a year later. Platinum was used for lead wires, and carbon filaments were connected first with clamps and after 1886 with carbon paste. Edison's bamboo filaments had first a rectangular and later a square cross section and a hairpin shape. Squirted cellulose filaments, which dominated the market after 1893, soon substituted a loop for a hairpin in a round cross section. Edison's bamboo filament was abandoned after Edison General Electric merged with Thomson-Houston in 1892 to form the still prosperous, and now highly diversified, General Electric Co. But modern lamps incorporate more than the

FIGURE 2.4. Different kinds of Edison's light bulbs from the 1880s. Reproduced from Figuier (1888).

great inventor's high-resistance, parallel-connection design; they also perpetuate his patented mounting.

Swan's lamps had a characteristic side-pin twist-lock base that was invented by his brother Alfred (G. B. Patent 9,185, June 19, 1884; U.S. Patent 313,965, March 17, 1885). This was a direct predecessor of the bayonet base and socket that is still used in the United Kingdom as well as by automakers for tail and parking lights of cars and trucks worldwide. Lamps made by Westinghouse, Sawyer and Man, and Thomson-Houston had a clip base, and more than a dozen different base styles still coexisted during the mid-1890s. But it was a simple screw base—whose initial inspiration came to Edison early in 1880 as he unscrewed the cover of a kerosene can, and whose design was done by two of Edison's long-time associates, Edward H. Johnson and John Ott, during the fall of 1880 (Friedel and Israel 1986)—that was borne by about 70% of all lamps sold in 1900 and that became an industry standard by 1902.

Nothing demonstrated better the state of incandescent lamps after more than a dozen years of their rapid evolution than did the lighting at Chicago's Columbian Exposition of 1893, the first such event solely illuminated by electricity (Bancroft 1893). Arc lights were still used for some outdoor locations, but the Westinghouse Electric & Manufacturing Co. was the principal lighting contractor, and it used 90,000 "Stopper" lamps operating at 105 V. These were basically Sawyer-Man lamps with a ground glass stopper as their base, nitrogen-filled bulb, and iron lead wires that were designed to circumvent various Edison's patents before their expiry. Inside the Electrical Building at the exposition were more than 25,000 incandescent lights, including 10,000 General Electric lamps as well as designs by Westinghouse, Western Electric, Brush Electric, Siemens & Halske Co., and five smaller companies.

But all of these lamps shared one undesirable property: very low efficiency of converting electric energy into visible light. The earliest designs produced just over one lm/W, which means (converting with the average of 1.46 mW/lm) that they turned no more than 0.15% of electric energy into light (figure 2.5). Although this was a very low value on the absolute scale, it was an order of magnitude better performance than for paraffin candles, whose combustion converted a mere 0.01% of energy in the solid hydrocarbon into light, and a nearly four times the rate for gas lights: using typical performance data from Paton (1890), I calculated that gas jets converted less than 0.04% of the manufactured fuel to visible light. Improved filaments raised the typical performance to 2.5 lm/W (0.37%) by the mid-1890s (figure 2.5), and it was obvious that the ongoing large-scale diffusion of electric lighting would greatly benefit from replacing inefficient, and fragile, bits of carbonized cellulose by more luminous, as well as sturdier, metallic filaments.

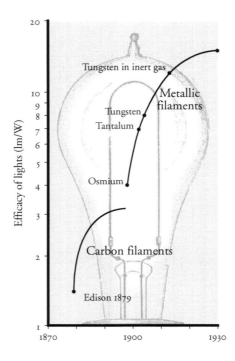

FIGURE 2.5. Efficacy of electric lights rose by nearly an order of magnitude between 1879 (Edison's first long-lasting carbon filament) and 1912 (Langmuir's tungsten filament in inert gases). Calculated and plotted from data in a variety of contemporary and retrospective publications.

Metallic Filaments

Standard efficacy of 3.5 lm/W (0.51%) was reached in 1898 by the deposition of carbon in the pores of cellulose filaments heated in the presence of petroleum vapor. In that year Carl Auer von Welsbach (1858–1929)—who in 1885 patented the incandescent mantle, a delicate gauze cylinder impregnated with thorium and cerium oxides whose greater luminosity delayed the demise of gas lighting (Welsbach 1902)—introduced the first working metal filament made of osmium. This metal's melting point (2,700°C) is 1,000°C above that of platinum, and osmium lights had efficacy of 5.5 lm/W, but the rarity, and hence the high cost of the element, prevented their commercial use.

Lamps with tantalum filaments (melting point of 2,996°C) made from a drawn wire reaching efficacies up to 7 lm/W were patented by Werner von Bolton and Otto Feuerlein in 1901 and 1902 and used until 1911. In 1904 Alexander Just and Franz Hanaman patented the production of tungsten filaments; the metal's high melting point (3,410°C) allowed efficacy of at least 8 lm/W, but preparing ductile (bendable) wire from this grayish white (hence its German name, *Wolfram*, and symbol W) lustrous metal was not easy (MTS 2002). Several methods were tried during the following years before William David Coolidge (1873–1975) demonstrated that the brittleness is due to the grain structure of the element. By using high temperature and mechanical treatment, tungsten becomes so ductile that it can be drawn into a fine wire even at ambient temperature; his patent application for making tungsten filaments was filed on June 19, 1912 (figure 2.6).

General Electric's first tungsten lamp, with the filament in vacuum, was introduced in 1910, and it produced no less than 10 lm/W (efficiency of about 1.5%). The metal became the dominant incandescent material by 1911, and the last cellulose lamps were made in 1918. When incandescing inside lamps, both filaments must be kept at temperatures far below their respective melting points (carbon just above 2,500°C and tungsten below 2,700°C) in order to avoid rapid vaporization. Consequently, their energy output peaks in the near infrared (at about 1 μm) and then falls through the visible spectrum to about 350 nm (into the ultraviolet A range), producing most of the visible light in red and yellow wavelengths, unlike daylight, which peaks at 550 nm and whose intensity declines in both infrared and ultraviolet directions.

One more step was needed to complete the evolution of standard incandescent light: to eliminate virtually all evaporation of the filament and thus to prevent any black film deposits inside a lamp. That step was taken in April 1913 by Irving Langmuir (1881–1957), who discovered that instead of maintaining a perfect vacuum, it is more effective to place tungsten filaments into a

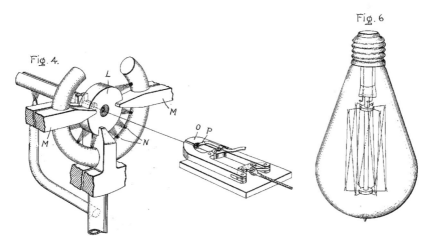

FIGURE 2.6. Two of six illustrations that accompanied Coolidge's U.S. Patent 1,082,933 for making tungsten filaments show his wire-drawing apparatus and an incandescent lamp made with the ductile metal.

mixture of nitrogen and argon and, in order to reduce the heat loss due to convection, to coil them (U.S. Patent 1,180,159; see figure 1.8). These measures raised the efficacy to 12 lm/W for common lightbulbs and more for the lamps of high power. Less than a quarter-century after Edison's 1879 experiments, the production of incandescent lamps was thus a mature technique. To be sure, gradual gains during the 20th century—improvements such as coiled tungsten coils (introduced in 1934) that further reduced heat loss from convection—brought further gains. Efficacy rose to more than 15 lm/W for 100-W lamps, and rated life spans increased to 1,000 hours for standard sizes, but no fundamental changes were ahead (figure 2.5).

By the 1990s more efficacious lights, based on entirely different principles, had captured large shares of the illumination markets. Although most of these lights also trace their origins to the pre-WWI period, they became commercially successful only after 1950. The first discharge lamps were built by Peter Cooper Hewitt (mercury vapor in 1900), and Georges Claude (1870–1960) demonstrated neon discharge (bright orange red) in 1910 (first neon lights were used in 1912 at a West End cinema in London). Further development of these ideas led to commercial discharge lamps: low-pressure sodium lamps were introduced in 1932, and low-pressure mercury vapor lamps, commonly known as fluorescent lights, were first patented in 1927 and came on the market during the late 1930s (Bowers 1998).

In fluorescent lamps, phosphorous compounds that coat the inside of their glass absorb ultraviolet rays generated by excitation of low-pressure mercury vapor and reradiate them as illumination approximating daylight. The best fluorescent lights are now producing nearly 110 lm/W, converting about 15% of electricity into visible radiation—and they also last about 25 times longer than does a tungsten filament. In 1912 Charles Steinmetz experimented with metal halide compounds in mercury lamps to correct their blue-green color— and exactly 50 years later General Electric revealed its first commercial metal halide light (whose successors now produce up to 110 lm/W).

Although the earliest lightbulbs were only as luminous as gas jets that they were designed to replace (typically equivalent to just 16 candles or about 200 lm), their light was safer, more reliable, quiet, and less expensive, which is why they rapidly displaced even Welsbach's greatly improved gas mantle. By 1914, 100-W tungsten lightbulbs were able to produce more than 1,000 lm, and during the coming decades they became the principal means of bringing daylight to billions of people. Remarkably, even the wide choice of nonincandescent lights that have been commercialized since the 1930s and that offer much higher efficacies and much longer life spans has not turned lightbulbs into rare relics of a simpler era.

Despite the inroads made by more efficient sources, incandescent lightbulbs continued to dominate the lighting market throughout the 20th century. The most detailed study of the U.S. residential energy consumption showed that during the early 1990s, 87% of all lights used one or more hours per day (453 million out of a total of 523 million) were incandescent (EIA 1996). And hundreds of millions of poor villagers in Asia, Africa, and Latin America whose dwellings have no connections to electric networks still wait for that soft light to change their lives as much as it has transformed the habits and opportunities of the world's more affluent minority. Those smallish fragile glass containers with an incandescing filament may not be around by the end of the 21st century—but they were the first source of convenient and flexible light that gave us, finally, an easy mastery over darkness, undoubtedly one the most common, most recognizable, and most beneficial inventions of the pre-WWI period that helped to create the 20th century and keep it brightly lit.

Edison's System

Invention of commercially viable lightbulbs had obviously a great practical and symbolic importance in ushering the electric era, but the bulbs were, both physically and figuratively, just the end of the line. Edison's lasting contribution is not that he set to invent a lightbulb (in that quest he was, as we have seen, preceded by a score of other inventors) or that he succeeded, relatively rapidly, in that effort. The fundamental importance of Edison's multifaceted and, even for him, frenzied activity that took place between 1879 and 1882 is that he put in place the world's first commercial system of electricity generation, transmission, and conversion. In Hughes's (1983:18) words,

> Edison was a holistic conceptualizer and determined solver of the problems associated with the growth of systems . . . Edison's concepts grew out of his need to find organizing principles that were powerful enough to integrate and give purposeful direction to diverse factors and components.

Few complex technical systems have seen such a rapid transformation of ideas into a working commercial enterprise. But the first step was taken reluctantly. One of the businessmen present at the Menlo Park demonstration on December 31, 1879, was Henry Villard, the chairman of Oregon Railway & Navigation Co., who became an instant convert and persuaded reluctant Edison, who preferred to concentrate on the development of a larger scale urban system, to install electric lights in his latest steamship. And so the first Edisonian system was put onboard the *Columbia* beginning in March 1880. The ship left for Portland in May 1880 and arrived in Oregon, after a 20-week journey around the South America, with its lighting in excellent condition. But this success was not followed immediately by any larger land-based installations.

This is understandable given the amount of technical, economic, and managerial challenges that had to be overcome. Electricity had to be cheaper than gas lighting, a mature, well-established industry with decades of experience, with extensive generation and transmission infrastructures in place in every major Western city and with profitable operations owned by some of the leading investors of that time. By the early 1880s, customers were used to the gently hissing sound of burning gas, but the experience was made less comfortable by the evolved heat and emissions of water vapor and carbonic acid. While electric light of 100 candles generated just 1.2 MJ of heat an hour and released no emissions, Argand gas burners of equivalent luminosity warmed the surroundings at the rate of about 20 MJ/hour and emitted nearly 1 kg of water vapor and nearly half a cubic meter of carbonic acid (Anonymous 1883; Paton 1890). Moreover, there were dangers of asphyxiation or explosion from leaking gas, and, obviously, the burner could be used only in an upright position.

Greater convenience of electric lights was obvious: steady, nonflickering, noiseless, nonpolluting, and flexible, ready to be used in all positions. But, as illustrated by Siemens's opinion cited in chapter 1, not everybody agreed that the days of gas lighting were numbered. And regardless of one's beliefs about the competitiveness of the two systems, it was very difficult to project the costs of electric system as there was no commercial production of most of the key components needed for its functioning and, naturally, no operational experience that would reveal the frequency of breakdowns and the overall system reliability to deliver electricity to customers. Planning far ahead, months before he had the first long-lasting lightbulb, Edison set aside his work on incandescent filaments and plunged not only into designs and construction of larger dynamos but also into detailed studies of operating costs, profits, and pricing of the coal gas industry. These studies were needed to set the goals that his system had to meet in order to prevail.

Dynamos, Engines, Fixtures

After his return from a memorable trip to California in August 1878, Edison began testing two of the most successful dynamos of the day, machines by Werner Siemens (1816–1892) and Zénobe-Théophile Gramme (1826–1901) that were used to power arc lights. By that time dynamo design advanced far beyond the first, hand-cranked, toylike generator built by Hypolite Pixii in 1831 (MacLaren 1943). By far, the most critical gain came in 1866–1867 with the realization that residual magnetism in electromagnets makes it possible for the generators to work even from a dead start, that is, without batteries and permanent magnets. In his memoirs, Werner Siemens (1893; figure 2.7) recalled his first presentation of the idea of what he named "dynamo-electric machine" to a group of Berlin physicists in December 1866 and its publication on January 17, 1867, as well as the fact that the priority of his invention was immediately questioned but confirmed later. But there is enough evidence to conclude that Charles Wheatstone (1802–1875), Henry Wilde, and Samuel Varley discovered the same phenomenon independently during the same time.

In an 1883 lecture, Siemens's brother William (1823–1883) noted that "the essential features involved in the dynamo-machine . . . were published by their authors for the pure scientific interest attached to them without being made subject matter of letters patent," and that this situation "retarded the introduction of this class of electrical machine" because nobody showed sufficient interest in the requisite commercial development (Siemens 1882:67). That is why Siemens gave a great credit to Gramme for his initiative in using Antonio Pacinotti's 1861 idea to build a ring armature wound with many individual coils

FIGURE 2.7. Werner Siemens, a founder of one of the world's leading makers of electric (and now also electronic) equipment, inventor of self-exciting dynamo and builder of first long-distance telegraph lines. Portrait reproduced from Figuier (1888).

of wire insulated with bitumen and also to introduce a new type of commutator.

Gramme's invention of a new *machine magnéto-electrique produisant de courant continu* was presented to the Academie des Sciences in Paris in July 1871 (Chauvois 1967). In contrast to Gramme's hollow cylinder, Siemens's improved dynamo had windings crossing near the center. Both of these machines could supply continuous current without overheating, were used to energize arc lights in an increasing number of European cities of the 1870s, and were later replaced by more powerful dynamos designed by Charles Brush and Elihu Thomson. Edison began his systematic work on producing better dynamos in February 1879, in parallel with the much better known work on incandescent lights, by examining closely the existing designs. He eventually adopted the drum armature, but the laboratory spent a great deal of time on devising new wiring patterns and designing new commutators and magnets.

The most distinguishing feature of dynamos that were actually used in the earliest lighting demonstrations was the use of two oversized, polelike, field magnets that earned them the nickname of "long-legged Mary-Ann" (Friedel and Israel 1986). Remarkably, these machines had efficiency as high as 82%, but a much larger dynamo was needed for the contemplated urban system capable of powering more than a thousand lights. Its design began in the summer of 1880, and its most distinct feature was the direct connection to a Porter-Allen steam engine, obviating the inefficient belting. The prototype could eventually energize 600 lamps of 18 candlepower, proving the basic viability of the concept.

The first installation of underground distribution network on the grounds of Menlo Park laboratory in spring and summer of 1880 uncovered problems with the existing insulation materials. The chosen wrapping consisted of layers of muslin with paraffin, linseed oil, and tar, and the first street lights supplied by underground conduits were lit on November 1, 1880. Edison's house was connected to the network a week later. Optimistic as ever, Edison now predicted he will have a working central station in the lower Manhattan before May 1, 1881. He needed to move fast, because by December 1880 no fewer than six companies were installing arc lights in the city, and because Hiram Maxim put in the first incandescent lights in the vaults and reading rooms of the Mercantile Safe Deposit Co. (Bowers 1998). But first, a new Edison Electric Illuminating Co. of New York, set up by Edison's attorney Grosvenor Lowrey, had to be formed (in December 1880 with the initial capital of $1 million) to conform to a state law that restricted the use of street lights to enterprises incorporated under gas statutes.

Then a deal had to be made between Edison and the company, controlled by the investors, for sharing the eventual profit, and all of the system's components had to be assembled. Armington & Sims supplied the steam engines, and Babcock & Wilcox provided aid on the boilers, but everything else— switch boxes, fittings, sockets, wall switches, safety fuses, fuse boxes, consumption meters, and insulated underground conductors—had to be designed, tested, improved, and redesigned. As Frank Lewis Dyer (the former general counsel for Edison's laboratory) recalled, the leading manufacturer of gas-lighting fixtures, Mitchel, Vance & Co.,

> had no faith in electric lighting and rejected all our overtures to induce them to take up the new business of making electric-light fixtures . . . Mr. Edison invited the cooperation of his leading stockholders. They lacked confidence or did not care to increase their investment. He was forced to go on alone. (Dyer and Martin 1929:718–719)

Once some of these new factories began turning out fairly large numbers of electric components, Edison abandoned his resistance to install small, isolated systems in individual plants or offices. In February 1881 the first installation of this kind was completed in the basement of Hinds, Ketcham & Co., a New

York lithography shop. Many similar installations followed during the rest of 1881, and in November of that year the Edison Company for Isolated Lighting was organized to handle new orders coming from universities, hotels, steamships, and textile mills, where the electric lights were particularly welcome in order to reduce the risk of fire. By the beginning of 1883, there were more than 150 isolated systems in the United States and Canada and more than 100 in Europe. But Edison's main goal was a large centralized system in Manhattan, whose realization was falling behind the previous year's expectation. Finally, the contract for this project was signed on March 23, 1881.

The First Central Plants

Edison initially wanted to supply the entire district from Canal Street on the north to Wall Street on the south but had to scale the coverage to about 2.5 km^2 between Wall, Nassau, Spruce, and Ferry streets, Peck Slip, and the East River. The location, the First District in lower Manhattan, was selected for a combination of reasons, above all its high density of lighting needs and its proximity to New York's financial and publishing establishments. Edison's crews surveyed the area's lighting needs, as well as its requirements for mechanical power, already during the year 1880 and then produced large maps annotated in color inks that showed Edison the exact number of gas jets in every building, their hours of operation, and their cost (Dyer and Martin 1929).

The laying of more than 24 km of underground conductors began soon after the contract was signed, and it proceeded according to Edison's feeder-and-main system for which he was granted a patent (U.S. Patent 239,147) in March 1881 (figure 2.8). This arrangement cost less than 20% than would have the tree system of circuits, reducing the cost of copper from the originally calculated $23.24 per lamp to just $3.72. Meanwhile, the work continued on a larger dynamo intended first for powering Edison's exhibits at the forthcoming Paris International Electrical Exposition (Beauchamp 1997).

Charles Clarke, Upton's college classmate and the first Chief Engineer of the New York Edison Co., had a major role in the development of a new design whose success was assured once the resistance of its armature was reduced to less than 0.01 Ω and the problems with air cooling were ironed out (figure 2.9). The massive machine (nearly 30 t) of unprecedented power (51.5 kW at 103 V) achieved public notoriety because of its nickname, Jumbo, derived from the fact that it was shipped to France on the same vessel that on a previous voyage transported P. T. Barnum's eponymous elephant from the London Zoo (Beauchamp 1997). By the end of 1881, the second Jumbo was installed at Edison's first operating central station in London, which began generating electricity on January 12, 1882.

The project was originally intended as a European demonstration of the

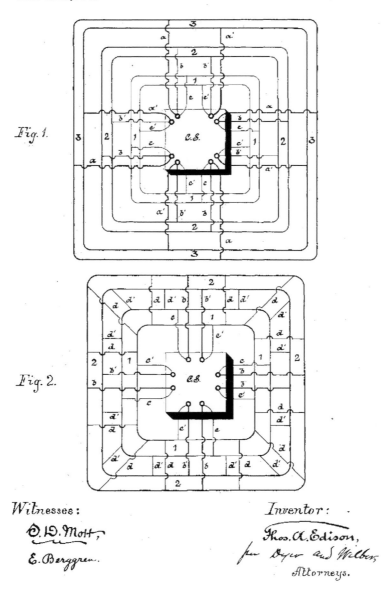

T. A. EDISON.
System of Electric Lighting.

No. 239,147. Patented March 22, 1881.

Fig. 1.

Fig. 2.

Witnesses:
Ø. ID. Mott,
E. Berggren.

Inventor:
Thos. A. Edison,
per Dyer and Wilber,
Attorneys.

FIGURE 2.8. Drawing attached to Edison's U.S. Patent 239,147 for a system of electric lighting (feeder and main arrangement). Reproduced from Edison (1881).

FIGURE 2.9. Edison's Jumbo, a massive dynamo that was first displayed in Paris in 1881. Reproduced from *Scientific American*, December 10, 1881.

new system, and its location was chosen by Edison's London representatives in order to circumvent the lengthy application process for digging up the streets by running electric lines under the Holborn Viaduct (Smithsonian Institution 2003). Machinery for this temporary installation was housed in a four-story row house (No. 57), with a 93-kW generator supplying initially 938 lamps, including 164 street lamps along 800 m of the viaduct and lights in the General Post Office, the City Temple, and businesses along the street, with direct current (DC) of 110 V. A second dynamo was added by April 1881, and the station remained in operation until 1884. But Edison's plans for a large central plant in London were thwarted by the Electric Lighting Act of the British Parliament, which made such installations impossible until it was revised in 1888.

Just as the Manhattan station was undergoing a series of final tests, Edison's first American small hydroelectric station, powered by a 107-cm waterwheel, was readied for service on the Fox River in Appleton, Wisconsin. Two small dynamos rated at a total of 25 kW were housed in a wooden shed, and they powered 280 weak lights (10 candle power), but the sturdy installation, ordered by the town's paper manufacturer H. F. Rogers, was in full operation by September 30, 1882, and it worked until 1899 (Dyer and Martin 1929). The first small English hydro-generating plant, opened during the previous year in Godalming by the Siemens brothers, whose company installed a small water turbine in the River Wey and laid cables in the gutters, operated only until 1884 (Electricity Council 1973).

FIGURE 2.10. Dynamo room of Edison's New York Pearl Street Station with six directly driven Jumbos. Reproduced from *Scientific American*, August 26, 1882.

The Manhattan station was to be in an entirely different class. Edison looked for a dilapidated building to site his generating station in a cheap property and ended up with two houses, 255 and 257 Pearl Street, both much more expensive than he anticipated. The station's heavy equipment was put in No. 257, coal (and ash) in the basement, four Babcock & Wilcox boilers (about 180 kW each) on the ground floor, and six Porter-Allen engines (each rated at 94 kW) and six large direct-connected Jumbo dynamos on the reinforced second floor (Martin 1922; figure 2.10). The neighboring house, No. 255, was used for material storage and repair shop and, frequently, as a dormitory for Edison and his crew working on the station. Edison devotion to the project included not only a close supervision of different tasks but actual work in the trenches (Israel 1998).

Even before all connections were completed, Edison turned on the first light in J. P. Morgan's office at 3 P.M. on September 4, 1882: just a single dynamo was online, and it supplied about 400 lights. When the time came to engage the second machine, the usually confident inventor admitted that he was extremely nervous as he scheduled a Sunday test—and, as he recalled later, his concerns were justified:

> One engine would stop and the other would run up to a thousand revolutions; and then they would seesaw. The trouble was with the governors . . . I grabbed the throttle of one engine and E. H. Johnson, who

was the only one present to keep his wits, caught hold of the other, and we shut them off. (Edison cited in Martin 1922:56)

To another witness,

it was a terrifying experience, as I didn't know what was going to happen. The engines and dynamos made a horrible racket, from loud and deep groans to a hideous shriek, and the place seemed to be filled with sparks and flames of all colors. It was as if the gates of the infernal regions had been suddenly opened. (Cited in Martin 1922:56–57)

By the end of 1882, three more Jumbos were added at the Pearl Street station, whose output was lighting more than 5,000 lamps, and in January 1883 the company began charging for its electricity using Edison's first ingenious but cumbersome electrolytic meters. Edison had every reason to be pleased. As he noted in 1904 in an article he wrote for *Electrical World and Engineer*, "As I now look back, I sometimes wonder at how much was done in so short a time" (cited in Jehl 1937:310). The scope of his early work during those years is perhaps best revealed by the patents he obtained between 1880 and 1882 in addition to nearly 90 patents on incandescent filaments and lamps: 60 patents dealing with "magneto or dynamo-electric machine" and its regulation, 14 patents for the system of electric lighting, a dozen patents concerning the distribution of electricity, and 10 patents for electric meters and motors (TAEP 2002).

Higher than anticipated costs and frequent outages of dynamos meant that the company lost money in 1883 ($4,457.50), but it made a good profit ($35,554.79) in 1884 (Martin 1922). Its operation provided a unique learning opportunity and served as an irreplaceable advertisement for the new system. One of the first adjustments Edison made soon after starting the Pearl Street station, on November 27, 1882, was to file a patent for the three-wire distribution system (U.S. Patent 274,290), an arrangement that was independently devised in England by John Hopkinson (1849–1898) a few months earlier and that remains the standard in our electric circuits.

Edison's first central station with the three-wire distribution was completed in October 1883 in Brockton, Massachusetts. This configuration saved about two-thirds of the copper mass compared to the two-wire conduit used in Manhattan (figure 2.11). While the feeder-and-main system is still with us, the three-wire DC transmission was not destined to be one of Edison's enduring innovations. In a few years it was superseded, both in its maximum spatial reach and its unit cost, by the transmission of three-phase and single-phase AC, which was initially shunned by Edison but strongly favored by Westinghouse, Tesla, Ferranti, and Steinmetz.

By 1884 the Pearl Street station was serving more than 10,000 lamps, and

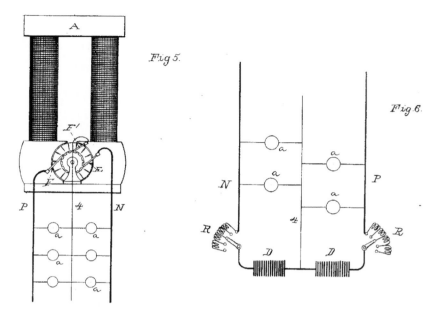

FIGURE 2.11. Two of seven drawings illustrating Edison's specification of his U.S. Patent 274,290 for three-wire electrical distribution. The left image shows wiring attached to a dynamo; the right, to secondary batteries. Reproduced from Edison (1883).

its success led to a rapid diffusion of similar installations for which the Edison Company had a virtual monopoly during most of the 1880s. By 1891 more than 1,300 central Edison plants were in operation in the United States, supplying about 3 million lights. Fire that damaged the Pearl Street station in 1890 destroyed all but one of the Jumbos, No. 9; in 1893 it was moved to the Columbian Exposition in Chicago, and it was eventually rebuilt and ended up in the Ford Museum in Greenfield Village near Dearborn, Michigan (Sinnott and Bowditch 1980; ASME 2003). The gutted station was reopened in just 11 days (Edison joined the repair crews in the round-the-clock effort to restart the operation), but it closed down in 1895, and the building was later demolished.

Edison's role in the genesis of the electric era cannot be overestimated. He was able to identify key technical challenges, resolve them by tenacious interdisciplinary research and development, and translate the resulting innovations into commercial use. There were other contemporary inventors of lightbulbs and dynamos, but only Edison had the vision of a complete system as well as the determination and organizational talent to make the entire system work, and he and his coworkers translated his bold ideas into realities in an astonishingly short period. One of the greatest tributes paid to this work came from

Emil Rathenau, one of the creators of Germany's electric industry and the founder of Allgemeine Elektrizitäts Gesselschaft, on the occasion of his 70th birthday on December 11, 1908.

Rathenau (quoted in Dyer and Martin 1929:318–319) recalled his impressions after seeing Edison's display at the Paris Electrical Exhibition of 1881:

> The Edison system of lighting was as beautifully conceived down to the very details, and as thoroughly worked out as if it had been tested for decades in various towns. Neither sockets, switches, fuses, lamp-holders, nor any of the other accessories necessary to complete the installation were wanting; and the generating of the current, the regulation, the wiring with distribution boxes, house connections, meters, etc., all showed signs of astonishing skill and incomparable genius.

This praise alone negates all the disparaging retelling of history and many derisory remarks that Edison's antagonists and critics have been producing since the very beginning of the electric era. Edison was, without any doubt and to resort to more of the modern parlance, an extraordinary systems thinker, but it seems to me that Friedel and Israel (1986:227) captured best the essence of his achievements by noting that "the completeness of that system was more the product of opportunities afforded by technical accomplishments and financial resources than the outcome of a purposeful systems approach." This more subtle interpretation does not change the basic facts. Edison was an exceptionally inventive, ambitious, and confident man who awed, motivated, inspired, and alienated his coworkers and financial backers. The combination of unusual insights, irrepressible enthusiasm, and the flair for (self)promotion led him often to voice exaggerated expectations and to make impossible promises.

Perhaps the most notable example of these attitudes was reflected in headlines a *New York Sun* reporter (cited in Friedel and Israel 1986:13) chose in September 1878 after talking to Edison at what was the very beginning of the search for practical incandescent light: "Edison's Newest Marvel. Sending Cheap Light, Heat, and Power by Electricity. 'I have it now.'" All that at a time when he actually had none of these capabilities! But obstacles and setbacks that derailed his overoptimistic schedules served only to fortify his determination to overcome them and to come up with better solutions. His devotion to the pursuit of invention was legendary, his physical stamina incredible. What is not perhaps stressed enough is that the financial support he received from his backers (including some of the richest men of his time) and the skills and talents of his craftsmen made it possible for him to explore fairly freely so many ideas and possibilities. In that sense, the Menlo Park laboratory was a precursor of the great corporate R&D institutions of the 20th century.

In retrospect, there is no doubt that the combination of Edison's inventions or radical improvements of several key components of the emerging electric system and his indefatigable push for its commercialization have been his greatest legacies. His were epoch-making, truly monumental, achievements, but I must hasten to note that while all of our electric networks are conceptual descendants of Edison's system of centralized electricity supply, their technical particulars as well as their operational arrangements had changed substantially before the end of the 19th century. Unlike the remarkably durable specifications and external features of Edison's incandescent lightbulbs, Edison's original electric system was not such a resilient survivor—and it did not deserve to be.

Above all, as advanced as it was for a pioneering design, Edison's first station was still a very inefficient generator of electricity: its heat rate was about 146 MJ/kWh, which means that it was converting less than 2.5% of the coal it burned into electricity. By 1900, the two key components of the electricity-generating system were different: steam turbines connected directly to alternators, rather than steam engines coupled with dynamos, became the preferred prime movers, and increasingly higher voltages of AC, rather than relatively low voltages of DC, distributed the generated electricity. On the consumption side, improved incandescent lights still used a large share of the generated electricity, but electric motors in factories and in urban transportation were rapidly becoming the largest consumers.

Turbines, Transformers, Motors

Incredibly, during the remainder of the 1880s the new electric industry saw advances that were no less fundamental than those made during the first three years of its development when it was dominated by Edison's race to introduce a profitable commercial system. These changes took place at every stage of the innovation process as they transformed the generation of electricity, its transmission, and its final uses. Edison remained an important, but increasingly a marginal, player, and the greatest acclaim must deservedly go to two engineers born in the mid-1850s: Charles Algernon Parsons (1854–1931) for his invention of steam turbogenerator, and Nikola Tesla (1856–1943) for his development of polyphase electric motor. Although William Stanley's (1858–1916) contributions to the design of transformers stand out, genesis of that device was much more a matter of gradual refinements that were introduced by nearly a dozen engineers in several countries.

Every one of these three great innovations made a fundamental difference for the future of large-scale electricity generation, and their combination amounted to a fundamental transformation in an industry that was still taking its very first commercial steps. By 1900, these advances came to define the performance of modern electric enterprises, determining possible size and efficiency of individual turbogenerating units, location and layout of stations, the choice of switchgear, and transmission and distribution arrangements. Without these innovations, it would have been impossible to reach the magnitude of generation and hence the economies of scale that made electricity one of the greatest bargains during the course of the 20th century.

All of these techniques were later transformed by series of incremental innovations whose pace had notably accelerated during the post-WWII economic expansion but then reached performance plateaus during the early 1970s. Consequently, all components and all processes of modern electric industry are now more efficient and more reliable than they were in 1914, yet they demand less material per unit of installed capacity and are more economical, and their operation has reduced impacts on the environment. At the same time, there is no mistaking their 1880s pedigree.

When Edison began to outline his bold plan for centralized electricity generation for large cities, he had only one kind of prime mover to consider: a steam engine. During the penultimate decade of the 19th century, design of these machines was a mature art (Thurston 1878; Ewing 1911; Dickinson 1939), but they were hardly the best prime movers for the intended large-scale electricity generation. After more than a century of development, their best efficiencies rose to about 25% by 1880, but typical sustained performances of large machines were much lower, closer to 15%. And although the mass/power ratio for the best designs fell from more than 500 g/W in 1800 to about 150 g/W by the 1880s, steam engines remained very heavy prime movers.

In addition to relatively low efficiency, high weight, and restricted capacity, top speeds of steam engines were inherently limited due to their reciprocating motion. Common piston speeds were below 100 m/min, and even the best marine engines of the 1880s, although they could reach maxima of up to 300 m/min, worked normally at 150–180 m/min (Ewing 1911). Moreover, this reciprocating motion had to be transformed into smooth rotation to drive dynamos. Edison's first generators installed in the experimental station in Menlo Park in 1880 were driven, most awkwardly and least reliably, by belts attached to steam engines. This arrangement was soon substituted by direct-connected high-speed engines, and Edison used the best available models in his first installations in 1881. A Porter-Allen machine he ordered was capable of 600 rpm but had problems with its speed regulator. An Armington & Sims engine for the Paris exhibition was rated at about 50 kW and had 350 rpm (Friedel and Israel 1986).

Simple calculations reveal the limitations inherent in these maxima. Imagine an urban system that does not serve any electric motors but only 1 million lights (merely six lights per household for 100,000 families and another 400,000 public, office, and factory lamps) whose average power is just 60 W. Such a system would require (assuming 10% distribution losses) 66 MW of electricity. With the best dynamos converting about 75% of mechanical energy into electricity, this would call for steam engines rated at 88 MW. Those at

Edison's Pearl Street station had the total capacity of nearly 3.4 MW, which means that our (dim and frugal) lighting system for a city of 400,000 people would have required 25 such generating stations. Their load factor would have been low (no more than 35%) as they had only one kind of final use, but even with relatively high efficiency of 20% they would have consumed annually about 250,000 t of steam coal.

This was, indeed, the inevitable setup of the earliest urban electric systems: they served just limited parts of large cities, and hence the generating companies had to bring large quantities of coal into densely populated areas and burn them there. Switching from DC to AC would have allowed for efficient transmission from peri-urban locations, but it would not have changed the required number of generating units. Moreover, this setup was so inefficient that were the entire U.S. population (about 57 million people in 1885) to be served by such low-intensity, systems they would have consumed the country's total annual production of bituminous coal during the mid-1880s! There was an obvious need for a more powerful, but a much less massive and more efficient, prime mover to turn the dynamos.

Steam Turbines: Laval, Parsons, Curtis

Water-driven turbines have been able to do this since the 1830s when Benoit Fourneyron introduced his effective designs, but before the 1880s there were not even any serious attempts to develop an analogical machine driven by steam. The first successful design was introduced by a Swedish engineer Carl Gustaf Patrick de Laval (1845–1913), whose most notable previous invention was a centrifugal cream separator that he patented in 1878 (Smith 1954). His impulse steam turbine, first introduced in 1882, extracted steam's kinetic energy by releasing it from divergent (trumpet-shaped) nozzles on a rotor with appropriately angled blades. Such a machine is subjected to high rotational speeds and huge centrifugal forces. Given the materials of the 1880s, it was possible to build only machines of limited capacity and to run them at no more than two-thirds of the speed needed for their best efficiency. Even then, a helical gearing was needed to reduce the excessive speed of rotation.

Blades in Laval's turbine are moved solely by impulse: there is no drop of pressure as the steam is passing the moving parts, and its velocity relative to the moving surfaces does not change except for the inevitable friction. Charles Parsons (figure 2.12) chose a very different approach in designing his successful steam turbine. As he explained in his Rede lecture,

> It seemed to me that moderate surface velocities and speeds of rotation were essential if the turbine motor was to receive general acceptance as a prime mover. I therefore decided to split up the fall in pressure of the

steam into small fractional expansion over a large number of turbines in series, so that the velocity of the steam nowhere should be great . . . I was also anxious to avoid the well-known cutting action on metal of steam at high velocity. (Parsons 1911:2)

Consequently, the steam moved in an almost nonexpansive manner through each one of many individual stages (there can be as many as 200), akin to water in hydraulic turbines whose high efficiency Parsons aspired to match. Starting with a small machine, he could not avoid high speeds: he filed the key British patent (G.B. Patent 6,735) on April 23, 1884 (the U.S. application for rotary motor was filed on November 14, and U.S. Patent 328,710 was granted on October 1885) and proceeded immediately to build his first compound turbine. This machine (now preserved in the lobby of the Parsons's Building at the Trinity College in Dublin) was rated at just 7.5 kW, produced DC at 100 V and 75 A, ran at 18,000 rpm, and had efficiency of a mere 1.6%. Many improvements followed during the decades before WWI (Parsons 1936).

Improvements concentrated above all on the design of blades and overcoming challenges of building a machine composed of thousands of parts that had to perform faultlessly while moving at high speeds in a high-pressure and high-temperature environment and do so within extremely narrow tolerances as clearances over the tips of moving blades are less than 0.5 mm (see figure 1.5). The first turbine had straight blading, for both rotating and guide wheels, which was later substituted by curved blades with thickened backs. Experiments showed that for the best efficiency the velocity of moving blades relative to the guide blades should be between 50% and 75% of the velocity of the steam passing through them.

Parsons also had to design a dynamo that could withstand high speeds and resist the great centrifugal force, and solve problems associated with lubrication and controls of the machine. The first small

FIGURE 2.12. There is no portrait of Charles Algernon Parsons dating from the early 1880s when he invented his steam turbine. This portrait (oil on canvas) was painted by William Orpen in 1922 and is reproduced here courtesy of the Tyne and Wear Museums, Newcastle upon Tyne.

single-phase turbo-alternators for public electricity generation were ordered in 1888, and the first two of the four 75-kW, 4,800-rpm machines began operating at Forth Banks station in Newcastle upon Tyne in January 1890. Their consumption was 25 kg of steam per kilowatt-hour at pressure of 0.4 MPa (or conversion efficiency of about just over 5%), and they supplied AC at 1 kV and 80 Hz. In 1889 Parsons lost the patent right to his parallel flow turbine when he left Clarke, Chapman & Co., established C. A. Parsons & Co. and turned to designs of radial flow machines that proved to be less efficient.

A 100-kW, 0.69-MPa turbo-alternator that his new company supplied to the Cambridge Electric Lighting in 1891 was Parsons's first condensing turbine. Parsons's small turbines of the 1880s were noncondensing, exhausting steam against atmospheric pressure and hence inherently less efficient. The condensing unit was also the first one to work with superheated steam (its temperature raised above that corresponding to saturation at the actual pressure), albeit only by 10°C. In 1894 Parsons recovered his original patents (by paying merely £1,500 to his former employers) and proceeded to design not only larger stationary units but also turbines for marine propulsion.

A rapid increase of capabilities then led to the world's first 1-MW units: two of them, with the first tandem turbine-alternator arrangement, were built by C. A. Parsons & Co. in 1899 for the Elberfeld station in Germany to generate single-phase current at 4 kV and 50 Hz (figure 2.13). The first 2-MW (6 kV, three-phase AC at 40 Hz) turbine was installed at the Neptune Bank station in 1903, and Parsons designs reached a new mark with a 5-MW turbine connected in 1907 in Newcastle upon Tyne (Parsons 1911). That turbine worked with steam superheated by nearly 50°C at 1.4 MPa, and it needed only 6 kg of steam per kilowatt-hour, converting coal to electricity with about 22%

FIGURE 2.13. An excellent engraving of Parsons's first 1-MW turbogenerator (steam turbine and alternator) installed in 1900 at Elberfeld plant in Germany. Reproduced from *Scientific American*, April 27, 1901.

efficiency. Parsons largest pre-WWI machine was a 25-MW turbo-alternator installed in 1912 at the Fisk Street station of the Commonwealth Edison Co. in Chicago. These large units converted about 75% of the steam's energy to the rotation of the shaft.

American development of steam turbines did not lag far behind the British advances (MacLaren 1943; Bannister and Silvestri 1989). George Westinghouse bought the rights to Parsons's machines in 1895, built its first turbine for Nichols Chemical Co. in 1897, and by 1900 delivered a 1.5-MW machine to the Hartford Electric Light Co. In 1897 General Electric made an agreement with Charles Curtis (1860–1953), who patented a new turbine concept a year earlier. This turbine could be seen as a hybrid of Laval's and Parsons's designs: it is a pure impulse turbine, but the kinetic energy of steam is not extracted by a single ring but by a multipulse action at several (commonly two or three but up to seven) stages (Ewing 1911). Curtis machines eventually reached capacities of up to 9 MW, and they were installed at many smaller plants throughout the United States during the first two decades of the 20th century.

General Electric's first turbine, a 500-kW unit installed in Schenectady, New York, in 1901, had a horizontal shaft, but its second machine of the same size was vertical, and the company soon began offering vertical Curtis turbines with capacities of up to 5 MW. Before GE abandoned the design in 1913, it shipped more than 20 MW of vertical-axis machines. The third American company that advanced the early development of steam turbine was Allis-Chalmers, originally also a licensee of Parsons's machines. By 1910 the company was producing turbines with capacities of up to 4.5 MW. And during the first decade of the 20th century, Swiss Brown Boveri Corporation became the leading producer of turbines on the European continent (Rinderknecht 1966).

Consequently, it took less than two decades after the first commercial installation to establish steam turbines as machines that were in every respect superior to steam engines. Their two most desirable features were the possibility of very large range of sizes and unprecedented generation efficiencies. In just 20 years, the size of the largest machines rose from 100 kW to 25 MW, a 250-fold increase, and by 1914 it was clear that it would be possible to build eventually sets of hundreds of megawatts. Performance trials of steam turbines were conducted from the very beginning of their design, and Parsons's (1911) figures show rapid gains of thermal efficiency. When his original numbers, expressed in terms of saturated or superheated steam per kilowatt-hour, are recalculated as percentages, they rise from about 2% for his 1884 model turbine to about 11% for the first low-superheat 100-kW machine in 1892 and to nearly 22% for the 5-MW, higher superheat design in 1907.

The efficiency for a 20-MW turbine installed in 1913 was just over 25% (Dalby 1920), which means that there was an order of magnitude gain in efficiency in three decades, clearly a very steep learning curve. In contrast, British marine engine trials of the best triple- and quadruple-expansion steam engines conducted between 1890 and 1904 showed maximum thermal efficiencies of 11–17% (Dalby 1920). This means that it made no engineering and economic sense to install steam engines in any large electricity-generating plant after 1904, if not after 1902. And so it was not surprising that America's largest new station completed in 1902, the New York Edison plant fronting the East River, still had 16 large Westinghouse-Corliss steam engines (Anonymous 1902; see the frontispiece to this chapter). But it was surprising when in 1905 London's County Council tramway power station in Greenwich began installing what Dickinson (1939:152) called a "megatherium of the engine world."

The "megatherium" was the first 3.5-MW vertical-horizontal compound steam engine, housed in a building of "cathedral-like proportions." This installation, so memorable for its poor engineering judgment, also provided the best contrast with inherently compact size and very low mass/power ratio of steam turbines. Greenwich machines were almost exactly as high as they were wide, about 14.5 m, and Parsons's company used to distribute a side-by-side drawing of these giants and of its turbogenerator of the same power (3.5 MW), whose width was just 3.35 m and height 4.45 m. Even Parsons's pioneering 100-kW turbine built in 1891 weighed only 40 g/W, or less than 20% than the best comparable steam engines. That ratio fell by 1914 below 10 g/W, and it was just over 1 W/g for the largest machines built after the mid-1960s (Hossli 1969).

Naturally, these declining power/mass ratios, especially pronounced with larger units (Hunter and Bryant 1991), translated into great savings in metal consumption and in much lower manufacturing costs per unit of installed power. Moreover, giant steam engines required additional construction costs for the enormous buildings needed to house them. Direct delivery of rotation power rather than awkward arrangements for converting the reciprocating motion was the most obvious mechanical advantage that was also largely responsible for lack of vibration that was often a problem with enormous piston engines.

The 20th century had witnessed a remarkable rise of turbine specifications very much along the lines that were set by Parsons during the first two decades of turbine development (Termuehlen 2001). Technical advances have been rather dramatic. Steam pressures rose by an order of magnitude by 1940, steam temperatures nearly tripled, and the highest ratings reached 1 GW by the late 1960s and eventually leveled off at 1.5 GW (figure 2.14). Three-phase alternators are now the world's largest continuous energy converters (their top ratings are an order of magnitude larger than those of the record-size gas turbines). Large modern turbines rotate at up to 3,600 rpm and work under pressures of 14–34 MPa and temperatures up to just above 600°C. The highest conversion efficiencies of modern fossil-fueled plants are now up to 40–43% (compared to just 4% in 1890) and can be raised further by resorting to combined generation cycles.

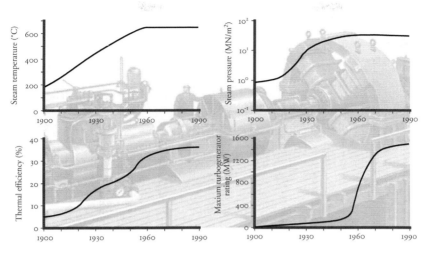

FIGURE 2.14. Long-term trends in the performance of U.S. steam turbogenerators show higher steam temperature and pressure, improved average efficiency, and rising unit power.

And there has been also some remarkable continuity in the manufacture of large steam turbogenerators. Both Westinghouse and General Electric remained their major designers and producers throughout the 20th century (Cox 1979). But unlike GE, Westinghouse has been dismembered: in 1998 Siemens bought Westinghouse Power Generation, and the following year BNFL bought West-inghouse Electric Co. Allis-Chalmers Division making the machines was even-tually bought by Siemens, and Brown Boveri joined with the Swedish ASEA. But, directly or indirectly, the earliest producers of these machines have re-mained, a century later, their major suppliers. Steam turbines and three-phase alternators produced by these companies dominate the world's electricity gen-eration, with some of the world's largest units installed in nuclear stations.

In the year 2000, about three-quarters of the world's electricity was gen-erated by steam turbines, and while the future growth of generating capacities in Western countries will be relatively slow, modernizing countries in Asia, particularly China and India, and in Latin America have been recently install-ing more than 30 GW of new turbine capacity every year (EIA 2002). Unmet electricity needs in these countries are such that high expansion rates would have to be sustained for several decades before the supply were to approach the levels prevailing in affluent countries. Consequently, even should we allow for relatively rapid advances in photovoltaics and wind-driven generation, there is little doubt that Parsons's machines will remain the leading producers of the world's electricity at least throughout the first half of the 21st century.

Before leaving the story of steam turbines, just a few paragraphs on their marine applications: superiority of Parsons's turbines for marine propulsion was demonstrated concurrently with increasingly more powerful stationary installations. Parsons's 1894 prospectus seeking the necessary capital for the new Marine Steam Turbine Co. claimed, quite correctly, that the new system will revolutionize the present method of using steam as a motive power. Experimental vessel *Turbinia* was 30 m long, displaced 40 t, and was driven by a single radial flow turbine capable of 715 kW (Osler 1981). But its first trials were disappointing as the propeller slippage was almost 50% and the speeds were far below the expectation.

Cavitation, formation of air bubbles around the propeller that was spinning too fast, was recognized as the key problem, and it was remedied by installing three parallel shafts driving the total of nine propellers. This configuration made the *Turbinia* the fastest vessel in the world, reaching 34.5 knots, 4 knots above the fastest British destroyers powered by compound steam engines. Public display of *Turbinia*'s speed at a grand Naval Review at Spithead on June 26, 1897, convinced the Royal Navy to order its first turbine driven destroyer in 1898. Six years later, 26 naval ships, including the massive *Dreadnought*, had Parsons direct-drive turbines, as did soon four of the world's largest passenger ships—*Mauretania, Lusitania, Olympic,* and *Titanic,* every one a great icon of the Golden Age of trans-Atlantic shipping of the early 20th century.

Diesel engines or gas turbines now power most freight and passenger ships, but more than a century after *Turbinia*'s triumph, steam turbines now capable of as much as 40 MW can still be found on vessels ranging from the largest aircraft carriers to the most modern tankers carrying liquefied natural gas. Vessels of the Nimitz class, the latest series of the U.S. nuclear carriers, have two pressurized water reactors that power four geared steam turbines. Large numbers of steam turbines also drive centrifugal pumps and compressors, including those employed during the Haber-Bosch synthesis of ammonia (Smil 2001). And, finally, commercial success of steam turbines led to the search for practical gas turbines whose remarkable rise will be detailed in this book's companion volume.

Transformers

I cannot think of another component of the electric system, indeed, of another device, whose ubiquitous service is so essential for continuous functioning of modern civilization, yet that would be so absent from public consciousness as the transformer. Coltman (1988) is correct when he ascribes this lack of recognition to the combination of transformer's key outward attributes: it does not move, it is almost completely silent, and it is usually hidden, be it underground, inside buildings, behind screens, or boxed in: plates protruding from large transformer boxes are external radiators used for cooling the device.

Yet without these ingenious devices, we would be stuck in the early Edisonian age of electricity generation and distribution where the prevailing limitations on the distance over which electricity can be transmitted would have required the engineers to sprinkle power stations with a rather high density throughout the urban areas, and would have forced smaller places to rely on isolated generating systems.

This prospect clearly worried the early proponents of electricity. William Siemens (1882:70) was concerned that "the extension of a district beyond the quarter of a square mile limit would necessitate an establishment of unwieldy dimensions." This would mean not only a large increase in the total cost of electric conductors per unit area but also that a "great public inconvenience would arise in consequence of the number and dimensions of the electric conductors, which could no longer be accommodated in narrow channels placed below the kerb stones, but would necessitate the construction of costly subways—veritable *cava electrica*."

Electricity is easiest to generate and most convenient to use at low voltages, but it is best transmitted over long distances with the least possible losses at high voltages because the power loss in transmission varies as the square of the current transmitted through the wires. Consequently, switching to higher voltages of AC reduced the dimensions of conductors, but such high voltages could not have been produced for the transmission and then reduced for the final use in households and offices without transformers. These devices do what their name implies: they convert one electric current into another, by either reducing or increasing the voltage of the input flow. What is most welcome from an engineering point of view is that transformers do it with virtually no loss of energy, and that the conversion works effectively across an enormous range of voltages (Coltman 1988; Calvert 2001).

Basic equations quantify the advantage of a system with transformers. The rate of transmitted electricity is equal to the product of its current and voltage ($W = AV$), and because voltage equals current multiplied by resistance (Ohm's law, $V = A\Omega$), power is the product of $A^2\Omega$. Consequently, to transmit the same amount of power with 100 times higher voltage will result in cutting the current by 99% and reducing the resistance losses also by 99%: the combination of low current and high voltage is always the best choice for long-distance transmission. Transformers provide the necessary interface to step up the generated current for transmission and then to step it down for local distribution and actual use. At that point, many gadgets may need further voltage reduction and current conversion; for example, 120 V AC has to be transformed into 16 V DC to run the IBM PC on which this book was written.

All transformers work by electromagnetic induction whose existence was demonstrated for the first time in 1831 by Michael Faraday. A loop of wire carrying AC (the primary winding) generates a fluctuating magnetic field that will induce a voltage in another loop (the secondary winding) placed in the field. In turn, the current now flowing in the secondary coil will induce a voltage in the primary loop, and both of the loops will also produce self-induction (figure 2.15). Because the total voltage induced in a loop is proportional to the total number of its turns (if the secondary has three times the number of turns of the primary, it will supply tripled voltage), and because the rate at which the energy is being transformed (W) must equal the product its voltage and its current (VA), the product of the secondary voltage and the secondary current must equal the product of the primary voltage and the primary current (leaving negligible losses aside). When Faraday wound two coils on an iron ring, rather than on a wooden one as in his initial induction experiments, he brought together the essential elements of the modern transformer.

Here is the other important reason why transformers do not occupy such a prominent position in the history of the 19th-century invention: there was no protean mind behind the device, no eureka-type discovery, just gradual improvements based on Faraday's experiments (Fleming 1901; MacLaren 1943). The need for transformers arose with the first tentative steps toward the AC system. Lucien H. Gaulard (1850–1888) and John D. Gibbs got the first patents for its realization in 1882 and displayed their transformer publicly in 1883 at the Electrical Exhibition in London. In order to add incandescent lightbulbs to an AC system that supplied arc lights connected in series (turned on and off at the same time), they used a two-coil transforming device that they called a secondary generator.

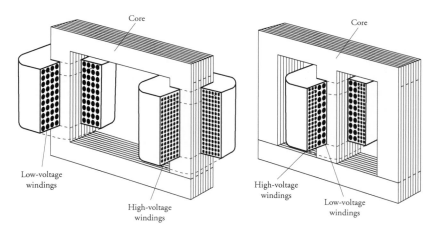

FIGURE 2.15. Two principal designs of modern transformers made with laminated iron sheets. In the core form, on the left, the two circuits enclose separate arms of the transformer, while in the E-shaped shell form both the primary and the secondary coils envelope the central bar.

The secondary generator's two windings were made of flat copper sheet rings inserted on cast iron wire core and insulated one from another by paper and varnish. A year later Gaulard and Gibbs staged another demonstration of this inefficient model at the international exhibition in Torino, Italy. In 1884 three Hungarian engineers employed at Ganz Works in Budapest improved the Gaulard-Gibbs device by designing two kinds of transformers for parallel connection to a generator and making them more efficient (Németh 1996). The key innovation—using closed iron cores that work much better than do the open-ended bundles—was suggested by Ottó Titusz Bláthy (1860–1939), the youngest engineer in the group. Shunt connection was the idea of Károly Zipernowsky (1853–1942), head of the electrical section of Ganz Works, and the experiments were performed by Miksa Déri (1854–1938). These transformers (later called ZBD) were demonstrated for the first time in May 1885 at the Hungarian National Exhibition in Budapest, where 75 of them were used to step down the current distributed at 1.35 kV to energize 1,067 incandescent lightbulbs.

George Westinghouse purchased the rights to the Gaulard-Gibbs design (for $50,000) and took options on ZBD transformers as well, but Edison's distrust of AC systems led him to ignore these critical developments, and transformers were the only fundamental electric device to whose development he did not make any significant contribution. In contrast, William Stanley, a young engineer working for Westinghouse, began designing improved transformers (he called them converters) in 1883, and by December 1885 he had a model that was much less expensive to make than the ZBD device and that became the prototype of devices we still use today (Stanley 1912; Hawkins 1951). In the first patented version (U.S. Patent 349,611 of September 21, 1886; three drawings on the left in figure 2.16), the soft-iron core, encircled by primary (b^1) and secondary (b^2) coil, was either annular or rectangular with curved angles and could consist either of a single piece of metal or wires or strips.

But, as shown in a patent drawing filed by George Westinghouse in December 1886 (U.S. Patent 366,362), soon the preferred composition of the core changed to thin plates of soft iron constructed with two rectangular openings and separated from each other, individually or in pairs, by insulating material (two drawings on the right in figure 2.16, right). From this design, it was a short step to stacked laminations that were stamped out from iron sheets in the form of letter E, which make it easy to slide prewound copper coils in place and then to lay straight iron pieces across the prongs to close the magnetic circuit; these became the most obvious marks of Stanley-type transformers (Coltman 1988; figure 2.15). Soon after the Westinghouse Electric Co. was incorporated in January of 1886, Westinghouse himself took out patents for improved designs of induction coils and transformers, including both the air-cooled and oil-insulated devices (both are still the standard practice today).

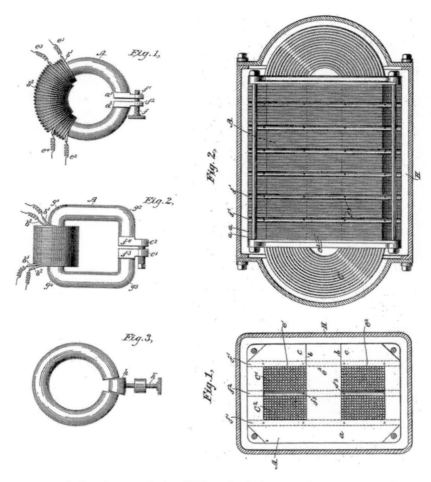

FIGURE 2.16. Drawings attached to William Stanley's 1886 U.S. Patent 349,611 for induction coil, a prototype of modern transformers (left), and to U.S. Patent 366,362 filed by George Westinghouse in December 1886. Stanley's device has an annular core; the later design has the core (A) made of thin insulated plates of soft iron with two rectangular openings.

Other patents for transformers were issued during the late 1880s to Elihu Thomson and Sebastian Ferranti, and in 1890 Ferranti designed the largest devices in operation for London's Deptford station. Electricity was generated at 2.5 kV and 83.5 Hz, stepped up to 10 kV for more than 11 km transmission by underground cables, and then stepped down by transformers in the central London area to 2.4 kV required for the distribution system (Electricity Council 1973). These were three-coil devices with the central primary separated from the two secondaries by sheets of ebonite, with coils made up of copper strips separated by vulcanized fibers, covered in varnished cloth, and designed to be cooled by either air or oil; their reliability can be best judged by the fact that they remained in service for more than three decades (Dunsheath 1962).

Also in 1891 William Stanley formed his owned electric manufacturing company, whose transformers were very similar to devices that he designed previously for Westinghouse. A big increase in transformer performance was needed by 1895 in order to accommodate transmission from the Niagara Falls hydroelectric station and the specifications required by the project's new industrial clients. Electricity generation, based on Tesla's polyphase AC concept, was at 5 kV and 25 Hz, and three-phase transmission to Buffalo carried the current at 11 kV. General Electric built two 200-kVA transformers for the aluminum plant of the Pittsburgh Reduction Co., and a 750-kVA transformer for the Carborundum Co. (MacLaren 1943). By 1900 the largest transformers were rated at 2 MVA and could handle inputs of 50 kV. But these devices operated with considerable hysteresis and eddy current losses.

Hysteresis is the memory effect of magnetic materials that weakens the transforming capacity by delaying the magnetic response, and eddy currents are induced in metals with low resistivity. Both of these phenomena dissipate large amounts of energy, and they made the early transformer cores, made of pure iron, relatively inefficient. Eddy currents can be reduced by laminating the core, and in 1903 English metallurgist Robert A. Hadfield (1858–1940) discovered that silicon steel greatly increased its resistivity while leaving the magnetic properties largely intact. Moreover, with the same maximum flux density silicon steel has hysteresis losses 75% lower than does sheet iron, and it is not subject to the phenomenon of aging whereby the hysteresis loss can double with time when the metal is exposed to higher temperatures ($> 65°C$) or mechanical fatigue (Calvert 2001). All transformer cores, as well as parts of rotating machinery subject to alternating fields, are made of silicon steel, constantly saving a great deal of electricity.

More than a century after its invention, the transformer remains little more than an artfully assembled and well-cooled bundle of iron sheets and copper coils. But as with so many other pre-WWI inventions, this conservation of a fundamentally ideal design has not prevented some very impressive gains in typical performance, mainly due to better production methods of silicon steel. Improved cooling and insulation were also introduced gradually beginning in the 1930s: low-power devices (up to 50 kVA) are still cooled by the natural air flow; larger ones have forced air circulation, and those above 200 kVA are usually immersed in mineral oil in a steel tank (Wildi 1981). Just before WWI, the largest transformers could work with inputs of up to 150 kV and power of 15 MW; today's largest transformers are rated at more than 1 GVA and can accommodate currents up to 765 kV, and the best units come very close to an ideal device.

But Stanley's (1912:573) appraisal is as correct today as it was when he confessed in 1912 to a meeting of American Institute of Electrical Engineers to

> a very personal affection for a transformer. It is such a complete and simple solution for a difficult problem. It so puts to shame all mechanical attempts at regulation. It handles with such ease, certainty, and economy vast loads of energy that are instantly given to or taken from it. It is so reliable, strong, and certain. In this mingled steel and copper, extraordinary forces are so nicely balanced as to be almost unsuspected.

First Electric Motors

History of electric motors resembles that of incandescent lights because in both cases the eventual introduction of commercially viable devices was preceded by a long period of experimental designs, and because in neither case did these unsystematic effort led directly to a successful solution of an engineering challenge. Tesla's contributions to the introduction and rapid diffusion of AC electric motors was no less important than Edison's efforts to commercialize incandescent light. And although the two most common converters of electricity, lights and motors, fill two very different final demands and hence it is not really appropriate to rank them as to their socioeconomic importance, there is no doubt that 20th-century economic productivity, industrial and agricultural, was revolutionized even more by electric motors than it was by electric lights.

The electric motor is fundamentally a generator working in reverse, and so it is not surprising that the first attempts to use the changing electromagnetic field for motion date to the 1830s, to the years immediately following Faraday's fundamental experiments. Given the eventual importance of motors, a great deal of historical research has been done to locate and to describe these earliest attempts as well as later developments preceding the first commercial introductions of DC motors during the 1870s and the invention of induction machines during the latter half of the 1880s (Hunter and Bryant 1991; Pohl and Müller 1984; Bailey 1911).

Certainly the most remarkable pioneering efforts were those of Thomas Davenport in Vermont in 1837 (U.S. Patent 132) and contemporaneous designs (beginning in 1834) by M. H. Jacobi in St. Petersburg. Davenport used his motors to drill iron and steel parts and to machine hardwood in his workshop, while Jacobi used his motors to power paddle wheels of a boat carrying 10–12 people on the Neva River. Both inventors had high hopes for their machines, but the cost and durability of batteries used to energize those motors led to early termination of their trials. A similar experience met Robert Davidson's light railway car that traveled on some British railroads in 1843, and heavy motors built during early 1850s (with considerable congressional funding) by Charles Page in Massachusetts. None of these battery-powered devices could even remotely compete with steam power: in 1850 their operating cost was, even under the most favorable assumptions, nearly 25 times higher.

Viewed from this perspective, it is not surprising that during the 1850s two of England's most prominent engineers held a very low opinion of such electric machines. Isambard K. Brunel (1806–1859), perhaps the most famous engineer, architect, and shipbuilder of his time, was against their inclusion in the Great Exhibition of 1851 as he considered them mere toys, and William Rankine (1820–1872), an outstanding student and popularizer of thermodynamics, wrote in 1859 that the true practical use of electromagnetism is not to drive machines but to make signals (Beauchamp 1997). Curiously, the first electric motor that was manufactured commercially and sold in quantity was Edison's small device mounted on top of a stylus and driving a needle in rapid (8,000 punctures a minute) up-and-down motion. This stencil-making electric pen, energized by a bulky two-cell battery, was the first device to allow large-scale mechanical duplication of documents (Pessaroff 2002). Edison obtained a patent for the pen in August 1876, and eventually thousands of these devices were sold to American and European offices (Edison 1876).

And it was also during the late 1870s that the first practical opportunities to deploy DC motors not dependent on batteries arose with the commercial introduction of Gramme's dynamo. This came about because of a fortuitous suggestion made in 1873 at the industrial exhibition in Vienna. Several incompatible versions of the story were published during subsequent decades, but the most authentic account comes from a letter sent in 1886 by Charles Félix, at that time the director of a sugar factory in Sermaize, to *Moniteur Industriel* and reproduced in full by Figuier (1888:281–282). Gramme Co. was exhibiting one dynamo powered by a gas engine and another one connected to a Voltaic pile.

During his visit of the exposition Félix suggested to Hippolyte Fontaine, who was in charge of Gramme's display, "Since you have the first machine that produces electricity and the second one that consumes it, why not let electricity from the first one pass directly to the second and thus dispense with the pile?" Fontaine did so, and the reversibility of machines and the means of retransforming electrical energy to mechanical energy were discovered. Electricity produced by the gas-engine-powered dynamo was led by a 100-m cable to the second dynamo, and this machine operating in reverse as a motor was used to run a small centrifugal pump.

Before the end of the decade came the first demonstration of a DC motor (at 130 V) in traction when Siemens & Halske built a short (300 m) circular narrow-gauge railway (*Bahn ohne Dampf und ohne Pferde*—railway without steam or horses—as an advertisement had it) at the Berlin Exhibition of 1879, with a miniature locomotive electrified from the third rail. A similar Siemens train was running in 1881 along Champs Elyséss during Paris Electrical Exposition in 1881, where Marcel Deprez displayed several motor-driven machines. As the first central electricity-generating stations were coming on line during the 1880s, such motor-driven devices began their transformation from display curiosities to commercial uses.

By 1887 the United States had 15 manufacturers of electric motors producing about 10,000 devices a year, most of them with ratings of only a small fraction of one horsepower, and many of them inefficient and unreliable (Hunter and Bryant 1991). Among the leading innovators in the United States were the Sprague Electric Railway & Motor Co. (the first one to demonstrate a powerful mine hoist in Aspen, Colorado, the precursor of electric elevators for tall buildings), the Thomson-Houston Co., and Eickemeyer Motors. Rudolf Eickemeyer's contribution was the winding of armature coils in a form in order to mold them to the exact shape, and to insulate them thoroughly before emplacing them in the armature; both of these procedures became the industry standard (MacLaren 1943). But the company that made the greatest difference was, thanks to its rights to Tesla's patents, Westinghouse Electric.

Nikola Tesla (1857–1943)—a largely self-educated Serbian engineer who

worked before his emigration to New York in 1884 with new electric systems in Budapest and Paris (figure 2.17)—was not actually the first inventor to reveal publicly a discovery of the principle of the rotating electromagnetic field and its use in an induction motor. Galileo Ferraris (1847–1897) presented the same insight to the Royal Academy of Science in Torino on March 18, 1888, and he published his findings in April, a month before Tesla's May 16, 1888, lecture at the American Institute of Electrical Engineers (Martin 1894; Popović, Horvat, and Nikić 1956). But Tesla, who claimed that he got the idea for his device one afternoon in 1882 while walking in a Budapest park with a friend and reciting a stanza from Goethe's *Faust*, was far ahead in developing the first practical machine (Ratzlaff and Anderson 1979; Cheney 1981; Seifer 1996).

FIGURE 2.17. Nikola Tesla: a portrait from the late 1880s when he worked on polyphase electric motors. Image available from Tesla Society.

He built its first models while working in Strasbourg in 1883, and after immigrating to the United States he hoped to develop the design for Edison's company. He was hired by Edison immediately after presenting the letter of introduction he got in Paris from Charles Batchelor, Edison's old associate, but the two incompatible men soon parted. Tesla, an outstanding scientist and a gifted mathematician, was appalled by what he felt to be Edison's brute force approach to problem solving and believed that a bit of theory and a few calculations would have eliminated a great deal of wasteful Edisonian searching for technical solutions. In addition, Edison, with his new central electric system built on low-voltage DC, was not ready to embrace the radical ideas of an avid high-voltage AC promoter.

After he secured generous financing by investors who appreciated the potential impact of his inventions, Tesla set up his electric company in April 1887, and he proceeded to build single-phase, two-phase, and three-phase AC motors and design appropriate transformers and a distribution system. Almost immediately, he also began a spate of patent filings (totaling 40 between 1887 and 1891), and the two key patents for his polyphase motor (U.S. Patents 391,968 and 391,969, filed on October 12, 1887) were granted on May 1, 1888. After many challenges, they were finally upheld by the courts in August 1900. The first patent specification illustrates the principle of a polyphase motor's operation (figure 2.18), and it gives clear indications of the simplicity and practicality of the device. That was Tesla's foremost goal, not an interesting gadget but

> a greater economy of conversion that has heretofore existed, to construct cheaper and more reliable and simple apparatus, and, lastly, the apparatus must be capable of easy management, and such that all danger from the use of currents of high tension, which are necessary to an economical transmission, may be avoided. (Tesla 1888:1)

Elegant design is one of the most highly valued achievements in engineering, and Tesla's AC motor was an outstanding example of this genre. By using the rotating magnetic field (with two or more alternating currents out step with each other) produced by induction, he had eliminated the need for a commutator (used to reverse the current's direction) and for contact brushes (allowing for the passage of the current). Westinghouse acquired all of Tesla's AC patents in July 1888 for $5,000 in cash and 150 shares of the company's stock, as well for the royalties for future electricity sales (the last being obviously the most rewarding deal). Tesla's motors were first shown publicly in 1888, and Westinghouse Co. produced its first electrical household gadget, a small fan powered by a 125 W AC motor, in 1889. This was a modest beginning of a universal conquest, but by 1900 there were nearly 100,000 such fans in American households (Hunter and Bryant 1991).

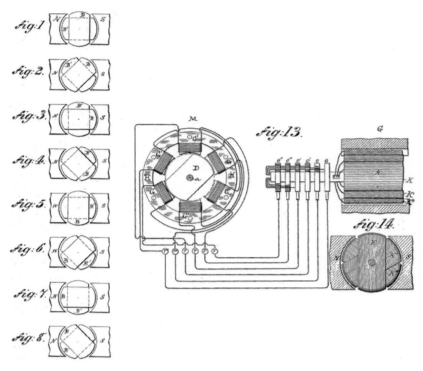

FIGURE 2.18. Illustrations attached to Tesla's U.S. Patent 391,968 for an electromagnetic motor. Drawings 1–8 show the principle of a two-phase motor's action, and drawing 13 shows the connections of a motor to a generator. Reproduced from Tesla (1888).

As with other inventions of the remarkable 1880s, subsequent improvements further raised the already high efficiencies of induction motors (the largest machines now convert more than 95% of electricity into kinetic energy), lowered their production costs, and vastly extended their uses without departing in any fundamental way from Tesla's basic designs. Tesla's first patent was for a two-phase machine. Mikhail Osipovich Dolivo-Dobrowolsky (1862–1919) built the first three-phase induction motor in Germany in 1889 while he was the chief electrician for the AEG, and three-phase machines soon became very common in industrial applications: the phase offset of 120° means that at any given moment one of the three phases is near or at its peak, and this assures a more even power output than with two phases, while a four-phase machine would improve the performance only marginally but would require another wire.

Adoption of Electric Motors and Their Impact

Only a few inventions have had such a transforming impact on industrial productivity and quality of life. In less than two decades after their commercialization, electric motors surpassed lights to become the single largest consumer of electricity in the United States. After 1900 they were rapidly adopted in manufacturing and for other industrial tasks. In 1899 only 22% of about 160,000 motors produced in the United States (only about 20% of them powered by AC) were for industrial uses; a decade later the share was nearly 50% of 243,000 (more than half of them run by AC), and before the end of 1920s electric motors became by far the most important prime mover in the Western industrial production (Schurr et al. 1990).

Many reasons explain this dominance. Induction motors are among the most rugged and most efficient of all energy-converting machines (Andreas 1992; Anderson and Miller 1983; Behrend 1901). Some of them are totally enclosed for operation in extremely dusty environments or work while completely submerged. All of them can be expected to deliver years of virtually maintenance-free operation and can be mass-produced at low cost, and their unit costs decline with higher capacities. They are also fairly compact, with large machines weighing less than 5 g/W (somewhat heavier than gasoline engines). By 1914 they were made in an impressive variety of sizes ranging from the smallest units driving dental drills to nearly 1,000 machines of 20–52 kW that were used to operate lock gates, rising-stem gate valves, and chain fenders of the newly built Panama Canal (Rushmore 1912; figure 2.19). Today's motors come in a much wider range of capacities, ranging from a fraction of a watt in electronic gadgets to more than 1 MW (and more than 300 A) for the largest synchronous motors that operate huge ore crushers.

And their scope of applications has increased to the point that everything we eat, wear, and use has been made with their help (Smil 2003). Electric motors mill our grain, weave our cloth, roll out our steel, and mold our plastics. They power diagnostic devices and are being installed every hour by tens of thousands aboard cars, planes, and ships. They do work ranging from frivolous (opening car sunroofs) to life saving (moving blood in heart-lung machines), from essential (distributing the heat from hydrocarbons burned by household furnaces) to entertaining (speeding small cars on their mad rides on roller-coasters). They also lift the increasingly urbanized humanity to high-rise destinations, move parts and products along assembly lines, and make it possible to micromachine millions of accurate components for devices ranging from giant turbofan jet engines to implantable heart pacemakers.

Perhaps the best way to appreciate their impact is to realize that modern civilization could have access to all of its fuels and even to generate all of its electricity—but it could not function so smoothly and conveniently without electric motors, these new alphas (in baby incubators) and omegas (powering

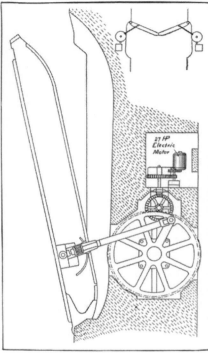

FIGURE 2.19. A set of lock gates of the Panama Canal and a plan view of mechanism for their operation. One 20-kW electric motor was needed for each leaf of 46 gate pairs, and 5.6-kW motors were used for the miter-forcing machines that made gates to come together with a tight seal and lock them in that position. Reproduced from *The Illustrated London News*, March 21, 1914, and from Rushmore (1912).

compressors in morgue coolers) of our world. Not surprisingly, electric motors now consume more than two-thirds of all electricity produced in the United States, and they are doing so with increasing efficiencies (Hoshide 1994). Also not surprisingly, one design cannot accommodate all of these demands, and induction motors have evolved to include several basic kinds of machines, with three-phase and single-phase motors being by far the most commonly encountered converters of electricity to kinetic energy.

Three-phase motors have the most straightforward construction. Their stators are made of a steel frame that encloses a hollow cylindrical core composed of stacked laminations. Two major types of three-phase induction motors are distinguished by their rotor wiring: squirrel-cage rotors (introduced by Westinghouse Co. during the early 1890s) are made up of bars and end rings; wound rotors resemble stators. Because the three-phase current produces a rotating magnetic field, there is no need for additional windings or switches within the motor to start it. And three-phase motors have also the highest efficiencies, are the least massive per the unit of installed power, and are the least expensive to make. Not surprisingly, three-phase induction motors are the norm for most of the permanently wired industrial machinery as well as for air- and liquid-moving systems.

Diffusion of three-phase induction motors in industrial production was a much more revolutionary step than the previous epochal transition to a new prime mover that saw waterwheels replaced by steam engines in factories. The first substitution did not change the basic mode of distributing mechanical energy that was required for countless processing, machining, and assembling tasks as factory ceilings continued to be clogged by complex arrangements of iron or steel line shafts that were connected by pulleys and belts to parallel countershafts and these, in turn, were belted to individual machines. With such an arrangement, a prime mover outage or an accident at any point along the line of power distribution (a cracked shaft, or just a slipped belt) stopped the whole assembly. Conversely, even if most of the machines were not needed (during the periods of slumping demand), the entire shaft assemblies were still running.

Electric motors were used initially to drive relatively short shafts for groups of machines, but after 1900 they were increasingly used as unit drives. Their eventual universal deployment changed the modern manufacturing by establishing the pattern of production that will be dominant for as long as electric motors will remain by far the most important prime movers in modern industrial production. Schurr et al. (1990) and Hunter and Bryant (1991) documented extensively the rapidity of this critical transition in the United States. While the total installed mechanical power in manufacturing roughly quadrupled between 1899 and 1929, capacity of electric motors grew nearly 60-fold and reached over 82% of the total available power compared to a mere 0.3% in 1890 and less than 5% at the end of the 19th century (USBC 1954; figure 2.20). Since then the electricity share has changed little: the substitution of steam and direct water-powered drive by motors was practically complete just three decades after it began during the late 1890s.

Benefits of three-phase motors as ubiquitous prime movers are manifold. There are no friction losses in energy distribution; individualized power supply allows optimal machine use and maximal productive efficiencies; they open the way for flexible plant design and easy expansion and enable precise control and any desired sequencing of tasks. Plant interiors look different because the unit electric drive did away with the overhead clutter, noise, and health risks brought by rotating shafts and tensioned belts, and the ceilings were freed for installation of better illumination and ventilation, steps that further aided in boosting labor and capital productivity.

But three-phase wiring is not normally supplied to homes and offices, and

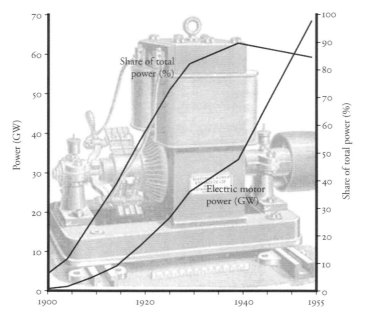

FIGURE 2.20. Capacity of electric motors and their share in the total power installed in the U.S. manufacturing. Plotted from data in USBC (1954).

hence a large variety of machines of smaller capacity draw on single-phase supply. Tesla filed his application for a single-phase motor on October 20, 1888, which was granted on August 14, 1894 (Tesla 1894), and Langdon Davies commercialized his single-phase design in 1893. Construction of these machines is very similar to that of three-phase motors: they have the same squirrel-cage rotor and a stator whose field alternates poles as the single-phase voltage swings from positive to negative. Efficiencies of single-phase motors are lower than in three-phase motors, but they, too, can deliver years of reliable work with minimal maintenance. Induction motors using single-phase 120-V AC are most common in household and office appliances, including ceiling and table fans, food mixers, refrigerators, and air conditioners.

The principle of synchronous motors was first demonstrated by Gramme's 1873 Vienna displays: they are simply alternators connected to a three-phase supply. Because its frequency is fixed, the motor speed does not vary no matter what the load or voltage of the line. Their principal use is at low speeds (below 600 rpm) where induction motors are heavy, expensive, and inefficient. West-inghouse deployed the first synchronous motor in 1893 in Telluride, Colorado, to run an ore-crushing machine. Modern electronic converters can produce very low frequencies, and synchronous motors can be thus run at very low speeds desirable in rotary kilns in cement plants and, as in its pioneering application, in ore or rock milling and crushing. And, naturally, synchronous motors are the best choice to power clocks and tape recorders.

But the ascendance of AC motors has not led to the total displacement of their predecessors, and DC motors, which flourished for the first time during the 1870s, remain ubiquitous. Some of them get their supply from batteries; many others receive DC from an electronic rectifier fed with AC. They have been entirely displaced by AC machines in nearly all common stationary industrial applications, but because of their very high starting torque, they have been always the motors of choice for electric trains, and between 1890 and 1910, before they were displaced by internal combustion engines, they were relatively popular as energizers of electric cars.

Despite many repeated promises of an imminent comeback of electric cars, those vehicles still have not returned, but there is no shortage of electric motors in road and off-road vehicles. Their total global number in the year 2000 was well over 2 billion: in addition to their starter motors, modern cars also have an increasing variety of small DC servomotors that operate windshield wipers, windows, locks, sunroofs, or mirrors by drawing the battery-supplied DC. DC motors are also common in chairs for invalids, as well as in treadmills, garage-door openers, heavy-duty hoists, punch presses, crushers, fans, and pumps used in steel mills and mines. Another advantage of DC motors is that their power tends to be constant as changes in torque automatically produce reverse change in speed; consequently, these motors slow down as a train starts going uphill, and in cranes and hoists they lift heavy loads slower than they do the light ones.

Systems Mature and (Not Quite) Standardize

Remarkable accomplishments of the 1880s refashioned every original component of new electrical system, and introduced machines, devices and converters whose performance had soon greatly surpassed the ratings, efficiencies, and reliabilities of initial designs. But this extraordinarily creative ferment was not conducive to standardization and optimization. This dissipative design is a phenomenon that recurs with all rapidly moving innovations. Before WWI it was also experienced by the nascent automotive industry. As a result, 10 years after the first central plants began operating in the early 1880s, one could find many permutations of prime movers, generators, currents, voltages, and frequencies among the operating coal-fired and hydro-powered stations.

Prime movers were still dominated by steam engines, but steam turbines were on the threshold of rapid commercial gains. Generators were mostly dynamos coupled indirectly or directly to reciprocating engines, but again, unit arrangements of turbines and alternators were ascendant. All of the earliest systems, isolated or central, produced and transmitted DC, commonly at 105–

110 V or at 200 V with a three-wire arrangement. But the first AC systems were installed within months after the commercial availability of AC transformers: Ferranti's small installation in London's Grosvenor Gallery in 1885, and Stanley's town plant in Barrington in Massachusetts, which led George Westinghouse, who came to see it on April 6, 1886, to enter the AC field (Stanley 1912). And so, inconspicuously, began the famous battle of the systems.

AC versus DC

Edison, the leading proponent of DC, was not initially concerned about the competition, but he soon changed his mind and embarked on what was certainly the most controversial, indeed bizarre, episode of his life. As he led an intensive anti-AC campaign, he did not base it firmly either on scientific and economic arguments (on both counts DC had its advantages in particular settings) or on pointing out some indisputable safety concerns with the high-voltage AC (HVAC). Instead, he deliberately exaggerated life-threatening dangers of AC, during 1887 was himself involved in cruel demonstrations designed to demonize it (electrocutions of stray dogs and cats coaxed onto a sheet of metal charged with 1 kV from an AC generator), and made repeated personal attacks on George Westinghouse (1846–1914), a fellow inventor (most famous for his railway air brake patented in 1869) and industrialist and, thanks to Stanley's influence, an early champion of AC (figure 2.21).

Perhaps worst of all, in 1888 Harold Brown, one of Edison's former employees, became a very active participant in a movement to convince the state of New York to replace hanging by electrocution with AC. Brown was eventually selected to be the state's electric chair technician, and he made sure that the AC was supplied by Westinghouse's alternators (MacLeod 2001). Edison's war on the HVAC lasted about three years. In 1889 he was still writing that laying HVAC lines underground would make them even more dangerous and that

> [m]y personal desire would be to prohibit entirely the use of alternating currents. They are as unnecessary as they are dangerous . . . and I can therefore see no justification for the introduction of a system which has no element of permanency and every element of danger to life and property. (Edison 1889:632)

Edison's vehement, dogmatic, and aggressive attitude was painful to accept and difficult to explain even by some of his most admiring biographers (Josephson 1959; Dyer and Martin 1929). But his opposition to HVAC stopped suddenly in 1890, and soon he was telling the Virginia legislature that debated an AC-related measure that "[y]ou want to allow high pressure wherever the conditions are such that by no possible accident could that pressure get into the houses of the consumers; you want to give them all the latitude you can" (quoted in Dyer and Martin 1929:418).

David (1991), in a reinterpretation of the entire affair, argues that Edison's apparently irrational opposition had a very rational strategic goal and that its achievement explains the sudden cessation of his anti-AC campaign. By 1886 Edison became concerned about the financial problems of the remaining electric businesses he still owned (by that time he had only a small stake in the original Edison Electric Light, the principal holding company), and hence he welcomed a suggestion to consolidate all Edison-related enterprises into a single new corporation. In return for cash ($1.75 million), 10% of the shares,

FIGURE 2.21. George Westinghouse had a critical role in the development of electric industry thanks to his inventiveness and entrepreneurial abilities, his unequivocal advocacy of alternating current, and his support of Tesla's work. Photograph courtesy of Westinghouse Electric.

and a place on the company's board, Edison's electric businesses were taken over by the Edison General Electric Co. organized in January 1889, and a year later Edison liquidated his shares and stopped taking any active role on the company's board. David (1991) argues that the real purpose of Edison's anti-AC campaign was to support the perceived value of those Edison enterprises that were entirely committed to produce components of DC-based system and thus to improve the terms on which he, and his associates, could be bought out. Once this was accomplished, the conflict was over.

The battle of the systems thus ended abruptly in 1890, although some accounts date it as expansively as 1886–1900. The latter delimitation is correct if one wants to include the period when AC became virtually the only choice for new central electric systems. And it must be also noted that, Edison's reasoning (no matter if guided mainly by stubbornness or strategic cunning) aside, the battle had its vigorous British version as the DC side—led by two luminaries of the new electric era, R.E.B. Crompton and John Hopkinson— was joined by one of the world's foremost physicists, Lord Kelvin. Sebastian Ziani de Ferranti, W. M. Mordey, and Sylvanus Thompson were the leading British experts behind HVAC.

While in 1886 one could see the DC–AC competition as something of a standoff, by 1888 the choice was fundamentally over thanks to the rapid advances of three technical innovations that assured the future AC dominance: the already described invention of induction motors, commercialization of an accurate and inexpensive AC meter, and the introduction of the rotary converter. No more needs to be said about the advantages of electric motors. In April 1888, Westinghouse engineer Oliver Shallenberger designed the first reliable meter for AC, and the very next year Westinghouse Electric began to manufacture it. But as David (1991) points out, neither the AC motors nor the AC meters could have dislodged the entrenched DC systems (by 1891 they accounted for more than half of all urban lighting in the United States) without the invention of rotary converter that made it possible to connect old DC central stations, as well as already fairly extensive DC transmission networks, to new long-distance HVAC transmission lines.

The first rotary converter was patented in 1888 by another of Edison's former employees, Charles S. Bradley. The device combined an AC induction motor with a DC dynamo to convert HVAC to low-voltage DC suitable for distribution to final users. When the original single-phase AC systems began to be replaced by more efficient polyphase transmission, new converters were eventually developed to take care of that transposition, and its reverse proved very useful during the 1890s when it made possible to use the existing DC-generating equipment to transmit polyphase HVAC over much larger areas than could be ever reached by a DC network. And so really the only matter that was unclear by 1890 was how fast will the AC systems progress, how soon would the superior polyphase AC generation and its high-voltage distribution become the industry standard.

A few bold projects accelerated this inevitable transition. The first one was the already mentioned Deptford station planned by the London Electric Supply Corporation, a new company registered in 1887, to provide reliable electricity supply to central London (figure 2.22). The site, about 11 km from the downtown on the south bank of the Thames, was chosen due to its location on the river and easy access to coal shipments by barges. Sebastian Ziani de Ferranti (1864–1930), the company's young chief electrical engineer, conceived the station on an unprecedented scale. Large steam engines were to drive record-size dynamos whose shafts, weighing almost 70 t, were the largest steel castings ever made in Scotland (Anonymous 1889b). Because of the magnitude of the service (more than 200,000 lights) and the plant's distance from the main load area in the downtown, Ferranti decided to use the unprecedented voltage of 10 kV. The company's financial difficulties eventually led to his dismissal from the project's leadership in 1891 and to the scaling-down of the Deptford operation, but Ferranti's unequivocal conviction about the superiority of high voltage was fully vindicated.

America's first AC transmission over a longer distance took place in 1890, when Willamette Falls in Oregon was linked by a 20-km, 3.3-kV line with

FIGURE 2.22. Massive exterior of London Electric Supply Corporation's Deptford station and one of its record-size Ferranti-designed dynamos under construction. Reproduced from *The Illustrated London News*, October 16, 1889.

Portland, where the current was stepped down first to 1.1 kV and then 100 and 500 V for distribution (MacLaren 1943). A more decisive demonstration of inherent advantages of long-distance HVAC transmission took place at the Frankfurt Electrotechnical Exhibition in 1891 (Beauchamp 1997). Oscar Müller built a 177-km-long line to a generator at Lauffen on the upper Neckar River in order to transmit 149 kW of three-phase AC, initially at 15 kV. The first small three-phase system in the United States was installed by the newly organized General Electric in 1893 in Concord, New Hampshire, and there is no doubt that the tide was completely turned when Westinghouse Co. and General Electric made designs for the world's largest AC project at the Niagara Falls (Hunter and Bryant 1991; Passer 1953).

On August 26, 1895, some of the water was diverted to two 3.7-MW turbines to generate electricity at 2.2 kV and use it at nearby plants for electrochemical production of aluminum and carborundum. In 1896, part of the output was stepped up to 11 kV and transmitted more than 30 km to Buffalo for lighting and streetcars. By 1900, 10 of Westinghouse two-phase 25-Hz generators of the first powerhouse (total of 37.3 MW) were in operation, and General Electric built 11 more units (total of 40.9 MW) for the second Niagara station completed by 1904. The project's aggregate rating of 78.2 MW accounted for 20% of the country's installed generating capacity (figure 2.23). Successful completion of the Niagara project confirmed the concept of large-scale HVAC generation and transmission as the basic paradigm of future electric systems.

General Electric's thrust into polyphase AC systems was greatly helped by insights and discoveries of Charles Proteus Steinmetz (1865–1923). Steinmetz was a brilliant mathematician and engineer who emigrated from Germany to New York in 1889, where he worked first for Rudolf Eickemeyer's company and was hired from there in 1893 by GE (Kline 1992). Steinmetz's work ranged from finding practical solutions to high transmission losses (his first patents at GE) to highly theoretical mathematical techniques used to analyze AC circuits with complex numbers. Between 1891 and 1923, he received more than 100 AC-related patents and published a dozen books. After 1900, Steinmetz had a critical role in setting up a new science-based laboratory at GE (one of the precursors of large-scale corporate R&D institutions) and guiding its early progress.

After the mid-1890s, it was only a matter of time before all of the necessary technical capabilities were put in place to support the expected expansion of AC systems. Parsons built the world's first three-phase turbo-alternator in 1900

FIGURE 2.23. Generator hall of Niagara Falls power plant. Reproduced from *Scientific American*, January 25, 1896.

for the Acton Hall Colliery in Yorkshire, where the 150 kW machine was to energize coal-cutting machinery. The first public supply of the three-phase current (at 6.6 kV and 40 Hz) came from the Neptune Bank station (still powered by steam engines) that was officially opened Lord Kelvin, the former opponent of AC, in 1901 (Electricity Council 1973). But long after the dominance of AC was assured, the commonly observed inertia of complex techniques made DC linger. In the United Kingdom, DC still had 40% of all generating capacity in 1906 (Fleming 1911). In the United States, DC's share in the total end-use capacity was 53% in 1902 and still 26% by 1917 (David 1991).

The trend toward increasing transmission voltages that was set during the first decade of AC developments continued, albeit at a slower pace, for nearly 100 years. Highest American ratings rose from 4 kV during the late 1880s to 60 kV by 1900 and to 150 kV in 1913; the first 300 kV lines were installed in 1934, and by the 1970s AC was transmitted by lines of 765 kV (Smil 1991). By the 1980s HVAC interconnections tied together most of the European continent as well as the eastern and western part of the United States and Canada, while east—west links across the continent remained very weak. And by the early 1970s, transmission techniques made a full circle by returning to DC, but this time at high voltages (HVDC).

HVDC, made possible by the use of mercury arc rectifiers that convert AC to DC and then reconvert it back to AC for distribution, was pioneered in Canada to transfer large blocks of electricity from hydro-projects in northern regions of Manitoba, British Columbia, and Quebec (Arrillaga 1983). This option becomes viable as soon as the cost of this conversion equipment is appreciably less than the money saved by building a cheaper two-conductor DC line. HVDC has been also used in submarine transmission cables, first pioneered on a smaller scale in 1913 when the British Columbia mainland was connected with Vancouver Island. In 1954 came the first large-scale application when the Sweden—Gotland cable carried 20 MW at 100 kV over 96 km, but a decade later New Zealand's two islands were connected in 1965 by a 250-kV tie that carries 600 MW.

Expanding Generation and Consumption

Proliferation of electric supply was very rapid: already in 1890 the United States had some 1,000 central electric systems and many more isolated generation units (Hunter and Bryant 1991). Many companies owned more than one station, and their service areas kept expanding. The Edison Electric Illumination Co. of New York, whose northernmost service boundary in 1883 was Nassau Street, reached 59th Street by 1890 and 95th Street by 1998, and by 1912 it served practically every street in Manhattan and the Bronx (Martin 1922).

In terms of 50-W lightbulb equivalents, its connections rose from 3,144 lights on December 1, 1882, to nearly 1.5 million lights by 1900 and to more than 12 million by the end of 1913. By 1900, every major city in Europe and North America, and many on other continents, had various systems of public electric supply from central stations. This electricity was used initially for lighting and transportation and soon also for a variety of household and industrial tasks (Wordingham 1901; Gay and Yeaman 1906).

During the first decade of the 20th century, many of these large cities were the first places where small stations that were located in or near downtowns or other densely populated areas were shut down and new, larger plants were built either on the outskirts of urban areas or directly in such coal-producing districts as the Midlands and Yorkshire in England and the Ruhr in Germany. In 1904 the Newcastle upon Tyne Electric Supply Co. commissioned its Carville station, whose design set the basic pattern for all subsequent large generating stations (Electricity Council 1973). The initial version had two 3.5-MW and two 1.5-MW units (combinations of boilers, turbines, and alternators that became later known as a turbogenerator) operated from a control room.

Large generating stations, mostly at mine-mouth locations when coal fired, have remained a preferred choice for Western utilities ever since. Plants of increasingly larger capacities were built until the 1960s in the United Kingdom, primarily in the Midlands, and in the United States concentrated in Pennsylvania, Ohio River Valley, and the Tennessee Valley Authority region. All of these stations burn finely pulverized coal that is blown into their boilers in order to maximize combustion efficiency. The first experiments with this technique began at the Willesden station in England in 1903. By 1914 the use of powdered coal was developed into a successful commercial practice (Holmes 1914), but most electricity-generating plants adopted it only after WWI.

The two pre-WWI generations also set the course for the construction of large-scale water power projects thanks to the introduction of two new turbines that eventually came to dominate hydroelectric generation. Several new water turbine designs appeared during the two decades after Benoit Fourneyron built his first reaction turbines during the late 1820s and the 1830s (Smith 1980). The most successful design was the one patented first in the United States in 1838 by Samuel B. Howd and then improved and commercially deployed by James B. Francis (1815–1892). This machine is known as the Francis turbine, but it is clearly a product of multiple contributions (Hunter 1979). Originally it had an inward flow, but during the 1860s it was developed into a mixed-flow turbine. During the 1880s Lester Allen Pelton (1829–1908) developed and patented an impulse turbine driven by water jets discharged into peripheral buckets, a design for very high water heads. In contrast, in 1913 Viktor Kaplan patented a reaction turbine with adjustable vertical-flow propellers that has become a preferred choice for low dams.

Before WWI more than 500 hydroelectric stations were in operation around the world (mostly in Europe and in North America), and the world's first pumped storage stations were also built. These stations use electricity produced during low-demand periods (mostly during the night) to pump water into a nearby high-lying reservoir, where it is available to be released into downpipes and converted almost instantly into hydroelectricity during periods of peak demand; a century later this is still the most economical way of an indirect electricity storage. But the period of hydroelectric megaprojects—exemplified in the United States by the Tennessee Valley Authority and by the Hoover and Grand Coulee dams—came only during the 1930s (ICOLD 1998).

Expanding availability of electricity led almost immediately to applications beyond household and street lighting. Few facts illustrate the rapid pace of post-1879 developments than the frequency of major national and international exhibitions that were devoted exclusively to that newly useful form of energy or where electricity played a major part (Beauchamp 1997). Of 38 international expositions that took place during the exceptionally inventive 1880s, seven were devoted solely to electricity, including the already noted Paris Electrical Exhibition of 1881 and similar displays in London in the same year, Munich (1882), Vienna (1883), and Philadelphia (1884). Electricity was a major attraction at British Jubilee Exhibitions in 1886–1887 and even more so at Chicago Columbian World Fair in 1893 (Bancroft 1893).

But already in Paris in 1881 several exhibitors showed, still as a curiosity rather than as a commercial proposition, sewing machines, lathes, and a printing press powered by electric motors. By 1890 General Electric was selling electric irons, fans, and an electric rapid cooking apparatus that could boil a pint of water in 12 minutes (Electricity Council 1973). Household appliances were introduced to the British market during the Crystal Palace Electrical Exhibition in 1891, and they included irons, cookers, and electric fires with the heating panels set behind wrought-iron screens. Two years later many more of these gadgets—including hot plates, fans, bells, bed warmers, radiators, and a complete Model Electric Kitchen with a small range, saucepan, water heater, broiler, and kettle—were featured at Chicago exhibition (Bancroft 1893; Giedion 1948). Early electric ranges resembled the well-established gas appliances, and the first toaster had only one set of resistance wires (figure 2.24).

Chicago displays also contained an enormous array of industrial applications of electricity ranging from ovens and furnaces to metal- and woodworking machines (lathes, drills, presses) and from railway signaling equipment to an electric chair. Among the notable pre-WWI commercial introductions, I should at least mention (with the dates of initial appearance in the United States) the electric washing machine (1907, featured soon afterward in the Sears & Roebuck Co.'s catalog), vacuum cleaners (originally called electric suction sweepers), and electric refrigerators (1912). Willis Haviland Carrier (1876–1950) designed his first air conditioning system in 1902 after he realized that air could be dried by saturating it with chilled water in order to induce conden-

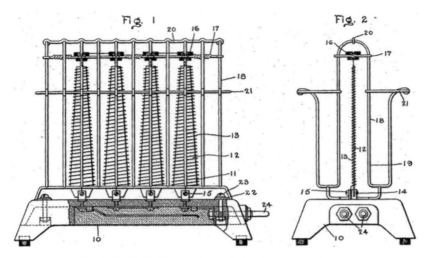

FIGURE 2.24. Frank Shailor's "electric heater," the first practical toaster whose U.S. Patent 950,058 was assigned to GE, which began its production in 1909. Its construction was made possible by a new nickel-chromium alloy; Shailor used John Dempster's (U.S. Patent 901,428) combination of 62% Ni, 20% Fe, 13% Cr, and 5% Mn for resistance wires.

sation. His first working system, with the cooling capacity of nearly 50 t of ice a day, was installed to control temperature and humidity for Brooklyn printing company Sackett-Williams. These controls eliminated slight fluctuations in the dimensions of their printing paper that caused misalignment of colored inks (Ingels 1952).

Carrier's first patent (U.S. Patent 808,897) for the "Apparatus for Treating Air" was granted in 1906, and in 1911 he completed *Rational Psychrometric Formulae*, which has become a lasting foundation of basic calculations required to design efficient air conditioning systems (ASHRAE 2001). Textile, film, and food processing industries were among the first commercial users of air conditioning, while the cooling for human comfort began only during the 1920s with department stores and movie theaters. The first air-conditioned American car was engineered in 1938, and in the same year Philco-York marketed the first successful model of a window air conditioner (NAE 2000). Widespread use of household air conditioning, and with it the large-scale migration to a suddenly comfortable Sunbelt, had to wait until the 1960s. By the end of the 20th century almost 70% of all U.S. households had either central or window air conditioning, and central units were being installed in virtually all new custom-built U.S. houses, regardless of their location.

While mechanical refrigeration for commercial operations (cold storages, ice making, meatpacking, dairies, breweries) was perfected during the 1880s, its large equipment, the need for manual operation by skilled personnel, and the use of potentially dangerous ammonia as the refrigerant did not make the technique readily transferable to household use. The first practical design for a kitchen unit, patented by Fred Wolf in 1913, used self-contained copper tubing with flared joints to reduce the risk of leaks—and it even produced ice cubes (Nagengast 2000). But the first affordable household refrigerator, General Motor's Frigidaire, became a common possession only during the late 1920s. Cost of refrigeration fell, and its safety rose, with the replacement of dangerous refrigerants by chlorofluorocarbons in 1931 (NAE 2000). But during the 1980s these gases were incontrovertibly implicated in the destruction of stratospheric ozone. Genesis, and solution, of this environmental concern will be covered in this book's companion volume. By the year 2000, only color televisions rivaled refrigerators as virtually universal household possessions in affluent countries.

The first portable electric vacuum cleaner, brought out in 1905 by Chapman & Skinner Co. in San Francisco, was an unwieldy machine weighing a little more than 40 kg with a fan nearly half a meter across to produce suction. The familiar upright machine (patented in 1908) was the invention of James W. Spangler. William Hoover's Suction Sweeper Co. began its production in 1907 and introduced its 0 model, weighing about 18 kg, a year later, and a high-speed universal motor for vacuum cleaners became available in 1909. The key features of the early Hoover arrangement are still with us: long handle, disposable cloth filter bag, and cleaning attachments. Electrolux introduced its first canister model in 1921. The first radical departure from the basic pre-1910 design came only with James Dyson's invention of nonclogging Dual Cyclone vacuum during the 1980s (Dyson 1998).

The age of electricity also brought new diagnostic devices and medical equipment whose basic modes of operation remain unchanged but whose later versions became more practical, or more reliable, to use (Davis 1981). The first dental drill powered by a bulky and expensive DC motor (replacing previous hand cranks, springs, and foot treadles) was introduced in 1874, and plug-in drills became available by 1908. A small battery was all that was needed to power the first endoscope, a well-lit tube with a lens for viewing the esophagus, designed in 1881 by a Polish surgeon Johann Mickulicz (1850–1905). At Buffalo exhibition in 1901, visitors could see for the first time an incubator for premature babies, and two years later Willem Einthoven (1860–1927) designed the first electrocardiograph (he was awarded the Nobel Prize for Physiology or Medicine in 1925), and the device has been manufactured for diagnostic use since 1909 (Davis 1981).

By far the most important diagnostic contribution that would have been impossible without electricity began with Wilhelm Conrad Röntgen's (1845–1923) experiments with electric discharges in highly evacuated glass tubes. That led, on November 8, 1895, to the discovery of x-rays (Brachner 1995). Their value as a diagnostic tool was appreciated immediately, and consequently, in 1901 when the President of the Royal Swedish Academy of Sciences was presenting Röntgen with the first Nobel Prize in Physics, he noted that they were already in extensive use in medical practice (Odhner 1901). By 1910 their use was standard in medical diagnostics in large cities of Western countries, and they also found many applications in science and industry.

Pre-WWI decades also set a number of basic management practices in the electric industry. Already in 1883 John Hopkinson (1849–1898) had introduced the principle of the two-part tariff method of charging for the consumed electricity, and the combination of a fixed (basic service) charge and a payment for the used quantity have been standard for more than a century. And almost immediately, the industry had to deal with the challenge of variable demand. As long as the electricity demand was dominated by lighting, the daily load factors of generating stations were very low, often no more than 8–10%, and in 1907 the average nationwide load for the United Kingdom was just 14.5%. That is why utilities were eager to see the installation of more electric motors, be they for lifting, in manufacturing, or for electric traction, and also looked for opportunities to reduce the large disparities between peak and average loads.

Electricity use for traction had begun on a small scale already in the early 1880s with the first electric tramway built by Werner Siemens in Berlin's Lichterfelde in 1881, and all major modes of electrically driven urban and intercity transport were introduced before 1913 (figure 2.25). The relevant first British dates were as follows (Electricity Council 1973): a short DC railway line in 1883 (on the beach at Brighton), tramway in 1885, underground railway 1890 (in London), main-line railway electrification 1895, AC railway line in 1908, and trolley buses in 1911 (but Werner Siemens had built the first trolley bus in Berlin already in 1882).

In the United States, the first electric streetcars were introduced by Frank J. Sprague (1857–1934) in Richmond, Virginia, in 1888 (cars lit with electric lights traveled at 24 km/h), and soon afterward in Boston (Cambridge). The subsequent diffusion of this innovation was very rapid: within five years 14 of the 16 U.S. cities with more than 200,000 people and 41 of 42 cities with populations between 50,000 and 200,000 had electric streetcars. By 1900, there were 1,200 such systems in operation or under construction; by 1902, 99% of the old horse-drawn systems were converted, and by 1908, more than 1,200 companies owned about 62,000 km of electrified track and elevated railways (Dyer and Martin 1929; Sharlin 1967).

As Hunter and Bryant (1991:206) put it, "it was a quick and complete revolution in urban transportation that touched lives of nearly all Americans living in cities," and, lighting excepted, it brought electricity's benefits to more people than any other application at that time. In 1892 Sprague introduced a

FIGURE 2.25. Electric streetcar between Frankfurt-am-Main and Offenbach had conductors of wrought iron (diameter of 3 cm) suspended by iron wires that were attached to telegraph poles. Identical arrangement was used in many other European cities. Reproduced from *Scientific American*, August 26, 1882.

multiple controller for operating several motors from the lead car at one time, opening up the applications for multicar trains and subways. Electric traction eventually spread from subways and commuter trains to high-speed rail links. Advances were rapid: already by 1902 Siemens & Halske was running an experimental four-motor polyphase car of 2.2 MW at speeds exceeding 160 km/h (Perkins 1902). All of today's fastest trains are electric. The French TGV, current speed record holder, draws 25 kV at 50 Hz from a fixed overhead wire; the current is then stepped down by massive transformers to 1,500 V, and it is supplied to synchronous 1.1 MW AC traction motors (TGVweb 2000).

In addition to transportation and industrial motors, electric welding, electrometallurgy, and electrochemical industries (particularly after Hall-Héroult process for producing aluminum was introduced in 1886) emerged as major industrial users of electricity. Electric arc welding with a carbon electrode was patented in the United Kingdom in 1885 by Nikolai Benardos and Stanislaw Olszewski, and seven years later Charles L. Coffin (1844–1926) got the U.S. patent for a metal electrode (Lebrun 1961). Advances in electrometallurgy were made possible by William Siemens's invention of electric furnace in 1878. A quantitatively small but qualitatively immensely important use of electricity is in the enormous variety of monitoring, and analytical devices in industry, science, health care, and, of course, millions of personal and mainframe computers.

As is always the case with technical innovations, diffusion of electricity was affected by national idiosyncrasies. Inertial reliance on the long-established steam power using cheap coal slowed down the British embrace of electricity, and the unfortunate Electric Lighting Act of 1882, which limited the operating licenses to seven years and gave the local authorities the right to take over the company assets after 21 years, created another disincentive that was remedied only by a new legislation in 1888. These two factors explain why the United States and Germany, rather than the United Kingdom, pioneered widespread applications of electricity. Statistics of electricity generation tell the story: by 1900, the national totals were (in GWh) 0.2 in the United Kingdom, 1.0 in Germany, and nearly 5 in the United States, and by 1914 the respective figures rose to 3.0, 8.8, and 24.8, which means that the United States per capita generation reached 250 Wh/year, compared to about 135 Wh in Germany and to just 70 Wh in the United Kingdom (USBC 1975; Mitchell 1998).

American and German cities were also far ahead of English urban areas in developing the symbiotic relationship between electricity and industrial production (Martin 1922; Hughes 1983; Platt 1991). By 1911 Chicago used more than 80% of its electricity for stationary and traction power; the analogical share was 66% in Berlin, but only 39% in London, where lighting still dominated the demand. And in both the United States and Germany, developments were disproportionately driven by just two companies: after the early 1890s it was General Electric and Westinghouse that introduced most of the technical advances in American electric industry, and AEG and Siemens had the same role in Germany (Strunk 1999; Feldenkirchen 1994).

American electrification efforts paid off rapidly in cities: by 1906 Boston's average per capita output of electricity was nearly 350 Wh/year, New York's almost 300 Wh/year, and Chicago's 200 Wh/year, compared to just over 40 Wh/year in London (Fleming 1911). But the country's large size delayed the access to electricity in smaller towns and particularly in rural areas: financial returns were either very low, or nil, given the high cost of extending the transmission lines from central stations. Only about 5% of potential customers had electricity in 1900; 10 years later still only 1 in 10 American homes was wired. The share surpassed 25% by 1918, and most of the urban households became connected only by the late 1920s (Nye 1990).

In technical terms, European and North American developments proceeded along very similar lines, as generating machinery was designed for more efficient performance with higher pressures and greater steam superheat, as established central stations grew by adding larger turbogenerators (often produced in large series of standard sizes), and as long-distance transmission steadily progressed toward higher AC voltages. Electric motors surpassed lights as the principal users of electricity in all industrializing countries (e.g., in the United Kingdom the crossover had taken place already before 1910), but Europe and North America retained distinct operating parameters for electricity distribution and its final use in households.

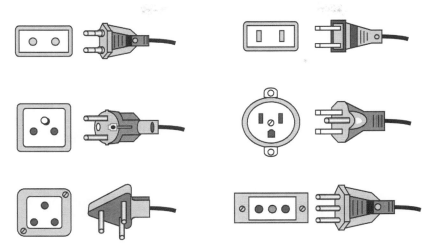

FIGURE 2.26. Major types of electric plugs used around the world. Simplified from a
more detailed illustration in USDC (2002).

Early polyphase systems operated with frequencies ranging from 25 to 133.33
Hz, but already by 1892 Westinghouse Co. had adopted just two rates, 25 Hz
for large-load supply and 60 Hz for general distribution. Accordingly, household and office supply in North America is now single-phase AC of 120 V
(maximum of 125 V) at 15 A and 60 Hz; the current oscillates 60 times every
second between ± 170 V, averaging 120 V with the high voltage supplied to
the smaller prong. In contrast, pre-WWI trend in the United Kingdom was
toward an increasing variation rather than standardization, and in 1914 London
had no less than 10 different frequencies and a multitude of voltages (Hughes
1983). The European standard eventually settled on single-phase AC of 220 V
at 50 Hz, but amperage is not uniform; for example, Italy uses 10 A or 16 A
(BIA 2001). Japanese ratings conform to the U.S. standard; Chinese, to the
European one.

But these disparities are nothing compared to the bewildering variety of
sockets (shaped as circles, rectangles, or squares), plugs (with flat or rounded
blades, and with or without grounding pins), and connectors used by households: worldwide there are 15 basic types of electric plugs and sockets with
different arrangements of flat and round pins (ITA 2002; figure 2.26). Unfortunately, there is little hope that this unhelpful diversity will be eliminated any
time soon; the only solution when using incompatible devices and plugs is to
buy appropriate transformers and adapters.

3

Internal Combustion Engines

> After supper my wife said, "Let's go over to the shop and try our luck once more . . ." My heart was pounding. I turned the crank. The engine started to go "put-put-put," and the music of the future sounded with regular rhythm. We both listened to it run for a full hour, fascinated, never tiring of the single tone of its song. The longer it played its note, the more sorrow and anxiety it conjured away from the heart.
>
> Karl Benz recalling the first successful run of his
> two-stroke engine on the New Year's Eve of 1879

Karl Benz (1844–1929) had a very good reason to be pleased as he listened to his two-stroke engine: his contract machine-building business was failing and a new gasoline-fueled engine, one that he designed and built in his free time, was to solve his difficulties. But neither he nor his wife Bertha (1849–1944)—who had more faith in his inventions that he did himself—could have foreseen the consequences of this experiment. By July 1886, Benz mounted a successor of that engine, still a low-powered but now a four-stroke machine, on a three-wheel chassis that was surmounted by two rather high-perched seats, an in-

FRONTISPIECE 3. Cover of the first catalog published by Benz & Cie. in 1888 shows a slightly modified version of the three-wheel vehicle patented and publicly driven for the first time in 1886. Courtesy of DaimlerChrysler Classic Konzernarchiv, Stuttgart.

FIGURE 3.1. Karl Benz. Photograph courtesy of DaimlerChrysler Classic Konzernarchiv, Stuttgart.

novation that made him eventually famous as the "inventor" of automobile (figure 3.1). Quotation marks are imperative because the motor car is one of the least appropriate artifacts to have its commercial introduction ascribed to a single inventor (Beaumont 1902; Walz and Niemann 1997).

But Benz was surely right about the music of the future. In a few generations its rhythm became ubiquitous, and there are now few other human creations that have been so admired and cherished and yet so disparaged and reviled as internal combustion engines. In these ingenious devices, steel pistons are tightly confined in cylinders as they are rapidly driven by mini-explosions of compressed fuel, and their frenzied reciprocating motion is then converted into smooth rotation that is transmitted to wheels or propellers. But whatever are our attitudes to these machines, their regular rhythm has come to define the tempo of modern civilization. Installed in cars, they bring choices and opportunities of personal freedom and mobility as well as environmental degradation and traffic congestion. They also enlarge the world of commerce as they power trucks, ships, and planes and, less visibly, as they energize many kinds of off-road machines that are now indispensable in mining and construction.

The process of introducing and diffusing internal combustion engines is an even better example of multifaceted, complex, synergistic origins of a major technical advance than is the formation of electric systems described in chapter 2. The new engines began their tentative careers as relatively inefficient coal-gas-fueled sources of stationary power, and as their performance improved they found many other useful applications besides eventually propelling machines of many kinds, the most demanding being civilian and military aircraft. Consequently, it is useful to separate the history of internal combustion engines into five stages, some being consecutive or partially overlapping, and others concurrent.

The first period embraces failed intermittent efforts to build explosively powered machines, a quest that goes back to the 17th century and whose pace picked up during the first half of the 19th century. This stage culminated with the first serious attempts to build steam-powered powered vehicles whose tests kept the idea of a self-propelled road machine in public consciousness: Walter Hancock's (1799–1852) series of quaintly named machines—Infant, Enterprise, Autopsy, Era, and Automaton—built during the early 1830s was particularly notable (Beaumont 1902). The steam engine has obvious disadvantages in road transportation, but it was not eliminated as a serious contender for powering passenger vehicles until the very end of the 19th century. Even some of the early proponents of powered flight tried to use it, an extraordinarily challenging task given its inherently high mass/power ratio.

The second stage in the development of internal combustion engines saw the construction of first commercially promising stationary machines powered by coal gas during the 1850s and their subsequent improvement, particularly thanks to the efforts of Nicolaus Otto and Eugen Langen. Their company, and its licensees, eventually produced thousands of relatively small horizontal engines for a variety of workshop and industrial uses. Initially, these were all noncompression and mostly double-acting machines, but Otto's pioneering 1876 design of a compression four-stroke engine proved immediately popular. This machine embodied key operating features of all modern engines, but it was not suitable for transportation. Although the design of Otto's engine was promptly patented, its priority, as I explain below, was later disputed.

Practical automotive designs emerged during the third stage with the development of gasoline-powered engine and with its use in the first carriagelike vehicles of the 1880s. This period is usually portrayed as a largely German invention of the automobile beginning with gasoline engines and motorized carriages built by Gottlieb Daimler and Wilhelm Maybach and, independently, by Karl Benz. But better engines are not the only precondition of a practical and reliable machine that is convenient and reasonably comfortable to use: a lengthy list of other requirements ranges from the overall machine design and good transmission of the generated power to cushioned wheels. These improvements began during the early 1890s with a fundamental redesign of motorized four-wheeled vehicles by Emile Levassor and continued with other important contributions made in France, Germany, the United States, and the United Kingdom. Another notable event of that intensively innovative period was Rudolf Diesel's (1858–1913) patenting of a new kind of engine, one that did not need either carburetor or a sparking device and that could use heavy oils rather than just the lightest liquid hydrocarbon fractions.

The fourth stage includes the rapid maturation of high-performance four-stroke engines and gradual emergence of road vehicles whose engine design and entire mechanical arrangement set the trends of automotive design for much of the 20th century. This stage also included the pioneering reciprocating engines for the first airplanes, whose subsequent development greatly advanced the performance, efficiency, and reliability of the new prime mover.

But if motorized vehicles were to claim a mass market, they could not be produced in the same way as they were during the first two decades of their existence, that is, in artisanal fashion either as unique items or in very small series by skilled (originally mostly bicycle) mechanics. Hence, the fifth key stage of the early development of internal combustion engines (and one that was largely concurrent with the fourth) was marked by the first steps toward highly efficient mass production of automotive engines and other car parts that eventually made the car industry the leading manufacturing activity of every large modern economy. Only a short lag separated these advances from a no less impressive large-scale production of aircraft.

Remarkably, nearly all of the basic challenges of durable design and affordable car manufacturing were resolved in a highly effective fashion before WWI. Today's cars do not obviously share outward features with their early 20th-century predecessors: low profile versus a tall, boxy appearance; enclosed versus open or semi-open bodies; curved versus angular body shapes; sunken versus free-standing headlights; and thick versus thin tires are just the most obvious differences. But as I show in this chapter, the evolution has been much more conservative under the hood. And while commercial long-distance flying ceased to rely on reciprocating engines during the 1960s, these efficient and highly reliable machines keep powering large numbers of smaller planes used for short-haul passenger transport, pastime flying, and tasks ranging from spraying crops to water-bombing of burning forests.

For decades we have taken internal combustion engines for granted, but they should still be a source of admiration if only because these affordable machines, now made up of many thousands of mostly precision-machined (or cast or extruded) parts, work so reliably for such long periods of time (Taylor 1984; Heywood 1988; Stone 1993). What is even more remarkable is that they do so requiring relatively little servicing while subject not only to various environmental extremes but also to abusive handling by users most of whom know nothing about their design or operating requirements. This combination of high performance and high reliability manifests even more admirably in aircraft engines, and the very reliability, durability, and affordability of internal combustion engines are major factors that work against their rapid replacement with alternatives that may be much less noisy and much less polluting but that do not have such a remarkable record of reliable service that is provided at such an acceptable cost.

Importance of internal combustion engines for smooth functioning of modern societies is taken entirely for granted in countries where their numbers match, and even surpass, the total numbers of population. There were about 275 million people in the United States in the late 1990s, but more than 210 million automotive and more than 100 million other internal combustion engines. The latter category includes more than 50 million engines in lawnmowers and other garden machines, about 17 million outboard and inboard engines in recreational boats, 4 million motorcycles, and nearly 2 million snowmobiles (NMMA 1999). About 20 million small internal combustion engines, ranging from units for ultralight

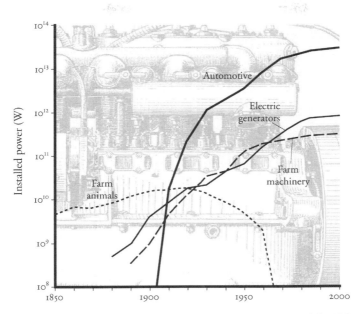

FIGURE 3.2. Long-term trends of the total power of main categories of the U.S. prime movers show that automotive internal combustion engines surpassed the aggregate power of both draft animals and electricity-generating equipment before 1910. Plotted from data in USBC (1975).

planes to emergency electricity generators, are now sold in the country every year. While the ubiquity of internal combustion engines is obvious, not too many people realize that in affluent countries their aggregate installed power is now considerably greater than that of any other prime mover.

Reliable U.S. data show that the total power of automotive engines had surpassed that of all prime movers used in electricity generation (steam engines, steam and water turbines) already before 1910 (USBC 1975). During the last decade of the 20th century, the total capacity of the U.S. vehicular internal combustion engines was more than 20 TW compared to less than 900 GW installed in steam and water turbines, or more than a 20-fold difference (figure 3.2). And a worldwide comparison shows that while in the year 2000 the total installed power of electricity-generating turbines was about 3.2 TW, the global fleet of road vehicles alone had installed power of at least 60 TW.

Multiple Beginnings

Cummins (1989:1) opens his meticulous history of internal combustion engines by noting, half seriously, that their development began with the invention of the cannon, a device whose key drawback was "that it threw away the piston on each power stroke." But this very action clearly demonstrated the potential utility of tamed explosion, and the first sketchy suggestions of engines exploiting that principle had appeared already in the late 17th century. But the first serious attempts to abandon steam as a working medium and to use hot gas instead date to the very end of the 18th century, and many noncompression (and hence inefficient) engines that used mixture of coal-gas and air were actually built before 1860. In this respect, the development of internal combustion engine resembles the early history of incandescent light, with more than a dozen inventors—from Philippe Lebon in 1801 to Eugenio Barsanti and Felice Mateucci in 1854—patenting and demonstrating their designs.

Two inventors should be singled out: William Barnett for proposing, in 1838, that the charged fuel should be compressed before ignition, and Isaac de Rivaz for actually installing his engine on a small wagon in 1807 and propelling it over a short distance at a speed of less than 5 km/h (Cummins 1989). But the first engine that caused a great deal of public interest and that was actually manufactured for sale was conceived only in 1858 and patented in 1860 by Jean Joseph Étienne Lenoir (1822–1900). Like a steam engine after which it was patterned and which it outwardly resembled, it was a horizontal, double-acting machine with slide valves to admit the mixture of illuminating gas and air and to release the burnt and expanded gas. Ignition was with an electric spark without any fuel compression.

A connecting rod joined the piston with a large flywheel to transmit the motion to its final use. This slow (about 200 rpm) engine had thermal efficiency of less than 4%, and fewer than 500 units, with typical rating of about 2 kW, were manufactured during the 1860s. They powered water pumps, printing presses, and other tasks requiring only limited amount of interruptible power. In 1862 a liquid-fuel version of the engine, mounted on a three-wheel cart and using a primitive carburetor, propelled the vehicle from Paris to Joinville-le-Point for a total distance of 18 km. This isolated trial had no follow-up, and it is not considered the real beginning of the automotive era. In fact, Lenoir did not pursue any further development of his engine after the 1860s.

Otto and Langen

Nicolaus August Otto (1832–1891; figure 3.3) was an unlikely candidate for revolutionizing the design of internal combustion engines (Langen 1919; Sittauer 1972). He was a traveling salesman for a wholesale food company (what Germans call *Kolonialwarengrosshandlung*)—but what he really wanted to do was to design a better engine. He became impressed by Lenoir's engine during its short spell of fame, and his experiments with its copy that a skilled mechanic built for him in 1861 led to the first Otto engine fueled with alcohol-air mixture. A better machine emerged only after Otto approached Eugen Langen (1833–1895), a well-off owner of a sugar-refining business, to invest in a newly established company (N. A. Otto & Cie.) that was formed in 1864.

The new company soon produced a two-stroke engine that resembled the Barsanti-Mateucci machine, but some of whose features were sufficiently different to be patented in 1866 in Germany and a year later in the United States. Its tall cylinder surmounted by a large five-spoked flywheel was heavy and noisy, but when it was displayed in

FIGURE 3.3. Nicolaus Otto, the inventor of eponymous two- and four-stroke gas-fueled internal combustion engines. Reproduced from Abbott (1934).

1867 at Paris Exhibition it was found to be more than twice as efficient as other featured gas engines. This earned it a Grand Prize, and the small company, suddenly beset with orders, began serial production in 1868. Once a new substantial investor was found, the whole enterprise was expanded and then in 1872 reorganized as Gasmotorenfabrik Deutz AG, named after a Cologne suburb where it was relocated. Today's Deutz AG remains one of the world's leading engine makers (Deutz 2003).

Gottlieb Daimler (1834–1900), an experienced engineer who worked previously also in France and England, was appointed as the production manager of N. A. Otto & Cie. Daimler brought along Wilhelm Maybach (1846–1929)—a draftsman whom he met for the first time in the machine workshop of Bruderhaus orphanage in Reutlingen where Maybach lived since the age of 10—as a new head of design department (LeMo 2003). The company's improved noncompression engine launched in 1874 had efficiency of about 10%, or as much as 75% higher than that of a comparably powerful Lenoir machine. But this performance was achieved by installing a nearly 4 m tall cylinder and letting the gas expand to 10 times its charged volume (compared to just 2.5 times in Lenoir's engine). Moreover, at about 900 g/W, the engine was much heavier than the best steam engines of the day.

Nevertheless, it was a commercial success, and hundreds of units were built every year during the latter half of the 1870s in Germany and under a license in France, the United Kingdom, and the United States. But as the company was engaged in numerous patent litigation trials and as many engineers, in Germany and abroad, were trying to come up with a design sufficiently different to warrant patenting, Otto felt that a much better machine was needed to preempt the competition. By the spring of 1876 he had such a design: a horizontal four-stroke compression engine that he conceived and had built not only against the prevailing wisdom of the time, but also without Daimler's help. The main objection appeared logical enough: why have just a single power stroke for four piston strokes and two full crankshaft revolutions when every stroke was a power stroke in double-acting engines?

Four-stroke operation was not the only innovation: Otto's first patents for the engine actually stressed more the novelty of stratified combustion, whereby gas and air mixture is preceded by air alone to create a lean charge near the piston and to produce a smoother, gradual burning of the fuel. Nothing came out of this proposition during Otto's time and for decades afterward, but the concept was revived during the 1970s as one of the means to lower emissions of combustion gases that act as precursors of photochemical smog. Otto's distrust of electric ignition led him to retain a small continuous ignition flame that was controlled by a sliding valve (figure 3.4).

This is how Otto's U.S. patent summarized the engine's key feature (Otto 1877:1):

> According to my present invention an intimate mixture of combustible gas or vapor and air is introduced into the cylinder, together with a separate charge of air or other gas, that may or may not support combustion, in such a manner and in such proportions that the particles of the combustible gaseous mixture are more or less dispersed in an isolated condition in the air or other gas, so that on ignition, instead of an explosion ensuing, the flame will be communicated gradually from one combustible particle to another, thereby effecting a gradual development of heat and a corresponding gradual expansion of the gases, which will enable the motive power so produced to be utilized in the most effective manner.

Under Maybach's direction the new design was improved and turned rapidly into a series of production models as both the company's tests and independent appraisals confirmed the engine's superior performance. Although its overall thermal efficiency was basically the same as for the noncompressing engine (about 17%), the new design reduced the piston displacement by 94% and mass/power ratio by nearly 70%. And although the engines were still quite heavy (at best, 200 g/W), they were now lighter than comparably sized small steam engines. Moreover, they could be built with much higher maximum capacities than the original Otto & Langen machines, and they were also much

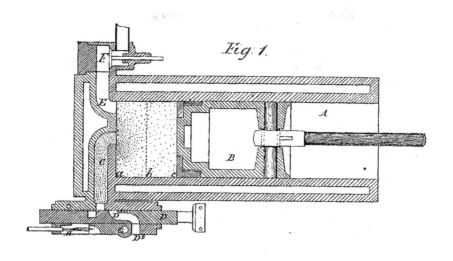

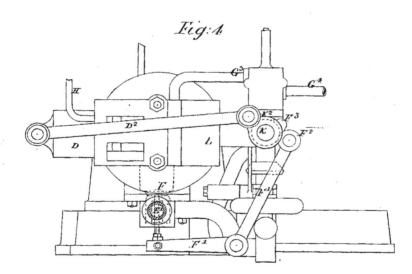

FIGURE 3.4. Cross section of a cylinder (with continuous ignition flame on the lower right) and overall arrangement of Otto's gas-fueled four-stroke engine patented in the United States in 1877. This illustration accompanies U.S. Patent 194,047.

quieter and vibrated much less. Otto's horizontal four-stroke engines were targeted for workshops too small for installing their own steam engine, and about 8,000 of these machines, with ratings ranging from 375 W to 12 kW, were sold by the end of the 1880s. Eventually, nearly 50,000 units with a combined capacity of about 150 MW (i.e., about 3 kW per engine) were made over the period of 17 years.

Specifications for Otto's typical early four-stroke machine were compression ratio about 2.6, air-to-fuel ratio 9:1, bore 210 mm, and stroke length 350 mm for a 6-kW engine running at 160 rpm, and they were still quite heavy at more than 250 g/W (Clerk 1909). The fundamental operating principle of Otto's invention has not changed, but the inventor would be impressed by much improved performance of today's identically powerful engines. A horizontal Honda, model GX240K1, commonly used as portable generator of electricity, provides an excellent example (Honda Engines 2003). This engine runs much faster (3,900 rpm), has a much higher compression ratio of 8.2:1 (hence a much smaller bore and stroke, 73 mm and 58 mm), and is much lighter (slightly more than 4 g/W). And, of course, it is fueled by gasoline and not by illuminating gas made from coal. But before I describe the rapid improvements of gasoline engines after their public demonstration on first clumsy carriages in 1886, I must take a short detour and clarify the origins of the four-stroke engine by looking at the role of Beau de Rochas.

Beau de Rochas and Otto: Ideas and Machines

Success of Otto's four-stroke engine led to many attempts to challenge its patent rights, but all of them were unsuccessful until a chance discovery of a forgotten patent. This claim was filed by a French engineer Alphonse Eugène Beau (1815–1893, who later in life styled himself Beau de Rochas) with the Société de Protection Industrielle on January 16, 1862 (No. 52,593) under an expansive title *Nouvelles recherches et perfectionnements sur les conditions pratiques de la plus grande utilisation de la chaleur et en général de la force motrice, avec application aux chemins de fer et à la navigation* (Payen 1993). The patent contained no drawings, and it was not based on any experimental model— but, undeniably, it detailed the principle of a four-stroke engine powered by a gas-air mixture that is compressed before its combustion.

Beau de Rochas did so by prescribing first the four parameters "for perfectly utilising the elastic gas in an engine" (citing from Donkin's [1896:432–433] translation):

1. The largest possible cylinder volume with the minimum boundary surface.
2. The greatest possible working speed.

3. Greatest possible number of expansions.
4. Greatest possible pressure at the beginning of expansion.

As for the sequence (shown in figure 3.5 by both piston movement and pressure-volume diagram), "the following operations must then take place on one side of the cylinder, during one period of four consecutive strokes:

1. Drawing in the charge during one whole piston stroke
2. Compression during the following stroke.
3. Inflammation at the dead point, and expansion during the third stroke.
4. Discharge of the burnt gases from the cylinder during the fourth and last stroke."

Because Beau de Rochas failed to pay the requisite fee, the patent was never published, and it circulated in just a few hundred copies distributed by the inventor. He also did not try to assert his rights when Otto's engine first appeared on the market, or for many years afterward. Only in 1884, after a patent attorney called attention to Beau's old private pamphlet did the Körting brothers, builders of large gas engines in Hannover, contest the originality of Otto's idea. Although it was obvious that Otto's engine was designed without any reference to Beau's obscure description, he eventually lost the legal fight in Germany in 1886, but his rights were upheld in the United Kingdom. More than a century later, the split persists: Deutz AG and German engineers, as well as most British and American automakers, talk about the Otto cycle, but

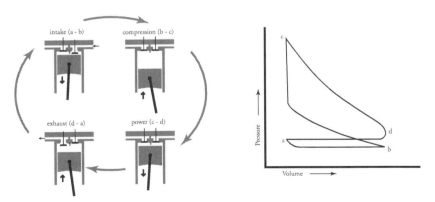

FIGURE 3.5. Idealized Otto cycle (with heat supplied and exhausted at constant volume) and piston movements corresponding to the four phases of the pressure-volume diagram.

in France it is *cycle de Beau de Rochas*. Whatever the name, the four strokes—*aspiration, compression, inflammation, refoulement* in Beau's language, *Ansaugen, Verdichten, Arbeiten, Ausschieben* in Otto's tongue—have become one of the ruling rhythms of modern civilization.

But the Beau-Otto affair goes far beyond the common patent interferences and usual complications and revisions in dating inventions, in this case pushing back the beginnings of four-stroke internal combustion engine to 1862 from 1876. Indeed, it goes to the very core of the concept of the Age of Synergy I advocate in this book. Beau de Rochas was content to set down his ideas on paper and then walk away from them, although he had opportunities to do otherwise: he was Lenoir's friend, since 1852 he lived and worked in Paris in the government's transportation department, and he was engaged in many projects that involved propulsion machinery.

I would argue that if Beau de Rochas were to get this priority, then Leonardo da Vinci might as well be regarded as the creator of helicopter on the basis of his famous sketch in *Manuscript B*. That brilliant Florentine did not have technical means (materials, prime mover) to convert this, and many other of his bold ideas, even into simple models, much less into real and practical machines. In contrast, Beau could have called on many technical advances available by the 1860s in order to construct at least a working prototype—but he simply chose not to do so. He made designs for railways and ship docks and planned the use of Lake Geneva for hydroelectric generation, but he did not try to convert his 1862 idea of a four-stroke engine even into a toylike model.

In contrast, Otto's approach was the very quintessence of the Age of Synergy. First, he developed a noncompression engine better than Lenoir's inefficient machine; then, working against the dominant consensus, and even against the judgment of his experienced production manager, he went on, entirely independently, to develop a much better four-stroke compression engine, patented it properly and promptly, and then committed resources to its further improvement, started its manufacturing in large series, and licensed it widely in his country and abroad. This is the very procedure whose numerous repetitions eventually created the civilization of the 20th century. Not surprisingly, Otto's inventive élan was sapped by the German denial of his four-stroke patent. His last innovation was the first low-voltage magneto ignition he designed in 1884, two years after the University of Würzburg awarded him, together with Alexander Graham Bell, an honorary doctorate. He died before his 59th birthday in Cologne, the town where he spent nearly all his adult life.

German Trio and the First Car Engines

The next critical stage in the history of internal combustion engines continued to unfold in Germany, and we have already met all three protagonists, Benz,

Daimler, and Maybach. Karl Friedrich Benz (1844–1929; figure 3.1), was born in Karlsruhe to a family whose men were smiths for several generations. After his technical studies, Benz set up an iron foundry and a mechanical workshop, and in 1877, with his main business struggling, he began to build stationary engines. Gottlieb Daimler (figure 3.6) worked for Otto & Langen (later Deutz AG) between 1872 and 1882 and left the company after disagreements with Otto. And Wilhelm Maybach, whom Daimler recruited to N. A. Otto & Cie. in 1873, improved designs of both Otto's two patented engines and brought them to commercial production. Independent efforts of these three men had eventually resulted in the first practical high-speed four-stroke gasoline-fueled engines.

FIGURE 3.6. Gottlieb Daimler. Photograph courtesy of DaimlerChrysler Classic Konzernarchiv, Stuttgart.

After leaving Deutz at the end of 1881, Daimler, soon joined by Maybach, set up a shop in the *Gartenhaus* on his comfortable property in Canstatt, Stuttgart's suburb, and set out to develop a powerful, yet light-weight engine fueled by gasoline (Walz and Niemann 1997). Gasoline, by that time readily available as the first volatile product of simple thermal refining of American and Russian crude oil, was an obvious choice. With about 33 MJ/L (nearly 44 MJ/kg), the fuel has about 1,600 times the energy density of the illuminating gas (which could be used in transportation only if highly compressed), and its low flashpoint ($-40°C$) makes it ideal for easy starting. At the same time, this very low flash point also makes it very hazardous to use. By 1883 the two engineers had a prototype of a high-revolution (about 600 rpm) gasoline-fueled engine with a surface carburetor and inflammation-prone hot-tube ignition (heated by an external gasoline flame) that caused repeated flare-ups.

In 1885 Daimler and Maybach fastened a small (about 350 W) and light, air-cooled version on a bicycle, creating a prototype of the motorcycle, which was driven for the first time on November 10, 1885, by Daimler's son Paul just 3 km from Canstatt to nearby Untertürkheim (Walz and Niemann 1997). Figure 3.7 shows this, still a rather unwieldy, machine as well as its subsequent transformation into a modern-looking motorcycle with electric engine starter that became fairly popular just before WWI (O'Connor 1913). Going a step

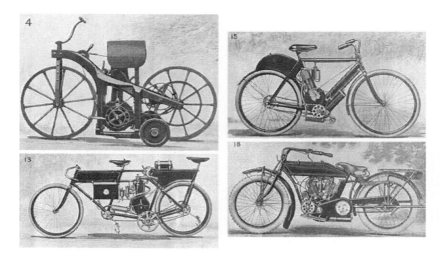

FIGURE 3.7. Daimler and Maybach's unwieldy 1885 motorcycle with subsidiary wheels compared to a tandem machine of 1900 and to standard motorcycles of 1902 and 1913. Reproduced from O'Connor (1913).

further, in March 1886 Daimler ordered a standard coach with four wooden wheels from a Stuttgart coachmaker and mounted on it a larger (0.462 L, 820 W), water-cooled version of the engine (figure 3.8).

Daimler and Maybach's inelegant vehicle, with passengers precariously perched high above the ground, made its first rides in Daimler's garden, a trip from Canstatt to Untertürkheim sometime during the fall of 1886, and the first extended journey in 1887 from Cannstatt to Stuttgart at about 18 km/h. Their early engine had only a single opening for inlet of the charge and the exhaustion of gases into the cylinder. Gasoline was supplied from the float chamber to carburetor, and the engine was cooled by water. Remarkably, their achievements were being independently anticipated and duplicated by Karl Benz working in Mannheim, about 120 km northwest from their Canstatt workshop (Walz and Niemann 1997).

On January 29, 1886, Karl Benz was granted the German patent (D.R. Patent 37,435) for a three-wheeled vehicle powered by a four-stroke, single-cylinder gasoline engine; this date is generally considered as the birth certificate of the first automobile. This was Benz's goal all along: once he decided in the late 1870s that engine construction is the best way to save his business, he concentrated, as already noted, on developing a two-stroke engine that he intended from the very beginning to use in vehicles. As described in this chapter's epigraph, he was successful just before the end of 1879. By 1882 he had fairly reliable two-stroke gasoline-fueled, water-cooled horizontal engine

Daimler-Motorwagen, 1886

FIGURE 3.8. Daimler and Maybach's first motor car. Their engine was installed in a coach ordered from Wimpf & Sohn in Stuttgart for the 43rd birthday of Daimler's wife Emma Pauline. Image courtesy of DaimlerChrysler Classic Konzernarchiv, Stuttgart.

with electric ignition, and in 1883, after securing an investor, a new company was finally set up. After the expiry of Otto's patent, Benz began designing four-stroke gasoline engines.

The first three-wheeled motorized carriage was driven publicly for the first time on July 3, 1886, along Mannheim's Friedrichsring. The light vehicle (total weight was just 263 kg, including the 96-kg single-cylinder four-stroke engine) had a less powerful (0.954 L, 500 W) and a slower-running (just 250 rpm) engine than did Daimler and Maybach's carriage, and it could go no faster than about 14.5 km/h. The vehicle's smaller front wheel was first steered by a small tiller and then by a horizontal wheel; the engine was placed over the main axle under the narrow double seat, and drive chains were connected to gears on both back wheels (see the frontispiece to this chapter). The car had to be started by turning clockwise a heavy horizontal flywheel behind the driver's seat, but water cooling, electric coil ignition (highly unreliable), spark plug (just two lengths of insulated platinum wire protruding into the combustion chamber), and differential gears were the key components of Benz's design that are still the standard features in modern automobiles.

The first intercity trip of Benz's three-wheeler took place two years later, in August 1888. Benz's wife Bertha took the couple's two sons and, without her husband's knowledge, drove the three-wheeler from Mannheim to Pforzheim to visit her mother, a distance of about 100 km (DaimlerChrysler 1999). After their arrival, they sent a telegram telling Benz about the completion of their pioneering journey. That trip was made in a vehicle with a backward-facing third seat that was placed above the front wheel, and by 1889 other additions included body panels, two lamps, and a folding top. But these machines were seen largely as curious, if not ridiculous, contrivances, and before 1890 only a few three-wheelers were actually made for sale, and the same was true about the 3-hp Victoria, Benz's first four-wheeler, which went on sale only in 1893.

Meanwhile, in 1889 Daimler and Maybach introduced a new two-cylinder V engine (angled at 17°) that displaced 0.565 L and ran at 920 rpm. Although they mounted the engine on a better chassis and used steel wire, rather than wooden spokes, the vehicle with larger back wheels still retained a decidedly carriagelike look. But the design pioneered the modern concept of power transmission through a friction clutch and sliding-pinion gears. A year later, in November 1890, when new partners were brought in and the Daimler Motoren Gesselschaft (DMG) was set up, they produced their first four-cylinder engine. At this point, their engines had much more power than needed to propel a carriage, but Daimler and Maybach did not take that step: after disagreements with new partners, they left the company and turned to a collaborative design of new engines that produced a new two-cylinder Phönix before they returned to the DMG in 1895.

And so it was a French engineer, Emile Levassor (1844–1897), who designed the first vehicle that was not merely a horseless carriage. As a partner in Panhard & Levassor, he was introduced to the German V engine by Daimler's representative in Paris, and in 1891, working for Armand Peugeot (1849–1915), he designed an entirely new chassis. Cars with Daimler and Maybach's engines took the four out of the first five places in the world's first car race in July 1894 (a steam-powered De Dion & Bouton tractor beat them). A year later—when Levassor himself drove his latest model to victory in the Paris—Bordeaux race (roundtrip of nearly 1,200 km) at average speed of 24 km/h (Beaumont 1902)—the car's Daimler-Maybach 4.5-kW engine weighed less than 30 g/W, an order of magnitude lighter than Otto's first four-stroke machines.

A momentous event in the company's history took place on April 2, 1900, when Emil Jellinek (1853–1918), a successful businessman and the Consul General of the Austro-Hungarian Empire in Monaco, set up a dealership for Daimler cars (Robson 1983). He placed the initial order for 36 vehicles worth 550,000 Goldmark, and then doubled it within two weeks. His condition for this big commitment: exclusive rights for selling the cars in the Austro-Hungarian Empire, France, Belgium, and the United States under the Mercédès (the name of Jellinek's daughter) trademark. Maybach responded to this opportunity by developing a design that Mercedes-Benz (2003) keeps describing as the first true automobile and a technical sensation and that Flink (1988:33) called the "first modern car in all essentials."

Being basically a race car, the Mercedes 35 had an unusually powerful, 5.9-L, 26-kW (35 hp) four-cylinder engine running at 950 rpm with two carburetors and with mechanically operated inlet valves; the vehicle had a lengthened profile and a very low center of gravity (DaimlerChrysler 2003). Shifting was done with a gear stick moving in a gate. What was under the hood was even more important: Maybach reduced the engine's overall weight by 30% (to 230 kg) by using an aluminum engine block, and the much greater cooling surface of his new honeycomb radiator (a standard design that is still with us) made it possible to reduce the coolant volume by half. Consequently, the mass/power ratio of this powerful engine was reduced to less than 9 g/W, 70% below the best DMG engine in 1895, and the vehicle's total body mass was kept to 1.2 t.

Within months the new car broke the world speed record by reaching 64.4 km/h, and a more powerful Mercedes 60, with unprecedented acceleration and more elegant bodywork, was introduced in 1903. This combination of performance and elegance, of speed and luxury, has proved to be the most endurable asset of the marque: the company advertised it as much in 1914 as it does today, and it continues to charge the buyers a significant premium for this renown (figure 3.9). Introduction of the Mercedes line could be seen as the beginning of the end of Germany's automotive *Gründerzeit*. During the first decade of the 20th century, the center of automotive development shifted to the United States as all three German protagonists left the automobile business. Daimler died in 1900, Benz left his company in 1906, and Maybach left DMG in 1907.

Before leaving this early history of German internal combustion engines, I should clarify the subsequent joining of Daimler's and Benz's names. In 1926 Benz & Cie. and DMG joined to form Daimler-Benz AG, uniting the names and traditions of the two great pioneers of internal combustion and automaking. Interestingly, the two men who lived for decades in cities that were just a bit more than 100 km apart never met during their long lives. And in 1998 Daimler-Benz AG bought Chrysler, America's fourth largest automaker (ranking behind General Motors, Ford, and Honda) to form DaimlerChrysler (Vlasic and Stertz 2000). That I have said so far nothing about the U.S. automakers has a simple explanation. As the leading British expert put it (Beaumont 1906:268), "progress in the design and manufacture of motor vehicles in America has not been distinguished by any noteworthy advance upon the practice obtaining in either this country or on the Continent." Matters changed just two years after this blunt and correct assessment with the arrival of Ford's Model T.

FIGURE 3.9. Ludwig Holwein's elegant pre-WWI advertisement for a Mercedes. Courtesy of DaimlerChrysler Classic Konzernarchiv, Stuttgart.

The Diesel Engine

Rudolf Diesel (1858–1913), the man who invented a different internal combustion engine, set out to do so as a 20-year-old student, eventually succeeded (although not in the way he initially envisaged) nearly 20 years later, but ended his life before he could witness his machine in all kinds of road and off-road vehicles as well as in all modes of shipping (figure 3.10). Diesel's account of his invention, written in 1913, explains how the idea of a better engine dominated his life since he was a university student (Diesel 1913:1–2):

> When my esteemed teacher, Professor Linde, explained to his audience at the Munich Polytechnic in 1878, during his lecture on thermodynamics, that the steam-engine transforms into effective work only 6–10% of the heat value of its fuel, when he explained Carnot's theorem, and pointed out that with isothermic changes of condition all heat conducted to a gas would be converted to work, I wrote in the margin of my college notebook: "Study whether it is not possible to realize in practice the isotherm?" Then and there I set myself the task! That was still no invention, not even the idea of one. Thereafter the wish to realize the ideal Carnot cycle dominated my existence.

Diesel's invention belongs to the same category as Charles Parsons's steam turbine: both had their genesis in understanding the laws of thermodynamics. Consequently, both of these developments are excellent examples of fundamental innovations that appeared for the first time only during the Age of Synergy: inventions driven and directed by practical applications of advanced scientific knowledge, not discoveries arrived at random, or by stubborn tinkering and experimenting. There is, however, an important difference: Parsons had realized his initial goals almost perfectly. Diesel's main objective, inspired by Carl Linde's (1842–1934) lectures, proved to be technically impossible, and he had to depart fundamentally from his initial idea of a near-ideal engine, but he still designed a prime mover of unprecedented efficiency.

Diesel also had an explicit social goal for his machine that he described in his famous programmatic publication (Diesel 1893): he wanted to change the fact that only large factories were able to invest in large efficient steam engines by in-

FIGURE 3.10. Rudolf Diesel, who did not succeed in designing a near-Carnot cycle engine, but whose new machine became the most efficient means of converting chemical energy of liquid fuels into motion through internal combustion. Reproduced from *Scientific American*, October 18, 1913.

troducing a prime mover whose affordable operation would make small businesses competitive. After his studies in Munich, Diesel became employed by Linde's ice-making company in Paris in 1880 and spent most of the 1880s in perfecting and developing various refrigeration techniques, including an ammonia absorption motor in 1887 (Diesel 1937). After several years of experiments, he filed a patent application for a new internal combustion engine in February 1892, and in December of the same year he was granted the first German patent (Patent 67,207) for the machine.

Thanks to the interest by Heinrich Buz (1833–1918), director of the Maschinenfabrik Augsburg, and with financial support by Friedrich Krupp, the difficult work of translating the idea into a working device began soon afterward. The engines were to follow the four-stroke Otto cycle, but their operating mode was to be unique. Air alone was to be drawn into the cylinder during the charging stroke, and then compressed to such an extent (Diesel's initial specifications called for as much as 30 MPa) that its temperature alone (between 800°C and 1,000°C) would ignite a liquid fuel as it entered the cylinder, where it would burn without exceeding the pressure of air and oil injection.

But, as Diesel was eager to stress, to claim that the self-ignition of the fuel is the most important attribute of his design would be an incorrect and superficial view: it would be absurd to build a much heavier engine merely for the sake of ignition when the machine is cold, because once the engine is warmed up self-ignition takes places at lower pressures. The engine's principal goal was the highest practical efficiency, and this necessitated the use of compressed air (Diesel 1913). Diesel's initial goal was to achieve isothermal expansion: after adiabatic compression of the air, the fuel should be introduced at such a rate that the heat of combustion would just replace the heat lost due to the gas expansion, and hence all thermal energy would be converted to kinetic energy and the temperature inside the cylinder would not surpass the maximum generated during the compression stroke.

As the conditions close to isothermal combustion could not be realized in practice, Diesel concluded in 1893 that combustion under constant pressure was the only way to proceed (Diesel 1893). But even then, it took more than three years of intensive technical development, and steady downward adjustments of operating pressure, before a new engine (German Patent 86,633, U.S. Patent 608,845) was ready for official acceptance testing. This took place in February 1897, and although the pressure—volume diagram bore little resemblance to Carnot's ideal cycle, the engine was still far more efficient than any combustion device (figure 3.11). Official testing results showed that the 13.5-kW engine, whose top pressure of 3.4 MPa was an order of magnitude below Diesel's initial maximum specification, had thermal efficiency of 34.7% at full load, corresponding to 26.2% of the brake thermal efficiency (Diesel 1913).

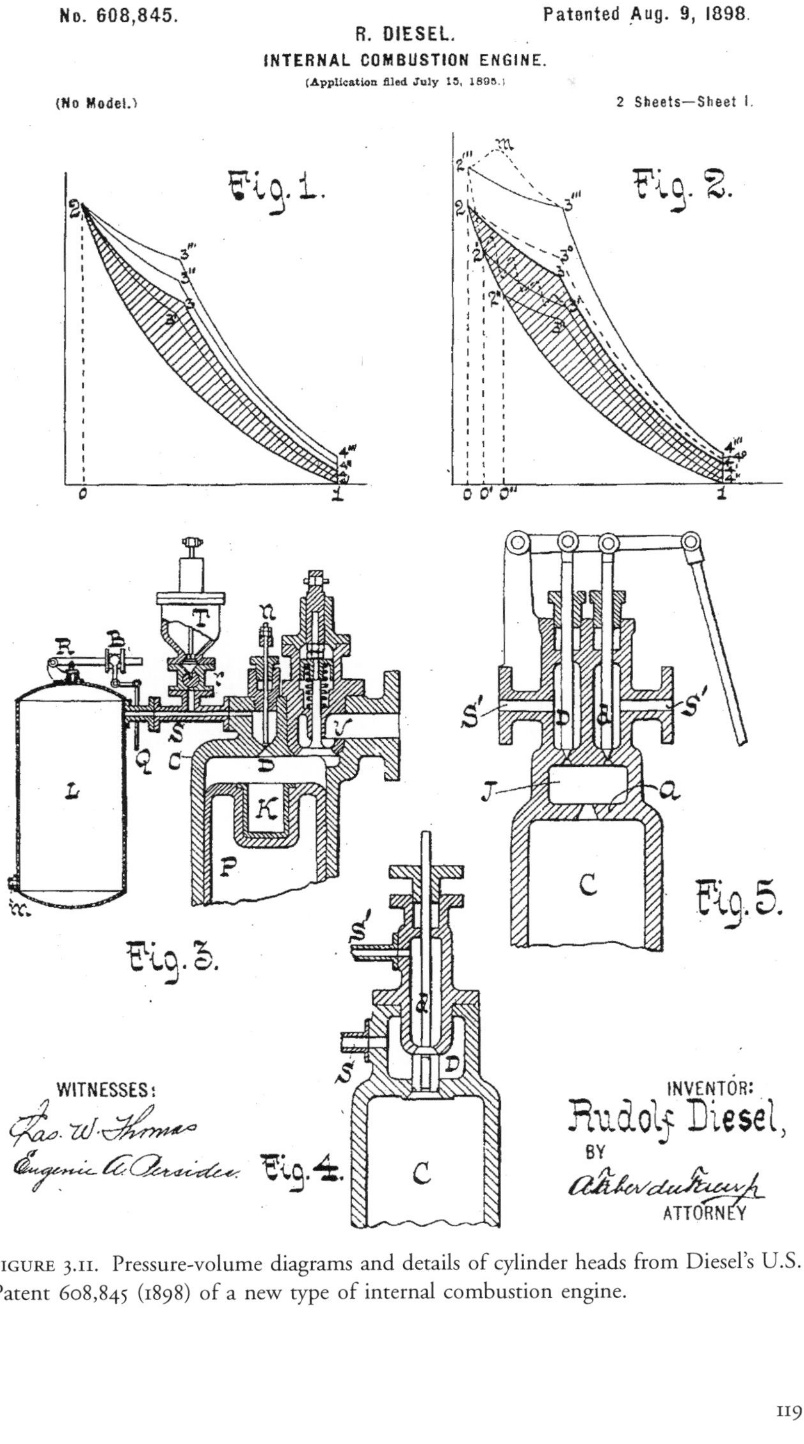

R. DIESEL.
INTERNAL COMBUSTION ENGINE.
(Application filed July 15, 1895.)

(No Model.)

2 Sheets—Sheet I.

Fig. 1.

Fig. 2.

Fig. 3.

Fig. 5.

Fig. 4.

WITNESSES:
Jas. W. Thomas
Eugenie A. Aersider.

INVENTOR:
Rudolf Diesel,
BY
Albert du Pury
ATTORNEY

FIGURE 3.11. Pressure-volume diagrams and details of cylinder heads from Diesel's U.S. Patent 608,845 (1898) of a new type of internal combustion engine.

By the end of 1897, brake thermal efficiency reached 30%, and 15 years later a much larger (445 kW) diesel engine had mechanical efficiency 77% and brake thermal efficiency 31.7% (Clerk 1911). In contrast, typical brake efficiencies of commonly deployed Otto engines at that time were between 14% and 17%. Diesel's dream of a Carnot-like engine was gone, but expectations for a machine whose efficiency was still twice as high as that of any other combustion device were high. Licensing deals with potential manufacturers were struck soon after the acceptance testing, but actual commercialization of diesel engines ran into problems associated above all with the maintenance of high pressure and with the timing of fuel injection. The latter challenge was solved satisfactorily only in 1927 when, after five years of development, Robert Bosch (1861–1942) introduced his high-precision injection for diesel engines.

Moreover, Diesel disliked to be engaged in the commercialization phase of his inventions, and he also judged many producers incapable of building his splendid machine properly. And like other inventors, he had to face the inevitable lawsuits contesting his invention, the first one already in 1897. Given the engine's inherently higher mass/power ratio, its first applications were in stationary uses (pumps, oil drills) and in heavy transport. A French canal boat was the first diesel-powered ship in 1903; a French submarine followed in 1904. A small electricity-generating plant to supply streetcars in Kiev was commissioned in 1906, and the first ocean-going vessel with diesel engines was the Danish *Selandia* in 1912. The first diesel locomotive, with engines by Sulzer, was built in 1913 for the Prussian Hessian State Railways (Anonymous 1913a). Four cylinders placed in the locomotive's center in 90° V configuration could sustain speed of 100 km/h. By that time there were more than 1,000 diesel engines in service, most of them rating 40–75 kW. A small diesel truck was built in 1908, but passenger diesel vehicles had to wait for Bosch's injection pump.

In 1913 Johannes Lüders, a professor at the Technische Hochschule in Aachen, published his *Dieselmythus*, a work highly critical of Diesel's accomplishments (Lüders 1913). That attack, initially less than enthusiastic commercial acceptance of his engines, slow pace of developing them for the automotive market, prolonged anxiety and illness (including a nervous breakdown), excessive family spending and resulting financial problems—any one of these factors makes it most likely that Diesel's disappearance was not an accident but a suicide. On September 29, 1913, he boarded a ship from Belgium to England, where he was to attend the opening of a new factory to produce his engines. Next morning he was not aboard the ship.

If Diesel had lived as long as Benz or Maybach (who died when 85 and 83 years old, respectively), he would have witnessed a universal triumph of his machine. Diesel's conviction that his engine will eventually become a leading prime mover of road vehicles began to turn into reality just 10 years after his death. But his accomplishment was indisputable even without that achievement. As Suplee (1913:306) noted in a generous obituary, Diesel's engine "is now known all over the world as the most efficient heat motor in existence, and the greatest advance in the generation of power from heat since the invention of the separate condenser by Watt."

Creating Car Industry and Car Culture

The early history of car industry has inevitable parallels with the creation of an entirely new system for the generation, distribution, and conversion of electricity. In both instances, the early progress was slow as there were well-established alternatives relying on extensive service infrastructures: gas lighting and steam power in the case of electricity, and steam-powered trains for transport over longer distance and horse-drawn urban vehicles and newly introduced electric streetcars and trains in cities in the case of automobiles. This slow pace of diffusion and acceptance of automobiles is reflected by small numbers of vehicles in use in the United States. There were mere 300 of them in 1895; in 1900, 17 years after Daimler and Maybach built their first gasoline engine, the total was just 8,000, which means that only one of out of every 9,500 Americans owned a motor vehicle; and by 1905 the nationwide registration total stood at just short of 78,000 cars (USBC 1975).

Another telling way to demonstrate this reality is to note that during the 1890s the *Scientific American*, a leading source of information on technical advances, kept paying a great deal of attention to what were by that time only marginal improvements in steam engine design while its coverage of emerging car engineering was rather unenthusiastic. Similarly, Byrn's (1900) systematic review of the 19th century inventions devoted only as many pages (7 out of 467) to automobiles as it did to bicycles. But there was also an obvious price difference: even relatively expensive early lights were much more affordable than the first-generation automobiles manufactured in small series by artisanal methods. And while electric lights and fans or irons were ready to use as soon as a household was connected to the central supply, driving on unpaved roads was difficult, and unreliability of early vehicle designs and a near complete absence of any emergency services turned these dusty (or muddy) trips into unpredictable adventures.

Three cartoons from *Punch* capture the essence of these experiences (figure 3.12). Travails of automotive pioneers of the 1890s were also vividly captured by Kipling's recollection of "agonies, shames, delays, rages, chills, parboilings, road-walkings, water-drawings, burns and starvations" (quoted in Richardson 1977:27). And there were still other, unprecedented, concerns. Matter of the public safety was addressed for the first time in 1865 in relation to heavy steam-

FIGURE 3.12. In the first year of the 20th century *Punch* carried many cartoons spoofing the new automotive experience. In the top image, a farmer calls, "Pull up, you fool! The mare's bolting!" "So's the car!" cries the motorist. The bottom left illustration shows "the only way to enjoy a motor-car ride through a dusty country," and the bottom right is not "a collection of tubercular microbes escaping from the congress but merely the Montgomery-Smiths in their motor-car, enjoying the beauties of the country." Reproduced from *Punch* issues of June 12, June 19, and July 31, 1901.

powered traction by the justly ridiculed British legislature, which required that three people driving any "locomotives" on highways and the vehicles, traveling at the maximum speed of 6.4 km/h, be preceded by a man with a red flag (Rolls 1911). Incredibly, these restrictions were extended in 1881 to every type of self-propelled vehicle, and they were repealed only in 1896. No other country had such retarding laws, but concerns about the speed of new machines and about accidents were seen everywhere. At the same time, nothing could erase a growing feeling that something profoundly important was getting underway.

In November 1895, the opening editorial of the inaugural issue of American *Horseless Age* had no doubts that "a giant industry is struggling into being here. All signs point to the motor vehicle as the necessary sequence of methods of locomotion already established and approved. The growing needs of our civilization demand it; the public believe in it" (cited in Flink 1975:13). But first the slow, fragile, unreliable, and uncomfortable vehicles with obvious carriage and bicycle pedigrees had to evolve into faster, sturdier, reliable, and much more comfortable means of passenger transportation. Although European engineers had accomplished much of this transformation by the beginning of the 20th century, without mass production these technically fairly advanced vehicles would have remained beyond the reach of all but a small segment of populations. Henry Ford took that critical step beginning in 1908, so by 1914 the two factors that were so important in shaping the 20th century—car making as a key component of modern economies and car ownership as a key factor of modern life—were firmly in place.

Technical Challenges

Virtually unchallenged dominance that the road vehicles powered by internal combustion engines had enjoyed during the 20th century was achieved within a single generation after their first demonstrations—but their fate appeared unclear as late as 1900. Decades of experience with high-pressure steam engines made it possible to build some relatively light-weight machines that won a number of car races during the 1890s. Among new steam-powered machines that were designed after 1900 was Leon Serpollet's beak-shaped racer that broke the speed record for 1 km with 29.8 s (equivalent to 120.8 km/h) in Nice in 1902, and Francis and Freelan Stanley's steam car that set a new world speed record for the fastest mile (205.4 km/h) in 1906. But after that steam-powered passenger cars went into a rapid decline, although heavy commercial vehicles stayed around for several more decades (Flink 1988).

And Edison was far from being alone in believing that electric cars will prevail. Late 1890s and the first years of a new century looked particularly promising for those vehicles. In 1896, at the first U.S. track race at Narangansett Park in Rhode Island a Riker electric car decisively defeated a Duryea vehicle, and three years later in France another electric car, a bullet-shaped *Jamais contente* driven by Camille Jenatzy, broke the 100 km/h barrier. And, of course, the electrics were clean and quiet, no high-pressure boilers and hissing hot steam, no dangerous cranking and refills with flammable gasoline.

Their commercial introduction began in 1897 with a dozen of Electric Carriage & Wagon Co.'s taxicabs in New York. In 1899 the U.S. production of electric cars surpassed 1,500 vehicles, compared to 936 gasoline-powered cars (Burwell 1990). Two years later Pope's Electric Vehicle Co. was both the largest maker and the largest owner and operator of motor vehicles in the United States (Kirsch 2000). But its operations were soon reduced to just occasional rides in and around the Central Park, and by the end of 1907 it was bankrupt. The combination of technical improvements, ease of use, and affordability shifted the outcome in favor of gasoline-powered vehicles—although Edison, still stubbornly searching for a high power-density car battery, kept predicting that electric cars will eventually cost less to run.

There was no single component of engine and car design that remained unimproved. While fundamental operating principles remain intact even today, the first quarter-century of automotive engineering was remarkable not only for the breadth of its innovations but, once the machines ceased to be a curiosity and gained the status of an indispensable commodity, also for the rapidity with which important substitutions and improvements took place. First to be resolved was the problem of convenient steering. In horse carriages, the front axle with fixed wheels swiveled on a center pivot, and while it could be turned easily by a horse, it was a challenge for a driver. Benz avoided the problem by opting for a single front wheel governed by a tiller. But there was an effective old solution to front-wheel steering, Rudolph Ackermann's 1818 invention that linked wheels by a rod so that they could move together while turning on independent pivots at the ends of a fixed front axle. England's first gasoline car, built by Edward Butler in 1888, was the first four-wheeler with Ackermann steering, setting a standard for nearly all future vehicles.

As already noted, in 1891 Emile Levassor of the Parisian company Panhard & Levassor, originally makers of wood-working machinery, began to reconfigure the motorized vehicles pioneered by Benz, Daimler, and Maybach in a radical and enduring fashion (figure 3.13). He moved the engine from under the seats and placed it in front of the driver, a shift that placed the crankshaft parallel with the car's principal axis rather than parallel with the axles and made it possible to install larger, more powerful engines, and that also led inevitably to the design of a protective, and later also aerodynamic, hood. The engine had a friction clutch and sliding-pinion gears, and he replaced the primitive leather drive belt with a chain drive.

As one of the founders of the most illustrious British car marque put it in his encyclopedic survey of motor vehicles, "with all the modifications of details, the combination of clutch, gear-box and transmission remains unaltered, so that to France, in the person of M. Levassor, must be given the honour of having led the development of the motor-car" (Rolls 1911:915–916). Subsequently, many particulars had to be improved or entirely redesigned in order to make this grand reconfiguration into a truly practical vehicle. Although the car had Ackermann steering, it was controlled by a long horizontal lever that was finally replaced by 1898 by modern wheel at the end of an inclined column.

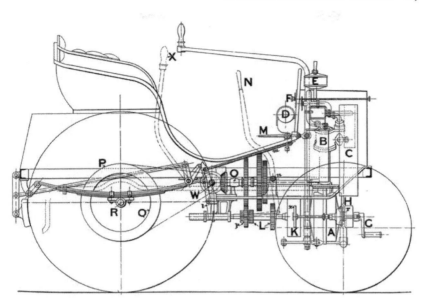

FIGURE 3.13. Section of Panhard & Levassor 1894 car equipped with Daimler's motor. *A* is the crankbox of the inclined cylinder, *B* the carburetor, and *C* the exhaust silencer. Gasoline flowed from *D* to *E* and *F* (ignition tube and mixture regulator). Three shifting wheels (*L*) were operated by the lever *N*. Reproduced from Beaumont (1902).

Even more important, ignition had to be made much less dangerous as well as much more reliable and efficient. The inherently risky open-flame hot tube was first replaced by low-voltage magnetos (generators producing a periodic electric pulse) whose design was improved in 1897 by Robert Bosch (1861–1942). Bosch's magneto was successfully used for the first time in a de Dion-Bouton three-wheeler that was then able to reach an unprecedented speed of 50 km/h (Bosch GmbH 2003). The device was also adopted by DMG, and so more than a decade after Bosch set up his workshop for precision mechanics and electrical engineering (in 1886, after he returned from the United States where he worked briefly for Edison), his business was finally prospering. But the real breakthrough came in December 1901, when Gottlob Honold showed his employer a new high-voltage magneto with a new type of spark plug.

The patent was granted on January 7, 1902, and the timing could not have been better as the high-voltage magneto and spark plug that made high-speed engines possible became available just as the serial car production was finally taking off. Fouling was a recurrent problem with the early designs, but gradual improvements raised the life span of spark plugs to more than 25,000 km. Nickel and chromium were used until the 1970s; copper core was introduced during the early 1980s, and shortly afterward a center electrode was made of

99.9% platinum whose use allows for the maintenance of optimal operating temperature and for reaching the self-cleaning temperature within seconds (Bosch GmbH 2002; Kanemitsu 2001). In 1902 Bosch produced the total of about 300 spark plugs; a century later the worldwide annual production topped 350 million.

Various carburetor designs of early years eventually gave way to the spray-nozzle device, developed by Maybach in 1893, which became the basis for modern devices that introduce a fine jet of gasoline into the air that is entering a cylinder. Mechanical, cam-operated, inlet valves replaced the automatic (atmospheric) devices that opened late on the induction stroke and closed prematurely, thus resulting in a weakened charge. Renault was the first car maker to introduce the transmission that was directly connected to the engine: it had three forward and one reverse gear selected with a gear shift. Engine control switched from cutting out impulses to throttling, that is, to reducing the volume of fuel charged into a cylinder at one stroke.

The overall engine design benefited from a greater variety of new high-performance steel alloys whose use made cylinders, pistons, valves, and connecting rods both lighter and more durable, and engine performance was greatly enhanced by improvements in lubrication. Final chain transmission to wheels running on a fixed axle was replaced by propeller drive on a rotating axle. Four- and even six-cylinder cars (the latter arranged in V) eliminated the need for a substantial flywheel, while the honeycomb radiator, patented by DMG in 1900, doubled the efficiency of cooling, and a single lever became the standard for engaging both forward and reverse gears.

As in all initial stages of a prime mover's development, installed power kept increasing not only in high-performance vehicles but also in cars designed for passenger transport. In 1904 Maybach completed the first six-cylinder Mercedes with 53 kW, and in 1906 he introduced a 90-kW race engine with overhead intake and exhaust valves and with dual ignition (DaimlerChrysler 2003). For comparison, today's small passenger cars (Honda Civic, Toyota Corolla) rate typically between 79 and 90 kW, and larger family sedans (Honda Accord, Toyota Camry) have engines capable of 112–119 kW.

There were also innovations that made cars easier to start, to drive, and to service in emergencies. All early vehicles were started by turning a hand crank, a demanding and, not infrequently, a fairly dangerous task: a premature engine firing due to an advanced spark setting would rotate the starting crank violently, and it could easily break a wrist or a thumb. Early automakers tried to design self-starting engines with acetylene, compressed air, gasoline vapors, springs, and other mechanical contrivances (Bannard 1914). The solution arrived finally in 1911 when Charles Kettering, at his Dayton Engineering Laboratories (the company, now Delco Remy, continues to make starters and other electric parts for cars), succeeded in designing the first practical electric starter, for which he received a U.S. patent in 1915.

After satisfactory tests, General Motors ordered 12,000 units for all of its 1912 Cadillacs. Laborious hand cranking remained common on cheaper models, but by 1920 nearly every U.S. car was available with at least optional electric starting. The three-speed and reverse gearbox (the H slot) that remains a standard in today's passenger cars was first offered with a Packard in 1900, the year when Frederick Simms also built the first car fender (NAE 2000). Electric lights and electric horns gradually replaced paraffin and acetylene lanterns and bulb horns, and mechanically operated windshield wipers kept clean the first windows in semi-enclosed automobiles. Detachable wheels and a spare wheel were first introduced in the United Kingdom, as was the first speedometer (made in 1902 by Thorpe and Salter with the range of 0–35 mph or 0–55 km/h). The inflatable rubber tire, patented in the United Kingdom 1888 (also U.S. Patent 435,995 in 1890) by Scottish veterinary surgeon John Boyd Dunlop (1840–1921) and almost immediately used on newly popular bicycles, was superior to the solid one, but as it was glued solidly to the wheel rim, it was not easily repaired.

The first detachable rubber tire was produced by Michelin brothers, Andre (1853–1931) and Edouard (1859–1940), in 1891. The older brother studied engineering and architecture, and the younger one painting, but both returned home to manage their small family business in Clermont-Ferrand, which they eventually transformed into a very successful multinational corporation (Michelin 2003). Their tire enabled Charles Terront to win the Paris–Brest–Paris bicycle race despite a puncture. As no manufacturer was ready to fit its cars with Michelin tires, the brothers used a Peugeot body and a Daimler engine to enter their own vehicle in the 1895 Paris–Bordeaux race: the *Éclair*'s tires had to be changed every 150 km, but the car finished the race in 9th place (Michelin 2003).

Four years later, Michelin tires were on Jenatzy's vehicle when it broke the 100 km/h barrier—and a year before that, the company adopted Bibendum, the tire-bulging Michelin Man, as its symbol. Once the early cars shed their carriage form and began to reach higher speeds, the detachable tire had a new huge market. The company's notable pre-WWI innovations included the dual wheel for buses and trucks that allowed heavier loads (in 1908), and the removable steel wheel that made it possible to use a ready-to-ride spare in 1913. That same year, Michelin released the first detailed map of France that was specifically designed with motorists in mind (today their maps cover all continents). Safer rides were made possible by replacing band brakes with drum brakes that were introduced by Renault in 1902. In December of the same year Frederick William Lanchester (1868–1946) patented disk brakes, whose clamps hold onto both sides of a disk that is attached to the wheel hub. But as these brakes required more pedal pressure to operate, they became common only with the power-assisted braking that became standard on many American cars during the 1960s.

Yet another important improvement to increase the safety of motoring arrived in 1913 with William Burton's (1865–1914) introduction of thermal cracking of crude oil, initially at the Standard Oil of Indiana's Whiting refinery (Sung 1945). This procedure was aimed at increasing the yield of gasoline in refining in order to meet the rapidly rising demand brought by mass-produced cars. Its product was the fuel that contained only about 40% volatile compounds, in contrast to almost purely volatile natural gasolines that were used since the beginning of the automotive era and that were a frequent cause of accidental flare-ups during hot summer months, particularly in early engines that relied on hot-tube ignition. Still, the less volatile fuel alone was not good enough to eliminate violent knocking that came with higher compression. That is why all pre-WWI engines worked with compression ratios no higher than 4.3:1 and why the ratio began to rise to modern levels (between 8 and 10) only after the introduction of leaded gasoline.

Perhaps no single achievement conveys more impressively the early advances in automotive design than the improvements in the key engine parameter, its mass/power ratio. I already noted that when Daimler and Maybach set out to develop a new engine, one of their key goals was to make it considerably lighter than Otto's four-stroke gas-fueled machine that weighed about 270 g/W. By 1890, the best DMG engine was down to about 40 g/W; in 1901, Maybach's famous first Mercedes weighed 8.5 g/W and Ford's Model T engine needed eventually less than 5 g/W (figure 3.14). As shown in figure 3.14, the mass/power ratio continued to decline after WWI before it stabilized around 1 g/W for typical passenger cars, which means that about 98% of the entire 1880–1960 decrease took place before 1913.

By the end of the first decade of the 20th century, the car was in many ways a rapidly maturing machine. Looking back, Flower and Jones (1981) noted that the performance of faster cars available by 1911 was comparable not just to that of a 1931 model but even to a typical car built immediately after WWII. There is a perfectly analogical case of tungsten-filament incandescent light bulbs, whose design has remained remarkably conserved ever since 1913. Being a much more complex artifact, a car engine had seen more improvements than a tungsten filament, but they all took place without any radical change to its basic design and to its mode of operation. But there had to be a fairly radical cut in price if car ownership were to diffuse among average urban and rural families.

In this respect one car will always have a very special place in the annals of automotive history: Ford's Model T deserves this position because of its low price, remarkably sturdy construction, and reliable service, the combination that had a profound socioeconomic impact. And the latter effect refers not only to the ways the car was promoted, marketed, and used but also to the novel way it was made. Ford's Model T was a key catalyst for the development

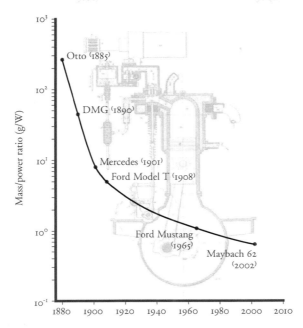

FIGURE 3.14. Rapid decline of mass/power ratios of automotive engines during the first three decades of their development was followed by a slower rate of improvement after WWI. Plotted from a wide variety of engine specifications.

of modern auto industry, and below I examine not only its technical parameters but also its wider impacts.

Ford's Model

The infatuation with machines that led Otto from grocery sales to four-stroke engines was also the beginning of Henry Ford's (1863–1947) remarkable achievement as the world's unrivaled producer of passenger cars (figure 3.15). Ford saw his first self-propelled vehicle, a wheeled steam engine moving to a new worksite, in 1876, and subsequently he admired the newly introduced Otto machines and experimented with small engines. While working at the Edison Illuminating Co. in Detroit, where he was the chief engineer, he build his first motorized vehicle, a small quadricycle, in 1896 (Ford 1922). But his entrepreneurial beginnings were not particularly auspicious. In 1898 he became the chief engineer and manager of Detroit Automobile Co., which was, after

FIGURE 3.15. Henry Ford's portrait taken in 1934. Library of Congress photograph (LC-USZ62-78374).

he left it, reconstituted in 1902 as Cadillac. His second, short-lived company did not produce any cars.

Third time lucky, Ford Motor Co. (FMC) was set up in 1903, with Alexander Malcolmson, a Detroit coal dealer, as the principal investor (Brinkley 2003). The company was at first quite successful with its models A and B, but its later upscale models—including Model K, Gentleman's Roadster, able to go 112 km/h but retailing for $2,800 (FMC 1908)—were not selling well. Model N introduced in 1906 was the first car aimed at a mass market, a goal that was finally achieved with Model T, which was launched on October 1, 1908. This was an unabashedly programmatic car, a vehicle that was intended, in Ford's own words, for the great multitude, but one that was built of good-quality materials and sold at a price affordable for anybody with a good salary. Unlike so many previous, and too many subsequent, advertising claims produced by the car industry, these statements were quite accurate. And unlike many new cars of the 20th century's first decade, this was a resolutely down-market vehicle designed to fill a large waiting market niche.

A still westward-moving America needed a car not just for urban residents but for settlers on vast expanses of newly cropped land, which is why Ford, himself a former farmer and the man with life-long interests in land, saw it as a farmer's car. Hence, he gave it a generous clearance to make it go not just along muddy roads (at that time fewer than 10% of all U.S. roads were surfaced) but if need be also across a ploughed field, and the combination of the simplest design (so any machine-minded farmer could fix it) and the best materials. Although the wide-tracked wheels, high seats, and small, initially wooden body (the combination that led to nicknaming the car a spider) conveyed a sense of overall fragility, this impression was false.

Heat-treated vanadium steel gave the car ruggedness that is amply attested by contemporaneous photographs of Model Ts carrying and pulling some heavy loads and running all kinds of attachments. Ford stressed this quality in his advertisements, claiming that "the Model T is built entirely of the best materials obtainable . . . in axles, shafts, springs, gears, in fact a vanadium steel car is one evidence of superiority" (FMC 1909). The first water-cooled, four-cylinder engines installed in Model Ts had capacity of 2,760 cm³, ran at 1,600 rpm, rated about 15 kW, and (including the transmission) weighed about 12 g/W. The touring version of the car weighed 545 kg and could reach speeds up to 65 km/h.

Of the five versions introduced during the first model year, the touring car for five passengers was, and subsequently remained, by far the most popular, while the coupe, looking much like a top hat, continued to be produced in only small numbers (figure 3.16). Ts were the first Fords with the left-hand drive (a logical feature for easy passing), and none of the early models had any battery: headlamps were lit by carbide gas, running and rear lights by oil, a magneto produced electricity for the four spark plugs, and starting was by a hand crank. Interestingly, the company's catalog reserved the greatest praise for springs that "no conceivable accident could possibly affect" and that made it "one of the easiest riding cars ever built" (FMC 1909).

Mass production was the company's strategy from the very beginning. The very first Model T catalog boasted that Ford had already worked the "quantity

FIGURE 3.16. Four Model Ts (clockwise starting from the upper left corner) as they appeared in the company's catalog (FMC 1909): touring car, by far the most popular version of the vehicle; roadster with a vertical windshield; a town car; and a coupe that looks much like a top hat.

production system" that allowed it to build "a hundred cars a day, a system whereby overhead expense is reduced to minimum . . . We know how to build a lot of cars at a minimum cost without sacrificing quality" (FMC 1909). But that was just a prelude. A much more efficient system, introduced for the first time at Highland Park factory in 1913, has been described, analyzed, praised, and condemned many times (Flink 1988; Ling 1990; Batchelor 1994; Brinkley 2003). Whatever the verdict, the system was widely copied, fine-tuned, and adjusted, and it remained the dominant means of large-scale automobile production until the combination of widespread robotization of many basic tasks and changes on assembly lines that were pioneered by Japanese automakers began its transformation during the 1970s (Womack, Jones, and Roos 1990).

The leading goal of Ford's assembly system was to do away with the artisanal mode of car manufacturing that most manufacturers relied on even after 1905. That arrangement saw small groups of workers assembling entire vehicles from parts they brought themselves (or which were carted or trucked to them) from assorted stores and whose fit they had to often adjust. As Ford expanded his production, this form of operation eventually required more than 100 chassis stations that were placed in two rows stretching over nearly 200 m, an arrangement that created logistic problems and floor congestion. The other option, whereby fitters and mechanics were walking around the assembly hall in order to perform their specialized tasks on each vehicle, was no more effective. As a result, productivity was low, with most European manufacturers requiring 1.5–2 workers to make one car per year, but Ford was able to make nearly 13 cars a year per worker already in 1904.

Ford's quest for high labor productivity relied on the use of identical parts to be used in identical ways in assembling identical-looking vehicles by performing repetitive identical tasks—and eventually doing so along endless moving belts whose speed could be adjusted to the point where rapid assembly did not compromise the desired quality of construction. This manufacturing process actually began at EMF Co. in 1911 when two of Ford's former employees, W. E. Flanders and Max Wollering, installed a mechanized assembly line to speed up chassis production. Chassis were first pushed, and then pulled, by cables actuated by electric motors, and Ford's first moving line, installed at Highland Park in 1913, was hardly more impressive: just 45 m long, with some 140 assemblers working on five or six chassis (Lewchuk 1989).

There was only a marginal increase in labor productivity at Ford during the early part of 1913, but then the idea of putting labor along a moving line began to bring the expected profit. The rope-drawn chassis assembly line cut the time from 12 hours and 30 minutes in October to 2 hours and 40 minutes in December (Flink 1988). The next year, the introduction of chain-driven conveyor cut the assembly time to just 90 minutes. The system was actually used for the first to assemble magnetos, where the time was cut from 20 to 5 minutes. By 1914 the new plant was producing thousand automobiles a day and

> the work has been analyzed to the minutest detail with a view to economizing time. Men are given tasks that are very simple in themselves, and, by dint of repetition day in and day out, acquire a knack that may cut the time of the operation in two. A man may become a specialist in so insignificant an operation, for instance, as putting in a certain bolt in the assembling of the machine (Bond 1914:8).

This method of production was undoubtedly exploitative, stressful, and, perhaps the strongest charge leveled against it, dehumanizing—and there is no simple explanation for why it was accepted by so many for so long. Opportunity for unskilled workers was an obvious draw, and relatively high wages had clearly an important effect. In 1914—in what *The Economist* called "the most dramatic event in the history of wages"—Ford more than doubled the assembly-line pay to $5/hour and cut the working day from nine to eight hours. This reduced high turnover and delayed the unionization of labor (Lewis 1986). The innovation opened the way toward dramatic rises in labor productivity, and Ford passed most of the savings achieved by his streamlined assembly to consumers. This decision initiated a period of impressive positive feedbacks as lower prices led to increasing sales, which led to higher production at lower costs.

Consequently, Ford's approach was not one of marketing but of true economies of scale: he did not start by offering the lowest priced car in order to generate subsequently large production volumes, but he kept increasing the volume of production and converting the higher productivity into lower prices. The introductory price (for red or gray vehicles) was $850 to $875, but by 1913 mass production of more than 200,000 vehicles (with bodies of sheet steel) lowered it by 35% to $550. A year later it was down to $440, and after WWI the car retailed for as little as $265 for the runabout model and $295 for the touring car (MTFCA 2003). Ford's great down-market move paid off. In 1903 his struggling company made about 15% of the country's 11,200 new cars. During the first fiscal year of Model T production (1908–1909), it shipped nearly 18,000 vehicles, about 15% of the U.S. passenger vehicles, but by 1914 Ford's share was about 44%, and three years later it rose to 48%.

Midnight blue became famous standard black by 1915, and no other colors were produced until 1925: the choice was dictated simply by the cost and durability of black varnishes, which had little resemblance to modern spray finishes (Boggess 2000). By the time the model was discontinued in 1927, the company's official data sheet shows that 14,689,520 Model Ts were made (Houston 1927), but other sources quote the total of just over 15 million. Durability of these vehicles and hardships of the deepest economic crisis of the 20th century combined to keep a few million of Model Ts on American roads throughout the 1930s. Nearly a century after their introduction, there are still many (rebuilt and reconditioned) Model Ts in perfect running order (MTFCA 2003).

The First Two Decades

Early car industry was much beholden to bicycle making. This is hardly surprising given the history of that simple machine (Whitt and Wilson 1982). Incredibly, the modern bicycle, an assembly that consists of nothing more than a two equally sized wheels, a sturdy frame, and a chain-driven back wheel, emerged only during the 1880s, a century after Watt's improved steam engines, half a century after the introduction of locomotives, and years after the first electric lights! Previous bicycles were not just clumsy but often very dangerous contrivances whose riding required either unusual dexterity and physical stamina or at least a great deal of foolhardiness. As such, they had no chance to be adopted as common means of personal transport (figure 3.17).

Only after John Kemp Starley and William Sutton introduced bicycles with

FIGURE 3.17. Evolution of bicycles from a dangerous vertical fork of 1879 and Starley's 1880 Rover (wheels are still of unequal size) to modern-looking "safety" models of 1890s with diamond frame of tubular steel and pneumatic tires. Images reproduced from *Scientific American*, April 19, 1892, and July 25, 1896.

equal-sized wheels, direct steering, and diamond frame of tubular steel—their *Rover* series progressed to direct steering and acquired a frame resembling standard diamond by 1886 (figure 3.17)—did the popularity of bicycles take off (Whitt and Wilson 1982). And they became soon even easier to ride with the addition of pneumatic tires and backpedal brake (U.S. Patent 418,142), both introduced in 1889. As with so many other techniques of the astonishing 1880s, the fundamental features of those designs have been closely followed by nearly all subsequent machines for most of the 20th century. Only since the late 1970s we have seen a number of high-tech designs that introduced expensive alloys, composite materials, and such unorthodox designs as upturned handlebars.

Early bicycle makers, building new metal-frame, metal-wheeled, rubber-tired machines destined for road trips, found the task of assembling small cars a naturally kindred enterprise. The bicycle contributed to the birth of the automotive era because it provided ordinary people with their first experience of individualized, long-distance transportation initiated at one's will and because these new speedy machines created demand for paved roads, road signs, traffic rules and controls, and service shops (Rae 1971; Flink 1975). Early cars benefited from these developments in terms of both innovative construction (steel-spoked wheels, welded tubes for bodies, rubber tires) and convenient roadside service, and the experience of free, unscheduled travel created new expectations and a huge pent-up demand for mass car ownership. But too many former bicycle mechanics switched to car building. Only some, including Opel, Peugeot, Morris, and Willys, prospered; most could not by producing either unique items or very small series of expensive yet often unreliable vehicles. That is why Ford's Model T was such a fundamental breakthrough.

Car races played an important role both in bringing the new machines to widespread public attention and in improving their design and performance. The first one was organized by *Le Petit Journal* in 1894 between Paris and Rouen (126 km), and the very next race in 1895 was ambitiously extended to a return trip between Paris and Bordeaux, a distance of nearly 1,200 km that the winning vehicle completed with average speed of 24 km/h (Flower and Jones 1981). In 1896 the Paris-Marseille-Paris race, beset by heavy rains, covered more than 1,600 km: Emile Levassor was among many injured drivers, and his injuries caused his death a few months later. The first decade of the 20th century saw more intercity races beginning with the Thousand Miles Trial in the United Kingdom in 1900, Paris-Berlin race in 1901, and Paris-Madrid race in 1903. The latter ended prematurely in Bordeaux after many accidents and 10 fatalities. Louis Renault, who won the first leg of the race in the light car category (figure 3.18), learned of the death of his brother Marcel only after arrival at Bordeaux.

But soon even longer races were held, such as the first long-distance rally from Paris to Constantinople in 1905, and, audaciously, the 14,000 km

FIGURE 3.18. Louis Renault winning the first leg of the ill-fated Paris-Madrid race in May 1903. Reproduced from the cover of *The Illustrated London News*, May 30, 1903.

Beijing-Paris run in 1907 when cars often had to be towed or even carried over rough, roadless terrain. This was followed by inauguration of the world's two most famous speedways at Le Mans in 1906 and Indianapolis in 1909 (Boddy 1977; Flower and Jones 1981). Today's variety of car races (from international to local, from monster drag car to dune buggy competitions) and the magnitude of annual spectator pilgrimages, particularly to the Formula 1 contests, demonstrate how enormously the interest in speed and performance has grown.

The international character of the emerging car industry was seen in both licensing agreements and sales. French winners of car races had German motors in their vehicles, DMG licenses launched the car-making industry in the United Kingdom (1893), and exports were important for most of Europe's pioneering car producers as well as Ford's company. For example, in 1900 when Benz & Cie. was still the world's largest automaker, it sold 341 out of the total of 603 vehicles abroad. As already noted, there were only some 8,000 cars in the United States in 1900, and fewer than 2,400 in France, the country that pioneered modern car design during the 1890s. In 1905, Germany and France each had fewer than 20,000 passenger vehicles, United Kingdom's total was about 30,000, and the U.S. registrations reached 77,000.

As is so often the case in early stages of a new industry, too many companies competed in the market. By 1899 there were more than 600 car manufacturers in France; nearly 500 car companies were launched in the United States before 1908, and about 250 of them were still in operation by the time Ford began selling his Model T (Byrn 1900; Flink 1988). Post-WWI consolidation reduced these numbers to fewer than 20 major companies, and by the end of the 20th century about 70% of all passenger cars made in North America were assembled by just three auto makers, General Motors, Ford, and Honda (Ward's Communications 2000).

Americans were late starters compared to Germans and the French. Although there were at least half a dozen experimental gasoline cars built in the United States between 1891 and 1893, the first machine that attracted public attention was designed by Charles and Frank Duryea in Springfield, Massachusetts. The two bicycle mechanics decided to build it after reading about Benz's latest vehicle in *Scientific American* in 1889. Their machine was ready in 1893; they won America's first (88 km) car race held in Chicago in November 1895 (beating a Benz, the only other car that finished the course) and sold the first car in February 1896. Other early automakers included Elwood Haynes and Alexander Winton. By the century's end, there were at least 30 American car makers, and they produced 2,500 vehicles in 1899 and just more than 4,000 in 1900 (Flink 1975).

All but one of the great enduring names in the American car industry appeared before 1905, with production conspicuously concentrated in Michigan (May 1975; Kennedy 1941). Ransom Olds (1864–1950), financed by S. L. Smith, began making his Oldsmobiles, which were nothing but buggies with an engine under the seat, in 1899. In 1901 Olds introduced America's first serially produced car, Curved Dash, with a single-cylinder 5.2 kW engine (May 1977). The Cadillac Automobile Co. began selling its vehicles in 1903, the same year when David D. Buick (1854–1929) sold his first car. In 1908 all of these brands became divisions of General Motors under William Durant (1861–1947). Only Walter Chrysler (1875–1940), who bought his first automobile in 1908 and became a Buick manager in 1912, did not begin making his cars before WWI.

Once the gasoline-powered engine prevailed, the speed of the post-1905 American automobilization was unparalleled. By 1913 Americans could choose among cars of more than 120 manufacturers, ranging from Ford's cheapest Model T (at $525 it was not the cheapest car available, as Metz and Raymond were selling inferior vehicles for, respectively, $395 and $445) to "America's Foremost Car," the Winton Six, whose "freedom-from-faults" had made it the leader with "up-to-the-minute in everything that makes a high-grade car worth having." France had the largest motor vehicle industry until 1904, when its output of nearly 17,000 cars was surpassed by the U.S. production of just over 22,000 vehicles (Kennedy 1941; Flink 1988). Then the gap widened rapidly: by 1908 U.S. output reached 60,000, and by 1913 it approached half a million (USBC 1975).

At that time U.S. car production was actually a quasi-monopolistic endeavor, a licensed enterprise with individual companies belonging to the Association of Licensed Automobile Manufacturers (ALAM) and paying a fee for every vehicle produced. This peculiar situation arose from an even more peculiar patent application. George B. Selden, a Rochester patent lawyer, filed the claim on May 8, 1879, but the patent was not issued until 16 years later, on November 5, 1895 (U.S. Patent 549,160), as Selden kept filing amendments and thus deliberately deferring the start of the 17-year protection period (Selden 1895; Kennedy 1941). By that time its design, shown in derisory drawings of an awkward looking carriage, was left far behind by the intervening engineering advances. Selden never built any vehicle, and the early automotive pioneers were not aware of the pending patent's existence until Albert A. Pope of the Electric Vehicle Co. bought the rights to Selden's patent in November 1899. He promptly filed and won an infringement suit against the Winton Co. (Flink 1975).

As the wording of Selden's claim was so broad—applying to any vehicle with "a liquid hydrocarbon gas-engine of the compression type comprising one or more cylinders" (Selden 1895:3)—a group of 32 automakers formed ALAM in 1903 and agreed to pay 1.25% of the retail value of their cars to the Electric Vehicle Co. (Kennedy 1941). Producers soon succeeded in reducing the rate, but ALAM continued legal fight against those manufacturers that refused to pay the fees as a means of regulating competition and keeping prices high. Henry Ford's new company was sued in October 1903 and, after a fight resembling the contested invention of incandescent lights, lost the case completely in 1909.

This put the Ford Co. in a very precarious position just as it was entering the mass-production stage with the Model T. Ford appealed, and maintained in his advertisements that the Selden patent does not cover a practical machine and that his company is the true pioneer of the gasoline automobile. The dubious patent was finally invalidated in January 1911, the year when other monopolies were outlawed with the antitrust verdicts that dissolved the Standard Oil Co. of New Jersey and the American Tobacco Co. (Kennedy 1941). And so the passenger car in America—though still adventuresome to drive, not particularly comfortable, and still rather expensive—was on the verge of becoming a mass property. After being first a marginal mechanical curiosity and then a pastime of some rich eccentrics, the car began to cast a widening spell.

Soon even the author of what is now seen as one of the great classics of

English children's literature, *The Wind in the Willows*, wrote admiringly of it. As Kenneth Grahame's animal friends, traveling in a horse-drawn gypsy, were swept into a deep ditch by a rapidly passing magnificent car, the Toad did not complain. Instead, they

> found him in a sort of a trance, a happy smile on his face, his eyes still fixed on the dusty wake of their destroyer . . . "Glorious, stirring sight!" murmured Toad, never offering to move. "The poetry of motion! The *real* way to travel! The *only* way to travel! Here to-day—in next week to-morrow! Villages skipped, towns and cities jumped—always somebody else's horizon! O bliss!" (Grahame 1908:40–41)

Toad's monologue ("O what a flowery track lies spread before me, henceforth! What dust-clouds shall spring up behind me as I speed on my reckless way!") prefigured the infatuation of hundreds of millions of drivers and passengers that followed during the 20th century. That some modernistic poets became early car enthusiasts is far less surprising. To give just one, and almost deliriously, overwrought example, William E. Henley (1849–1903), one of young Kipling's mentors, thought in his *Song of Speed* (written in 1903) that a Mercedes touring car was nothing less than a divine gift (Henley 1919:221):

> Thus has he slackened
> His grasp, and this Thing,
> This marvellous Mercédes,
> This triumphing contrivance,
> Comes to make other
> Man's life than she found it:
> The Earth for her tyres
> As the Sea for his keels.

Apparently little has changed since 1903, as many drivers still think of a Mercedes as a "triumphing contrivance," and some are even willing to pay for its latest embodiment—the overhyped ("well-pedigreed contender . . . ultra smooth ride") Maybach 62 Large Sedan—no less than $350,000 (yes, that was the starting price in 2003). Besides the gargantuan engine and electrohydraulic brakes this gives them also "private jet levels of luxury," with Grand Nappa leather, fully reclining airplane-style rear seats ("actively ventilated"), TV and DVD, a refrigerator and compartments for champagne flutes, choice of various woods for the inside finish, and (what a bargain!) a sunroof, two-tone paint, and a refrigerator as no-cost options.

Expanding Applications

Not surprisingly, it did not take long for European and American engineers to begin mounting new gasoline engines on chassis other than flimsy buggies or sedate carriages. Some of these applications, including small delivery vehicles and taxicabs, did not call for any special engines or body designs, but many other uses required more powerful engines and sturdier bodies. Heavier designs began appearing during the late 1890s, and by 1913 just about every conceivable type of a heavy-duty commercial vehicle that is now in service was present in forms ranging from embryonic to fairly accomplished. A century later we rely on a great variety of internal combustion engines whose ratings span five orders of magnitude. The smallest and lightest engines (<1 kW) are installed in lawnmowers, outboard motors, and ultralight airplanes. Most passenger cars do not have engines more powerful than 100 kW, off-road machines can rate above 500 kW, airplane engines more than 1 MW, and the largest marine engines and stationary machines for electricity generation can surpass 20 MW.

After 1903 another process began unfolding in parallel with the development of specialized engines for buses, trucks, off-road vehicles, locomotives, and ships: increasingly more sophisticated efforts to use new engines in flight, to become finally airborne with a machine heavier than air. A different type of engine was needed to meet this challenge, a light and powerful one with very low mass/power ratio. As is so well known, Wright brothers achieved that goal on December 17, 1903, when their airplane had briefly lifted above the dunes at North Carolina's Kitty Hawk by an engine that, like its wooden body, they built themselves (Wright 1953). The new aeroengines had seen an even faster rate of improvement during after 1905 than did the automotive ones. These impressive early advances put in place all the key features and basic design trends that dominated the world of flight until the gas turbines (jet engines) took over as the principal prime movers in both military and commercial planes about half a century later.

But before turning to these machines I should mention small engines that are used wherever low weight, small size, and low cost operation are needed. Their development began during the late 1870s with Dugald Clerk's (1854–1932) two-stroke gas-fueled compression engine designed in 1879 and Benz's first two-stroke gasoline-fueled machines. In North American homes, they can be found in more than 50 million lawn mowers, grass trimmers, and snow and leaf blowers, in many millions of chain saws, in outboard engines for motorboats, and in motorcycles. But in 2002 Honda, the world's leading maker of motorcycles, announced plans to use only four-stroke engines because they produce fewer emissions and consume less fuel than do two-stroke models (Honda 2002). Small two-stroke engines have another disadvantage: most of them have only one cylinder, which means that there is only one power stroke per revolution, and the resulting widely spaced torque pulses cause much more perceptible vibrations than do multicylinder engines.

Heavy-Duty Engines

After 1905 it was obvious that steam-powered wagons, trucks, and tractors faced an inevitable extinction as more powerful internal combustion engines were being installed in a widening variety of commercial vehicles. Benz & Cie. built the world's first gasoline-powered bus able to carry just eight passengers in 1895. London got its first famous red double-deckers, built on a chassis virtually identical with that of a 3-t truck, in 1904, and by 1908 there were more than 1,000 of them in service. They displaced some 25,000 horses and 2,200 horse-drawn omnibuses, and Smith (1911) concluded that their higher speed had an effect of more than doubling the width of city's main thoroughfares. Horse-drawn omnibuses disappeared from London by 1911; Paris saw the last one in January 1913. As much as we complain about congestion caused by motor vehicles, the equivalent amount of traffic powered by horses would be at least twice or three times more frustrating (see figure 1.7).

DMG built its first gasoline-powered truck, a 3-kW vehicle with two forward and one reverse speed, in 1896; it could carry up to 5 t at speeds not surpassing 12 km/h. In the United States Alexander Winton began manufacturing his first delivery wagons, powered by a single-cylinder, 4.5-kW engine, two years later. Two-ton and 3-t trucks were the first common sizes, with smaller machines popular for city and suburban delivery (figure 3.19). Fire engines were the most frequently encountered special truck applications before WWI (Smith 1911), and heavier, 5-t trucks that could haul gross trailer loads of up to 40 t became standard in the United States for long-distance transport before WWI (Anonymous 1913b). The preferred U.S. practice that emerged before WWI is still with us. The semi-trailer is made up of the tractor unit (now normally with two-wheel steer axle and two rear drive axles with a pair of double wheels on each side) that carries a turntable on which is mounted the forward end of the trailer (initially just a two-wheeled, and now typically eight-wheeled), hence named a standard 18-wheeler.

By 1913 the area served daily by a truck was more than six times larger than the one served by horse-drawn wagons, and total work nearly was four times greater, with less than 15% of space required for garaging as opposed to stabling, and all that at an overall cost of motorized hauling that was only about two-fifths of the expense for animal draft (Perry 1913). Massive deployment of trucks began only with the U.S. involvement in WWI: by 1918 the country's annual truck production was nearly 230,000 units, more than nine times the 1914 total (Basalla 1988). Postwar production continued at a high rate, but

The ¾ Ton Utility Truck–$1250

FIGURE 3.19. A 1913 advertisement for a small (four-cylinder) utility truck (maximum carrying capacity of 0.9 t) made by the Gramm Motor Truck Co. This "most practical and serviceable truck of its size ever built" was designed for city and suburban delivery. Reproduced from *Scientific American*, January 18, 1913.

truck capacities remained relatively low until after WWII, and eventually they have reached maximum gross weights of as much as 58 t (truck tractor and two trailing units). But most U.S. states limit the 18-wheelers (with engines of about 250 kW) used in highway transport to about 36 t. Where these restrictions do not apply, trucks have grown much bigger: a fully loaded Caterpillar 797B, the world's largest truck powered by a 2.65-MW engine and used in surface mining, weighs 635 t (Caterpillar 2003).

Use of internal combustion engines in field machinery began almost as soon as the first small-scale sales of passenger cars (Dieffenbach and Gray 1960). The first, and excessively heavy, gasoline-powered tractor was built by John Froehlich in Iowa in 1892 and sold to Langford in South Dakota, where it was used just in threshing for a few months. Froelich formed the Waterloo Gasoline Tractor Engine Co., which was later acquired by the John Deere Plow Co., whose successor remains a leader in producing tractors and other self-propelled agricultural machinery. By 1907 there were no more than 600 agricultural tractors in the United States, and most of these early models resembled more the massive steam-powered machine rather than today's tractors: their mass/power ratios were between 450 and 500 g/W, compared to just 70–100 g/W for modern machines. Only Henry Ford experimented with very light (his 1907 test model weighed less than 700 kg) and small machines.

By 1912 the number of tractor makers rose above 50, the first international trials had taken place (Ellis 1912), and smaller, more practical, and more affordable machines began to appear (Williams 1982). Given the climate and terrain of California's Central Valley, it is not surprising that this region pioneered the widespread use of gasoline-fueled tractors and combines (Olmstead and Rhode 1988). By 1912 the United States had some 13,000 working tractors, and 80 companies produced more than 20,000 new machines in 1913 as the massive replacement of draft animals by internal combustion engines was finally underway (Rose 1913). But it was only during the 1920s when the power of agricultural machinery surpassed that of draft animals (see figure 3.2).

And shortly after WWI diesel engines, helped by their combination of advantageous characteristics, finally began their conquest of heavy automotive market. Besides the engine's inherently higher conversion efficiency, diesel fuel is also cheaper than gasoline, yet it is not dangerously flammable. The last attribute makes it ideal for applications where fire could be particularly dangerous (on vessels) as well for use in the tropics, where high temperatures will cause little evaporation from truck and bus tanks. Moreover, high engine efficiency and low fuel volatility mean that these diesel-powered vehicles have a much longer range per tankful than do equally powerful gasoline-fueled machines. Additional mechanical advantages include diesel engine's high torque, its resistance to stalling when the speed drops, and its sturdiness (well-maintained engines can go 500,000 km without an overhaul).

In 1924 Maschinenfabrik Augsburg-Nürenberg (MAN, the successor of the company where Diesel worked between 1893 and 1897) was the first engine maker to use direct-injection diesel engine, and eight years later the company stopped producing any gasoline-fueled vehicles and concentrated on diesel engines, which remain one its major products (MAN 2003). Both Benz and Daimler produced their first diesel trucks in 1924, and after they merged in 1926 they began developing a diesel engine for passenger cars; it was ready in 1936 when model 260-D, a rather heavy 45-hp saloon that became a favorite taxicab, was displayed at the Berlin car show (Williams 1972; Mercedes-Benz 2003). By the late 1930s, most of the new trucks and buses built in Europe had diesel engines, a dominance that was extended after the WWII to North America and Asia and, with exported vehicles, to every continent. Mass/power ratios of automotive diesels eventually declined to less than 5 g/W, and today's lightest units in passenger cars are not that different from gasoline-fueled engines.

The use of steam engines in shipping continued well after WWI, while steam turbines were gaining an increasing share of the market. But by 1939 25% of the world's merchant marine ships were propelled by diesel engines. Almost a century after the first vessel was equipped with a small diesel engine, that triumph is even more evident. Today some 90% of the world's largest freight ships, including the crude oil supertankers, are powered by diesel engines. MAN, Mitsui, and Hyundai are their leading producers. Maximum size of these large machines still keeps increasing: in 1996 the world's largest engine rated about 56 MW, but just five years later Hyundai built a 69.3 MW engine (Hyundai 2003). Dimensions of the largest machines are best illustrated by their cylinder bores and piston strokes: these are, respectively, almost 1 m and more than 2.5 m, dimension an order of magnitude larger than those of automotive diesels (MES 2001).

Low-rpm diesel engines of different sizes are also deployed in tens of thousands of localities far away from centralized electrical supply—be it in low-income countries of Asia, Africa, and Latin America or in isolated places in North America and Australia—in order to provide light and mechanical energy for refrigeration and crop processing. By 1998 the world's largest diesel generator, a Hyundai design for India, reached 200 MW. Other market niches where diesel engines are either dominant or claim large shares of installed power include heavy construction (cranes, and earth-moving, excavating, and drilling machines), tractors and self-propelled harvesters, locomotives (both for freight and passenger service, but the fastest ones are electric), and main battle tanks (but the best one, the U.S. Abrams M1/A1, is powered by a gas turbine).

FIGURE 3.20. Ayres's bizarre new aerial machine as illustrated in *Scientific American* of May 9, 1885.

Engines in Flight

At the close of the 19th century, the old ambition to fly in a heavier-than-air machine appeared as distant as ever. Utterly impractical designs of bizarre flying contraptions were presented as serious ideas even in technical publications. One of my favorites is a new aerial machine that was pictured and described in all seriousness in May 9, 1885, issue of *Scientific American* (figure 3.20), but I could have selected other preposterous designs from the last third of the 19th century. The goal was elusive, but some innovators were very determined: Otto Lilienthal (1848–1896), the most prominent German aviation pioneer, completed more than 2,500 short flights with various gliders before he died in 1896 when his glider stalled. So did (fiction anticipating reality) a man obsessively driven to invent a flying machines in Wells's *Argonauts of the Air* that was published a year before Lilienthal's death.

Building the flying machines heavier than air during the last decade of the Age of Synergy was about much more than just mounting a reciprocating engine on a winged fuselage. Airplanes fit perfectly into the class of achievements that distinguish that period from anything that preceded it: they came about only because of the combination of science, experimentation, testing, and the quest for patenting and commercialization. The two men who succeeded where so many other failed—Wilbur (1867–1912) and Orville (1871–1948) Wright (figure 3.21)—traced their interest in flying to a toy, a small helicopter powered by a rubber string they were given by their father as children when they lived in Iowa. This toy was invented by Alphonse Pénaud in 1871, and the two boys built a number of copies that flew well before they tried to construct a much larger toy.

Years later, their interest in flying was rekindled by Lilienthal's gliding experiments and his accidental death. Finally, after reading a book on ornithology in the spring

FIGURE 3.21. Wilbur and Orville Wright seated on the rear porch steps of their house, 7 Hawthorne Street, in Dayton, Ohio, in 1909. Library of Congress portrait (LOT 11512-A).

of 1899, Wilbur sent a letter to the Smithsonian Institution inquiring about publications on flight. The letter was answered in a matter of days by a package that contained reprints of works by Lilienthal and Langley. The brothers also ordered Octave Chanute's (1832–1910) book that described the progress in flying machines (Chanute 1894) and began a correspondence with this theoretical pioneer of manned flight. Soon they augmented the engineering experience from their bicycle business with numerous tests as they embarked on building and flying a series of gliders (figure 3.22). Beginning in the fall of 1901, these

FIGURE 3.22. Wilbur Wright piloting one of the experimental gliders above the dunes of Kill Devil Hills at Kitty Hawk, North Carolina, in October 1902. Library of Congress images (LC-W861-11 and LC-W861-7).

experiments also included a lengthy series of airfoils and wing shapes in a wind tunnel (Culick 1979).

While gliding attempts have a fairly long history, the first successful trial of an unmanned powered plane took place on October 9, 1890, when Clément Ader's *Eole*, a bat-winged steam-driven monoplane, was the first full-sized airplane that lifted off under its own power. In 1896 Samuel Pierpoint Langley (1834–1906), astronomer, physicist, and the secretary of the Smithsonian Institution, received a generous U.S. government grant ($50,000) to build the gasoline-powered aircraft (NASM 2000). By 1903 his assistant, Charles M. Manly, designed a powerful (39 kW, 950 rpm) five-cylinder radial engine. This engine was mounted on *Aerodrome A*, and the plane was launched, with Manly at the controls, near Widewater in Virginia by a catapult from a barge on October 7, 1903. But the plane immediately plunged into the river and did so once more during the December 8 test on the Potomac River in Washington, DC, when it reared and collapsed unto itself. Manly was pulled unhurt from icy water, but the collapse spelled the end of Langley's project. But the pressure was clearly on the Wrights if they were to claim the primacy of flying a motorized aircraft.

Nine days after Manly's mishap, on December 17, 1903, Orville Wright piloted the first successful flight at Kitty Hawk near Kill Devil Hills in North Carolina: just more of a jump of 37 m with the *Flyer* pitching up and down, staying airborne for just 12 seconds, and damaging a skid on landing (Wright 1953; USCFC 2003; see figure 1.3). Their second flight, after repairing the skid, covered about 53 m, and the third one 61 m. During the fourth flight, the machine began pitching up and down as it did in previous attempts, but Wilbur got it eventually under control, covered nearly 250 m in level flight, and then the plane began plunging and crash-landed with a broken front rudder frame, but in 59 seconds it traveled 260 m. When it was carried back to its starting point, a sudden wind gust lifted the plane, turned it over, and destroyed it. The first *Flyer*, a fragile machine with a wingspan of 12 m and weighing just 283 kg, thus made only those four flights totaling less than 2 minutes. The brothers telegraphed their father in Ohio with their news and took their broken *Flyer* to Dayton.

Why did the Wrights, bicycle makers from an Ohio town, succeed in less than five years after they ordered information on flight from the Smithsonian? Because they started from the first principles and studied in detail the accomplishments and mistakes of their immediate predecessors. Because they combined good understanding of what was known at that time about the aerodynamics with practical tests and continuous adjustments of this knowledge. Because they aimed to produce a machine that would not only lift off by its own power but that would be also properly balanced and could be flown in a controlled manner. There is no doubt that the matters of aerodynamic design and flight control were a much greater challenge than coming up with an engine to power the flight, although they did that, too (Culick 1979; Wright 1953).

What is so remarkable about the Wrights' accomplishment is that they designed the plane's every key component—wings, balanced body, propellers, engine—and that in order to do so they prepared themselves by several years of intensive theoretical studies, calculations, and painstaking experiments with gliders that were conducted near Kill Devil Hills between 1901 and 1903. And after getting negative replies from engine manufacturers whom they contacted with specifications required for their machine, the Wrights designed the engine themselves, and it was built by their mechanic, Charles Taylor, in just six weeks. This was by no means an exceptional engine. Indeed, as Taylor noted later, it "had no carburettor, no spark plugs, not much of anything . . . but it worked" (cited in Gunston 1986:172).

The four-cylinder 3.29-L engine lied on its side, its body cast of aluminum and its square (10 × 10 cm) steel cylinders were surrounded by water jackets, but the heads were not cooled, so the engine got progressively hotter and began losing its power. The crankshaft was made from a single piece of steel; one of its ends was driving the camshaft sprocket, and the other one the flywheel and two propeller chain sprockets. The Wrights' initial aim was to get 6 kW, but they did better: the finished engine weighed 91 kg; at first it developed 9.7 kW, and at Kitty Hawk it was rated as high as 12 kW—this unexpected performance gave it a mass/power ratio as low as 7.6 g/W. The brothers applied for a patent on March 23, 1903, nine months before the *Flyer* took off, and received a standard reply that the U.S. Patent Office was sending to many similar applications: an automatic rejection of any design that had not already flown. Their patent (U.S. Patent 821,393) was granted in May 1906, and, expectedly, it was commonly infringed and ignored. The key patent drawing shows no engine, just the detailed construction of wings, canard, and the tail (figure 3.23).

Afterward, as with the development of automobiles, there was a pause in development. The Wrights did not do actually any flying in 1906 and 1907 while there were building six or seven new machines. But starting in 1908, the tempo of aeronautic advances speeded up noticeably, with new records set after just short intervals. In September 1908, Wilbur Wright stayed over Le Mans for 91 minutes and 25 seconds, covering nearly 100 km, and during that year's last day he extended the record to 140 minutes. Louis Blériot (1872–1936)—an engineer whose first unsuccessful flying machine built in 1900 was an engine-powered ornithopter designed to fly by flapping its wings—crossed the English Channel by flying from Le Borques to Dover on July 25, 1909, in 37 minutes. This flight, worth £1,000 from London's *Daily Mail*, was done by a monoplane of his own design, the fourth such machine since he built the world's first single-winged aircraft in 1907 (see figure 7.1).

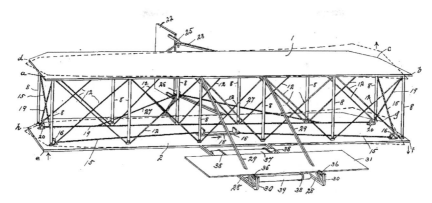

FIGURE 3.23. Drawing of the Wrights' flying machine that accompanied their U.S. Patent 821,393 filed on March 23, 1903. This drawing is available at http://www.uspto .gov.

Everything of importance was improving at the same time. The Wrights' first plane was a canard (tail-first) and a pusher (rear engine) with skids, and its engine was stationary with in-line cylinders. But soon there was a variety of propulsion, tail, and landing gear designs as well as the tendency toward a dominant type of these arrangements, and in-line stationary engines were replaced by radial machines (Gunston 1999). The Antoinette—an engine designed by Léon Levavasseur and named after the daughter of his partner—was originally built for a speedboat, but between 1905 and 1910 it became the most popular radial (eight cylinders arranged in 90° V) engine, and its very low mass/power ratio (less than 1.4 g/W) remained unsurpassed for 25 years.

The first successful rotary engine (with the crankshaft rigidly attached to the fuselage, and cylinders, crankcase, and propeller rotating around it as one unit) was the Gnome, a 38-kW machine designed in France by Séguin brothers in 1908 and whose improved versions powered many fighter planes during WWI (Gunston 1986). Ovington (1912:218) called it "theoretically one of the worst designed motors imaginable, and practically the most reliable aeroplane engine I know of." Other rotary engines followed soon afterward, including the Liberty, the most popular engine of WWI whose mass/power ratio was just above 1 g/W. Before 1913, there were also planes with shock absorbers and with simple retracting landing gear, as well as the first machines (in 1912) with monocoque (single-shell) fuselage that was required for effective streamlining and was made of wood, steel, or aluminum. Hoff (1946:215) captured the speed of these advances, and the driving force behind them, by noting that in the early 1900s "the only purpose of the designer was to build an airplane that would *fly*, but by 1910 military considerations became paramount."

France—where Alberto Santos Dumont (1873–1932), an avid and famously flamboyant Brazilian airship pioneer, made his first flight in a biplane of his own design in 1906—led this military development. By 1914 the country had about 1,500 military and 500 private planes, ahead of Germany (1,000 military and 450 private). Much like with cars, early development of American military airpower was slow. The U.S. Army bought a Wright biplane in 1908, but little progress was achieved even by April of 1917 when the United States declared war on Germany: at that time the country had only two small airfields, 56 pilots (and 51 students), and fewer than 300 second-rate planes, none of which could carry machine guns or bombs on a combat mission (USAF Museum 2003). Only in July 1917 did the Congress approved the largest ever appropriation sum ($640 million, or about $8 billion in 2000 US$) to build 22,500 Liberty engines; by October 1918 more than 13,000 of them were built.

But it was an American pilot who demonstrated for the first time a capability that eventually became one of the most important tools in projecting military power. On November 14, 1910, Eugene Ely (1886–1911) took off with a Curtiss biplane from the *USS Birmingham* to complete the first ship-to-shore flight, and less than 10 weeks later, on January 18, 1911, he took off from the Tanforan racetrack and landed on an improvised deck of cruiser *Pennsylvania* anchored in the San Francisco Bay (see figure 6.8). Thirteen years later the Japanese built the world's first aircraft carrier, and it was the destruction of their carrier task force by three American carrier air groups during the battle of Midway between June 4 and 7, 1942, that is generally considered the beginning of the end of Japan's Pacific empire (Smith 1966).

Of course, both the pre-WWI and the dominant WWI (wood-steel-fabric) airplane designs were constrained by many limitations. Most obviously, in order to minimize the drag (an essential consideration given the limited power of early aeroengines), structural weight had to be kept to minimum and thin wings (maximum thickness to length ratio of 1:30), be they on mono- or biplanes, had to be braced. This requirement disappeared only with the adoption of light alloys (Hoff 1946), and today's largest passenger planes have unbraced cantilevered wings, some as long as 50 m. These structures are designed to withstand considerable vertical movements: even seasoned air travelers may get uncomfortable as they watch a wingtip rising and dipping during the spells of high turbulence.

American airpower lagged in WWI, but an unprecedented mobilization of the country's resources made it globally dominant during WWII. Some P-51 Mustangs, perhaps the best combat aircraft of WWII with engines that rated as high as 1.1 MW, had maximum level speed just over 700 km/h (Taylor 1989). That was the pinnacle of reciprocating engines. Jet propulsion eliminated them first from combat and long-distance passenger routes and soon afterward from all longer (generally more than two hours) flights. But reciprocating engines still power not only those nimble, short-winged machines that perform amazing feats of aerial acrobatics at air shows but also airplanes that provide indispensable commercial and public safety services (Gunston 2002). These range from suppressing forest fires (Canadian water bombers are the best) and searching for missing vessels to seeding Californian or Spanish rice fields, from carrying passengers from thousands of small airports (Canadian Bombardier Industries and Brazilian Embraer are now the main contenders in this growing market) to applying pesticides to food and feed crops.

And before leaving the world of engines for the world of new materials and new syntheses that follows in chapter 4, I should mention that even Diesel's inherently heavier machine did eventually get light enough to power airplanes. The first success came in 1928 with Packard DR-980, a 168-kW radial engine that weighed only 1.4 g/W, that set a new world record of keeping a plane aloft (circling above Jacksonville) without refueling for 84 hours (Meyer 1964). Diesel engines were eventually installed on German WWII bombers, and recently lightweight diesels have become finally available to the general aviation market. In 2002 the Société de Motorisations Aéronautiques (SMA) received the FAA certification for its SR305 diesel engine, which weighs only 1.29 g/W (i.e., no more than typical modern gasoline-fueled aeroengines) and is suitable for installation in such small single-engine airplanes as Cessna, Piper, or Socata (SMA 2003). In the U.S. Delta Hawk and Continental Motors are working on similar engines.

4

New Materials and New Syntheses

The careful text-books measure
(Let all who build beware!)
The load, the shock, the pressure
Material can bear.
So, when the buckled girder
Lets down the grinding span,
The blame of loss, or murder,
Is laid upon the man.
Not on the Stuff—the Man!

Rudyard Kipling, *Hymn of Breaking Strain*

Motion and speed—rotation of steam turbines and electric motors, recipro-
cation of internal combustion engines, travel by new automobiles and bicycles,
promise of airplanes—were the obvious markers of the epoch-making pre-
WWI advances. At the same time, another kind of less flamboyant innovation
was changing the societies of Western Europe and North America. None of
those new speedy machines would have been possible without new superior
materials whose applications also revolutionized construction. The bold cantilever of the Firth of Forth bridge, shown in the frontispiece to this chapter,

FRONTISPIECE 4. The foreground of this dramatic photograph, taken in April 1889,
shows the bottom member of the unfinished North Queensferry cantilever of the Forth
Bridge and Garvie main pier in the background. Reproduced from the cover of *The
Illustrated London News*, October 12, 1889.

was thus no less a new departure and an admirable symbol of the 1880s than were Benz's engines or Tesla's motors. The bridge was the first structure of its kind to be built largely of steel: only that alloy could make this defiant design possible as only that material could bear the loads put on those massive cantilevers.

This was not the first use of metal in bridge building: wrought-iron chains were used for centuries to suspend walkways across deep chasms in China, and the era of cast iron structures began in 1779 when Abraham Darby III completed the world's first cast iron bridge across the Severn's Shropshire Gorge. But these were expensive exceptions as wood continued to rule even the most advanced pre-industrial societies, be it Qianlong's China or the Enlightenment France. Wood was the dominant material in construction, farm implements, artisanal machinery, and household objects. If these items, and other objects made of natural fibers, were to be excluded from a census of average family possessions, then the total mass of materials that were not derived from biomass amounted to as little as a few kilograms and typically to no more than 10 kg per capita. Moreover, those artifacts were fashioned from low-quality materials, with cast iron pots and a few simple clay and ceramic objects dominating the small mass of such possessions.

In contrast, by the end of the 20th century an average North American family owned directly more than 2 t of high-performance metal alloys in its cars, appliances, furniture, tools, and kitchen items. In addition, two-thirds of these families owned their house, and its construction needed metals (iron, steel, copper, aluminum), plastics, glass, ceramics, paper, and various minerals. Any reader who has never tried to visualize this massive increase of material possessions will enjoy looking at Menzel's (1994) photographs that picture everything that one family owned in 30 different countries (from Iceland to Mongolia) during the early 1990s. Differences between possessions in some of the world's richest and poorest countries that are so graphically captured in Menzel's book are good indicators of the material gain that distinguished modern economies from pre-industrial societies.

And, of course, total per capita mobilization of materials in modern economies, including a large mass of by-products and hidden flows (e.g., coal mining overburden, ore and crop wastes, or agricultural soil erosion), is vastly greater than direct household consumption or possessions. Studies of such resource flows show that annual per capita rates during the mid-1990s added up to 84 t in the United States, 76 t in Germany, and 67 t in the Netherlands (Adriaanse et al. 1997). When the U.S. total is limited only to materials actually used in construction and to industrial minerals, metals, and forestry products, its annual rate shows increase from about 2 t per capita in 1900 to about 10 t per capita after 1970.

This great mobilization of nonbiomass materials began during the two pre-WWI generations. In 1800 steel was a relatively scarce commodity, aluminum was not even known to be an element, and there were no plastics. But by 1870 the world's annual output of steel still prorated to less than 300 g per capita (yes, grams); there was no mass production of concrete, and aluminum was a rarity used to make jewelry. And then, as with electricity and internal combustion, the great discontinuity took place. By 1913 the world was producing more than 40 kg of steel per capita, a jump of two orders of magnitude in half a century, and the American average was more than 200 kg/person. The cost of aluminum fell by more than 90% compared to 1890, and concrete was a common, although an oft-reviled, presence in the built environment.

But the material revolution did not end with these newly ubiquitous items. Maturing synthetic inorganic chemistry began producing unprecedented volumes of basic chemicals (led by sulfuric acid), and by 1909 Fritz Haber succeeded to do what generations of 19th-century chemists had tried to accomplish as he synthesized ammonia from its elements, an achievement that opened the way for feeding 6 billion people by the end of the 20th century (Smil 2001). On the destructive side of the technical ledger, chemistry moved from gunpowder to nitroglycerine and dynamite and beyond to even more powerful explosives. Dynamism of the Age of Synergy thus did not come only from new energies and new prime movers; it was also fashioned by new materials whose quantity and affordability combined with their unprecedented qualities to set the patterns whose repetition and elaboration still shape our world.

Steel Ascendant

Steel is, of course, an old material, one with a long and intricate history that is so antithetically embodied in elegant shapes and destructive power of the highest quality swords crafted in elaborate ways in such far-flung Old World locations as Damascus and Kyoto (Verhoeven, Pendray, and Dauksch 1998). But it was only during the course of the two pre-WWI generations that steel became inexpensive, its output truly massive, and its use ubiquitous. Unlike in the case of electricity, where fundamental technical inventions created an entire industry *de novo*, steelmaking was a well-established business long before the 1860s—but one dominated by artisanal skills and hence not suited to high-volume, low-price production that was to become a hallmark of the late 19th century.

Lowthian Bell (1884:435–436), one of the century's leading metallurgists, stressed that by 1850 "steel was known in commerce in comparatively very limited quantities; and a short time anterior to that period its use was chiefly confined to those purposes, such as engineering tools and cutlery, for which high prices could be paid without inconvenience to the customer." But by

1850 at least one cause of the relative rarity of steel was removed as the production of pig (cast) iron, its necessary precursor, expanded thanks to taller,
and hence much more voluminous and more productive, coke-fueled blast
furnaces. Bell concluded that the typical furnaces were too low and too narrow
for the efficient reduction of iron ore, and his design increased the overall
height by 66%, the top opening by more than 80%, and the hearth diameter
by 33% (Bell 1884; figure 4.1).

Other important pre-1860s innovations were the introduction of hot blast
(this increased combustion efficiency), the recovery and reuse (for heating of
the blast air) of CO-rich hot gases escaping from the furnace's open top, freeing
of the hearth (support of the furnace's outer casing by cast iron pillars made
the hearth accessible from every direction and allowed the placing of a maximum number of tuyeres, or blast-delivering nozzles), and capping of the furnace with cup-and-cone apparatus introduced by George Parry in 1850 (it also
enabled a more even distribution of the charge inside the furnace). During the
latter half of the 19th century, the technical primacy in ironmaking passed
from the British to the Americans (Hogan 1971). Before 1870, the world's most
advanced group of tall blast furnaces operated in England's Northeast, using
Cleveland ores discovered in 1851 (Allen 1981). Pennsylvanian ironmakers then
took the lead, especially with the furnaces built at Carnegie's Edgar Thomson
Works: their hearth areas were more than 50% larger; their blast pressures and
rates were twice as high.

These advances resulted in unprecedented outputs of more than 300 t/day
by 1890. But by 1900 nearly all U.S. coke was still made in inefficient beehive
ovens rather than in by-product coking batteries, which nearly halved the
coking cycle, recovered valuable gases and chemicals, produced higher coke
yield, and could use a greater variety of coals (Porter 1924). Abundance of

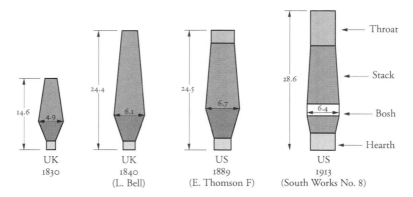

FIGURE 4.1. Increasing height and volume of blast furnaces, 1830–1913. Based on data
in Bell (1884) and Boylston (1936).

excellent coking coals and cheapness of the beehive ovens are the best explanations of this lag. As a result of these cumulative advances, the global output of pig iron roughly doubled to 10 Mt between 1850 and 1870, and then, as Americans and Germans took the technical lead away from the United Kingdom, it jumped to more than 30 Mt by 1900 and reached 79 Mt by 1913 (Kelly and Fenton 2003; figure 4.2).

This growth was not driven primarily by new technical capabilities that made such outputs of pig iron possible, but by rapidly rising demand for steel whose inexpensive production became a reality thanks to new processes that could convert pig iron into an alloy with superior qualities and hence with a much larger scope of applications than cast iron. Both pig iron and steel are alloys whose composition is dominated by the most common of all metals, but whose differences in carbon content translate into special physical, and hence also structural, properties. Cast iron contains between 2% and 4.3% carbon, while steel has merely 0.05–2% of the element (Bolton 1989).

Cast iron has very poor tensile strength (much less than bronze or brass), low impact resistance, and very low ductility. Its only advantage is good strength in compression, and its widest uses have been in such common artifacts as water pipes, motor cylinders, pistons, and manhole covers. In contrast, the best steels have tensile strengths an order of magnitude higher than

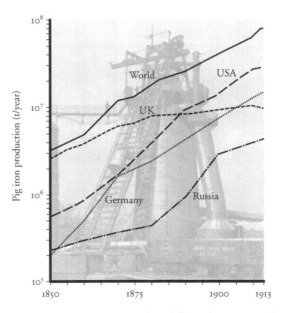

FIGURE 4.2. Pig iron smelting, 1850–1913. Plotted from data in Campbell (1907) and Kelly and Fenton (2003).

does pig iron, and they can tolerate impacts more than six times greater. They remain structurally intact at up to 750°C, compared to less than 350°C for cast iron. Addition of other elements—including, singly or in combination, Al, Cr, Co, Mn, Mo, Ni, Ti, V, and W, in amounts ranging from less than 2% to more than 10% of the mass—produces alloys with a variety of desirable properties. Low-carbon sheet steel goes into car bodies; highly tensile and hardened steels go into axles, shafts, connecting rods, and gear. Stainless steels are indispensable for medical devices as well as for chemical and food-processing equipment, and tool steels are made into thousands of devices ranging from chisels to extrusion dies.

Pre-industrial societies were producing limited amounts of steel primarily by a process of cementation, which added desired amount of carbon to practically carbon-free wrought iron. The alloy remained a commodity of restricted supply even by the mid—18th century, when cast iron production began rising substantially. Benjamin Huntsman (1704–1776) began producing his crucible cast steel by carburizing wrought iron during the late 1740s, but that metal was destined only for such specialized, limited-volume applications as razors, cutlery, watch springs, and metal-cutting tools. As the advancing industrialization needed more tensile metal, particularly for the fast growing railways, wrought iron filled the need.

This iron had to be produced in puddling furnaces by the sequence of reheating the brittle pig iron in shallow coal-fired hearths, and pushing and turning it manually with long rods in order to expose it to oxygen and to produce an alloy with a mere trace (0.1%) of carbon. Later this extraordinarily hard labor—which involved manhandling iron chunks of nearly 200 kg for longer than an hour in the proximity of very high heat—was done mechanically. The puddled material was then, after another reheating, rolled or hammered into desired shapes (rails, beams, plates). The first innovation that made large-scale steelmaking possible was introduced independently and concurrently by Henry Bessemer (1813–1898) in England and by William Kelly (1811–1888) in the United States. Bessemer revealed his process publicly in August 1856 and patented it in England (G.B. Patent 2219) in 1856 and in the United States (U.S. Patent 16,083) in 1857 (Bessemer 1905).

Molten pig iron was poured into a large pear-shaped tilting converter lined by siliceous (acid) material, and subsequent blasting of cold air through tuyeres would decarburize the molten metal and drive off impurities. Kelly rushed to file his patent (U.S. Patent 17,628 in 1857) only after he learned that Bessemer filed his in the United Kingdom, but was declared the inventor of the process by U.S. courts. Hogan (1971) compares the key sentences from both patent applications in order to demonstrate how similar was the reasoning behind them. The blowing period lasted as little as 15 and usually less than 30 minutes per batch, and it produced spectacular displays of flames and smoke issuing from the converter's mouth (figure 4.3). But the process proved to be a great

FIGURE 4.3. Turning gears and a cross section of Bessemer converter pouring the molten metal (top; reproduced from Byrn 1900) and a converter in operation at John Brown & Co. Foundry (bottom; reproduced from *The Illustrated London News*, July 20, 1889).

disappointment: while the air forced into molten iron would burn off carbon and silicon, phosphorus and sulfur were left behind.

This drawback was not discovered during Bessemer's pioneering work because, by chance, he did his experimental converting with pig iron made from Blaenavon ore that was virtually phosphorus-free. The simplest solution to the phosphorus problem was to use iron ores of great purity, but their supply was obviously limited. A technical fix for the sulfur problem that made the Bessemer process much more widely applicable was discovered by Robert Forester Mushet (1811–1891). This experienced metallurgist added small amounts of spiegel iron, a bright crystalline iron ore with about 8% Mn and 5% C, to the decarbonized iron in order to partially deoxidize the metal (Mn has a great affinity for O_2) as well as to combine with some of the sulfur and remove the resulting compounds in slag. Patent specifications for this process were filed in 1856 (G.B. Patent 2219), with most of the rights assigned to the Ebbw Vale Iron Co., which financed Mushet's experiments (Osborn 1952).

When the company failed to pay a required patent stamp duty, the process became a public property, and it could be freely used by Bessemer in licensing his technique. By 1861 Bessemer steel was being rolled into rails in a number of mills around England, and it was first produced in the United States in 1864. One more essential step was needed to make the process universal: to remove phosphorus from pig iron in order to be able to use many iron ores that contain the element. Many engineers tried for several decades to solve this challenge, and the solution was found by two young metallurgists, Sidney Gilchrist Thomas (1850–1885) and his cousin Percy Carlyle Gilchrist (1851–1935). Their reasoning was not new: to use a basic material that would react with the acidic phosphorus oxides present in the liquid iron and remove them in slag.

Practical realization of the idea had to overcome a number of problems, beginning with the preparation of durable basic linings and ending with the challenge of dumping large volumes of slag (Almond 1981). The cousins persevered, and after several years of experiments with different linings, they produced hard, dense, and durable blocks from impure limestone and sealed the joints with a mixture of tar and burned limestone (or dolomite). As the basic lining would be insufficient to neutralize the phosphorus compounds, they also added lime to the charged ore. In order to safeguard their process, Thomas and Gilchrist took out a dozen British and foreign patents between 1878 and 1879, and their innovation was finally acclaimed at the Iron and Steel Institute meeting in May 1879.

Soon afterward, the production of basic Bessemer steel took off throughout the continental Europe, where poor ores were widely used, and Germans turned the phosphoric slag into a fertilizer by simply grinding it. By 1890, nearly two-thirds of all European steel was smelted by the basic Bessemer process. This technique produced most of the world's steel between 1870 and 1910. Its share in the American output peaked at 86% of the total in 1890 (Hogan 1971). But this success was short-lived as it was a different procedure, one whose first practical application was one of the markers of the beginning of the Age of Synergy, that proved to be an epoch-making innovation and produced most of the world's steel for more than two-thirds of the 20th century.

Open-Hearth Furnace

The story of open-hearth steelmaking is one of the best illustrations of difficulties with attributing the origins of technical inventions and dating them accurately. The principal inventor of the furnace was one of four brothers of the German family that has no equals as far as the inventiveness of siblings is concerned. We have already met Werner, the founder of Siemens & Halske Co. and one of the inventors of self-excited dynamos (see figure 2.8). The adaptation of open-hearth furnace for more efficient steelmaking was the idea of Carl Wilhelm, known as William after he became a British subject in 1859 (figure 4.4). Wilhelm, an inventor and promoter in the fields of thermal and electrical engineering, elaborated the idea in cooperation with his younger brother Friedrich (1826–1904).

The idea of open-hearth melting itself was not new; it was the heat economy introduced into the process by Siemens's regenerative furnace that made the commercial difference. The furnace was very simple: a rectangular brick-lined chamber with a wide, saucer-shaped and shallow hearth whose one end was used to charge pig iron as well a small quantity of iron ore or steel scrap, and to remove the finished steel (Riedel 1994). Unlike in ordinary furnaces, where much of the heat generated by fuel combustion escaped with hot gases through a chimney, Siemens's furnace led the hot gases first through a regenerator, a chamber stacked with a honeycomb mass of bricks (chequers) that absorbed a large share of the outgoing heat.

As soon as the bricks were sufficiently heated, the hot gases were diverted into another regenerating chamber, while the air required for combustion was preheated by passing through the first, heated, chamber. After its temperature declined to a predetermined level, the air flow was reversed, and this alternating operation guaranteed the maximum recovery of waste heat. Moreover, the

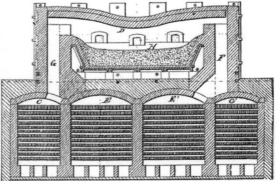

FIGURE 4.4. William Siemens (reproduced from *Scientific American*, December 22, 1883) and a section through his open-hearth steelmaking furnace. Gas and air are forced through chambers *C* and *E*, ascend separately through *G*, ignite in *D*, and melt the metal in hearth (*H*); hot combustion gases are led through *F* to preheat chambers *E'* and *C'*. Once these are hot, they begin receiving gas and air flows and the entire operation is reversed as combustion gases leave through *G* to preheat *C* and *E*. Reproduced from Byrn (1900).

gaseous fuel used to heat the open-hearth furnace (usually produced by incomplete combustion of coal) was also led into a regenerative furnace that then required four brick-stacked chambers (figure 4.4). This energy-conserving innovation (fuel savings amounted to as much as 70%) was first used in 1861 at glassworks in Birmingham.

In 1867, after two years of experiments, Siemens was satisfied that high temperatures (between 1,600°C and 1,700°C) generated by this process will easily remove any impurities from mixtures of wrought iron scrap and cast iron charged into the furnaces. As Bell (1884:426) put it, "the application of this invention to such a purpose ... is so obvious, that its aid was speedily brought into requisition in what is now generally known as the Siemens-Martin or open-hearth process." The double name is due to the fact that although the Siemens brothers did their first tests in 1857, and then patented the process in 1861, it took some time to perfect the technique, and a French metallurgist Emile Martin (1814–1915) succeeded in doing so first and filed definitive patents in summer 1865. Concurrent trials by Siemens were also promising, and by November 1866 Siemens and Martin agreed to share the rights: Siemens-Martin furnace was born, and the new process was finally commercialized for the first time in 1869 by three British steelworks.

Unlike in the Bessemer converter, where the blowing process was over in less than half an hour, open-hearth furnaces needed commonly half a day to finish the purification of the metal. Tapping of the furnace was done by running a crowbar through the clay stopper and pouring the hot metal into a giant ladle, which was then lifted by a crane and its contents poured into molds to form desired ingots. Experiments with open-hearth furnaces that were lined with basic refractories began during the early 1880s, and between 1886 and 1890 the process spread to the U.S. steelmaking (Almond 1981). Moreover, in 1890 Benjamin Talbot (1874–1947) devised a tilting furnace that made it possible to tap slag and steel alternatively and hence to turn the basic open-hearth steelmaking from a batch into a virtually continuous process.

As in so many other cases of technical advances, the initial design was kept largely intact, but the typical size and average productivities grew. During the late 1890s, the largest plant of the U.S. Steel Corporation had open-hearths with areas of about 30 m²; by 1914 the size was up to 55 m², and during WWII it reached almost 85 m² (King 1948). Heat sizes increased from just more than 40 t during the late 1890s to 200 t after 1940. At the beginning of the 20th century, an observer of an open-hearth operation at the Carnegie Steel Co.'s Homestead plant in Pennsylvania (Bridge 1903:149) noted that this way of steelmaking had

> none of the picturesque aspects of the Bessemer converter. The most interesting thing about it to a layman is to see, through colored glasses, how the steel boils and bubbles as if it were so much milk ... but the

gentle boiling of steel for hours without any fireworks or poetry, in a huge shed empty of workmen as church on weekdays, is not a very interesting sight.

But it was this placid, slow-working, improved, tiltable, basic Siemens-Martin furnace that came to dominate American steelmaking during the first two-thirds the 20th century. The steel that built most of New York's skyline, that went to reinforce the country's largest hydroelectric plants that were built between the 1930s and 1960s, steel that protected aircraft carriers at Midway in 1942 and made the armor of Patton's 5th Army tanks as they raced toward Paris in 1944, steel that is embedded in long stretches of the U.S. interstate highways and large airport runways—virtually all of that ferrous alloy came from open hearths. In the United States, its share rose from just 9% of all steel production in 1880 to 73% by 1914, and it peaked at about 90% in 1940; similarly, in the United Kingdom its share peaked just above 80% in 1960 (figure 4.5). These peaks were followed by rapid declines of the technique's importance as basic oxygen furnace became the dominant means of modern steelmaking during the last third of the 20th century, with electric arc furnaces not far behind.

The last smelting innovation that was introduced before 1914 was the electric arc furnace. William Siemens built the first experimental furnaces with electrodes placed at the top and the bottom of a crucible or opposed horizontally (Anonymous 1913c). Paul Héroult commercialized the process in 1902 in order to produce high-quality steel from scrap metal. These furnaces operate on the same principles as the arc light: the current passing between large carbon

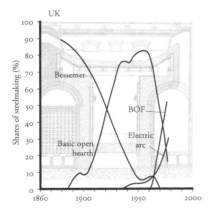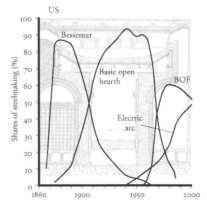

FIGURE 4.5. Shares of leading methods of steelmaking in the United Kingdom (1880–1980) and the United States (1860–2000). Plotted from data in Almond (1981) for the United Kingdom and in Campbell (1907) and USBC (1975) for the United States.

electrodes in the roof melts the scrap, and the metal is produced in batches. Other electric arc furnace designs were introduced shortly afterward, and by 1910 there were more than 100 units of different designs operating worldwide, with Germany in the lead far ahead of the United States (Anonymous 1913c). Subsequent combination of decreased costs of electricity and increased demand for steel during WWI transformed the electric arc furnaces into major producers of high-quality metal, with nearly 1,000 of them at work by 1920 (Boylston 1936). Their rise to dominance in modern steelmaking will be described in the companion volume.

Steel in Modern Society

All of these metallurgical innovations meant that the last quarter of the 19th century was the first time in history when steel could be produced not only in unprecedented quantities, but also to satisfy a number of specific demands and to be available in large batches needed to make very large parts. Global steel output rose from just half a million tons in 1870 to 28 Mt by 1900 and to about 70 Mt by 1913. Between 1867 and 1913, U.S. pig iron smelting rose 21-fold while steel production increased from about 20,000 t to nearly 31 Mt, a more than 1,500-fold rise (figure 4.4). Consequently, steel's share in the final output of the metal rose precipitously. In 1868 about 1.3% of the U.S. pig iron were converted to steel; by 1880 the share was slightly less than a third, and it reached about 40% by 1890, almost 75% by 1900, and 100% by 1914 (figure 4.6; Hogan 1971; Kelly and Fenton 2003).

As the charging of scrap metal increased with the use of open-hearth and, later, electric and basic oxygen furnaces, steel production began surpassing the pig iron output: by 1950 the share was 150%, and by the year 2000 U.S. steel production was 2.1 times greater than the country's pig iron smelting (Kelly and Fenton 2003). Inexpensive steel began to sustain industrial societies in countless ways. Initially, nearly all steel produced by the early open-hearth furnaces was rolled into rails. But soon the metal's final uses became much more diversified as it filled many niches that were previously occupied by cast and wrought iron and, much more important, as it found entirely new markets with the rise of many new industries that were created during the two pre-WWI generations.

Substitutions that required increasing amounts of steel included energy conversions (boilers, steam engines), land transportation (locomotives, rolling stock), and after 1877 when the Lloyd's Register of Shipping accepted steel as an insurable material, the metal rapidly conquered the shipping market. Steel substitutions also took place in the production of agricultural implements and machinery as well as in textile and food industries, in industrial machines and tools, and building of bridges. Undoubtedly the most remarkable, elegant,

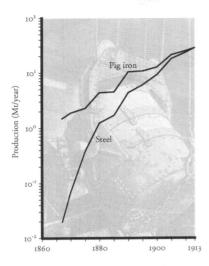
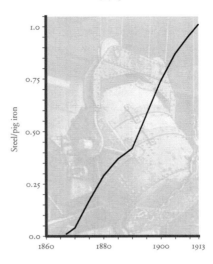

FIGURE 4.6. U.S. steel production and steel/pig iron shares, 1867–1913. Plotted and calculated from data in Temin (1964) and Kelly and Fenton (2003).

and daring use of steel in bridge building was the spanning of Scotland's Firth of Forth to carry the two tracks of the North British Railway (see frontispiece to this chapter).

This pioneering design by John Fowler and Benjamin Baker was built between 1883 and 1890, and it required 55,000 t of steel for its 104-m-tall towers and massive cantilevered arms: the central span extends more than 105 m, and the bridge's total length is 2,483 m (Hammond 1964). The structure—both reviled (William Morris called it "the supremest specimen of all ugliness") and admired for its size and form—has been in continuous use since its opening. Although a clear engineering success, this expensive cantilever was not the model for other similarly sized or larger bridges. Today's longest steel bridges use the metal much more sparingly by hanging the transport surfaces on steel cables: the central span of Japan's Akashi Kaikyo, the world's longest suspension bridge by 2000 that links Honshu and Shikoku, extends 1,958 m.

New markets that were created by the activities that came into existence during the last third of the 19th century, and that had enormous demands for steel, included most prominently the electrical industry with its heavy generating machinery, and the oil and gas industry dependent on drilling pipes, well casings, pipelines, and complex equipment of refineries. An innovation that was particularly important for the future of the oil and gas industry was the introduction of seamless steel pipes. The pierce rolling process for their production was invented in 1885 by Reinhard and Max Mannesmann at their father's file factory in Remscheid (Mannesmannröhren-Werke AG 2003). Several years later they added pilger rolling, which reduces the diameter and wall thickness while increasing the tube length. The combination of these two techniques, known as the Mannesmann process, produces all modern pipes.

The first carriagelike cars had wooden bodies, but a large-scale switch to sheet steel was made with the onset of mass automobile production that took place before 1910. Steel consumed in U.S. automaking went from a few thousand tons in 1900 to more than 70,000 t in 1910, and then to 1 Mt by 1920 (Hogan 1971). Expanding production of motorized agricultural machinery (tractors, combines) created another new market for that versatile alloy. Beginning in the 1880s, significant shares of steel were also used to make weapons and heavy armaments for use on both land and sea, including heavily armored battleships.

The pre-WWI period also saw the emergence of many new markets for specialty steels, whose introduction coincides almost perfectly with the two generations investigated in this book (Law 1914). Reenter Bessemer, this time in the role of a personal savior rather than an inventor. After Mushet lost his patent rights for perfecting the Bessemer process, his mounting debts and poor health led his 16-year-old daughter Mary to make a bold decision: in 1866 she traveled to London and confronted Bessemer in his home (Osborn 1952). We will never know what combination of guilt and charity led Bessemer to pay Mushet's entire debt (£377 14s 10d) by a personal check and later to grant him an allowance of £300 a year for the rest of his life. Mushet could thus return to his metallurgical experiments, and in 1868 he produced the first special steel alloy by adding a small amount of tungsten during the melt.

Mushet called the alloy self-hardening steel because the tools made from it did not require any quenching in order to harden. This precursor of modern tool steels soon became known commercially as RMS (Robert Mushet's Special Steel), and it was made, without revealing the process through patenting, by a local company and later, in Sheffield, under close supervision of Mushet and his sons. This alloy was superior to hardened and tempered carbon steel, the previous standard choice for metal-working tools. One of its most notable new applications was in the mass production of steel ball bearings, which were put first into the newly introduced safety bicycles and later into an expanding variety of machines; eventually (in 1907) the Svenska Kullager Fabriken (SKF) introduced the modern method of ball-bearing production.

Manganese—an excellent deoxidizer of Bessemer steels when added in minor quantities (less than 1%)—made brittle alloys when used in large amounts. But in 1882 when Robert Abbot Hadfield (1858–1940) added more of the metal (about 13%) to steel than did any previous metallurgist, he obtained a hard but highly wear-resistant, and also nonmagnetic, alloy that found its most important use in toolmaking. During the following decades, metallurgists began alloying with several other metals. Between 3% and 4% molybdenum was added for tool steels and for permanent magnets; as little as 2–3% nickel sufficed to make the best steel for cranks and shafts steel and, following James Riley's 1889 studies, also for armor plating of naval ships. Between 0.8% and 2% silicon was used for the manufacture of springs, and niobium and vanadium were added in small amounts to make deep-hardening steels (Zapffe 1948; Smith 1967).

Nickel and chromium alloy (Nichrome), patented by Albert Marsh in 1905, provided the best solution for the heating element for electric toasters: dozens of designs of that small household appliances appeared on the U.S. market in 1909 just two months after the alloy became available (NAE 2000; see figure 2.25). Finally, in August 1913 Harry Brearley (1871–1948) produced the first batch of what is now an extended group of stainless steels by adding 12.7% chromium, an innovation that was almost immediately used by Sheffield cutlers. Some modern corrosion-resistant steels contain as much as 26% chromium and more than 6% nickel (Bolton 1989), and their use ranges from household utensils to acid-resistant reaction vessels in chemical industries. Introduction of new alloys was accompanied by fundamental advances in understanding their microcrystalline structure and their behavior under extreme stress. Discovery of x-rays (in 1895) and the introduction of x-ray diffraction (in 1912) added new valuable techniques for their analysis.

During the 20th century, steel's annual worldwide production rose 30-fold to nearly 850 Mt (Kelly and Fenton 2003). Annual per capita consumption of finished steel products is a good surrogate measure of economic advancement. The global average in 2002 was just more than 130 kg, and the rate ranged from highs of between 250 and 600 kg for the affluent Western economies to about 160 kg in China, less than 40 kg in India, and just a few kilograms in the poorest countries of the sub-Saharan Africa (IISI 2003). Exceptionally high steel consumption rates for South Korea and Taiwan (around 1 t per capita) are anomalies that do not reflect actual domestic consumption but rather the extensive use of the metal in building large cargo ships for export.

I would be remiss if I were to leave this segment, devoted to the most useful of all alloys, without describing in some detail its most important use in a hidden and, literally, supporting role. Although normally painted, steel that forms sleek car bodies, massive offshore drilling rigs, oversize earth-moving machines, or elegant kitchen appliances is constantly visible, and naked steel that makes myriads of households utensils and industrial tools is touched every day by billions of hands. But one of the world's largest uses of steel, and one whose origins also dates to the eventful pre-WWI era, is normally hidden from us and is visible only in unfinished products: steel in buildings, where it is either alone in forming the structure's skeleton and bearing its enormous load or reinforces concrete in monolithic assemblies.

Steel in Construction

Buildings of unprecedented height were made possible because of the combination of two different advances: production of high-tensile structural steel, and the use of steel reinforcement in concrete. Structural steel in the form of I beams, which had to be riveted together from a number of smaller pieces, began to form skeletons of the world's first skyscrapers (less than 20 stories tall) during the 1880s. Its advantages are obvious. When tall buildings are built with solid masonry, their foundation must be substantial, and load-bearing walls in the lower stories must be very thick. Thinner walls were made possible by using cast iron (good in compression) with masonry, and later cage construction used iron frames to support the floors. The 10-story (42 m high) Chicago's Home Insurance Building, designed by William Le Baron Jenney (1832–1907) and finished in 1885 (the Field Building now occupies its site), was the first structure with load-carrying frame of steel columns and beams. The total mass of this structural steel was only a third of the mass of a masonry building and resulted in increased floor space and larger windows (figure 4.7).

As electric elevators became available (America's first one was used in Baltimore in 1887, and the first Otis installation was done in New York in 1889), and as central heating, electric plumbing pumps, and telephones made it more convenient to build taller structures, skyscrapers soon followed, both in Chicago and in New York. The early projects included such memorable, and no more standing, Manhattan structures as the World Building (20 stories, 94 m in 1890) and the Singer Building (47 stories, 187 m in 1908). Henry Grey's (1849–1913) invention (in 1897) of the universal beam mill made it possible to roll sturdy H beams. In the United States they were made for the first time by the Bethlehem Steel's new mill at Saucon, Pennsylvania, in 1907, and their availability made the second generation of skyscrapers possible.

Rolled in a single piece directly from an ingot, H beams had a substantially higher tensile strength (Cotter 1916; Hogan 1971). An undisputed paragon of structures using this new material is New York's Woolworth Building at 233 Broadway, designed by Cass Gilbert (1859–1934) and finished in 1913. The building used for the first time all of the key techniques that still characterize the construction of skyscrapers. Its 57 stories (241 m high) are founded on concrete piers that reach Manhattan's bedrock; sophisticated wind bracing minimizes its swaying, and high-speed elevators provide local and express service. A new, elite group of fearless construction ironworkers emerged to assemble these massive and tall steel skeletons—and a hundred years later they still form a special professional caste (figure 4.7).

The Woolworth building remained the city's tallest structure until 1930, when it was surpassed by the Bank of Manhattan and Chrysler buildings and, a year later, by the Empire State Building (Landau and Condit 1996). All of

FIGURE 4.7. The beginnings of the skyscraper era. The Home Insurance Building in Chicago, completed in 1885 (and demolished in 1931), was the first structure supported by steel beams and columns. The engraving of New York's ironworkers building an early skyscraper and taking "coolly the most hazardous chances" is reproduced from *The Illustrated London News* of December 26, 1903. A century later, safety harnesses are compulsory, but the basic assembly procedure and the need for casual tolerance of great heights and uncommon sense of balance and physical dexterity have not changed.

these structures, and most of their post-1970 companions—in the United States peaking with 443 m of Sears Tower in Chicago, and globally, as of the year 2000, with 452 m of Petronas Towers in Kuala Lumpur (figure 4.8)—share the same fundamental structural property, as all of the world's tallest skyscrapers hide variably shaped steel skeletons adorned with ornamental cladding of stone, metal, plastic materials, or glass.

The other major use of steel in construction, burying it inside concrete, also emerged during the pre-WWI period. This technique, whereby concrete gets its tensile strength from embedded steel bars or steel lattices, created much of the built environment of the 20th century. Concrete—a mixture of cement, sand, gravel, and water—is of course an ancient material, the greatest structural invention of the Roman civilization. The Pantheon, the pinnacle of its concrete architecture, built between 126 and 118 B.C.E., still stands in the heart of Rome's old city: its bold dome, spanning 43.2 m, consists of five rows of square coffers

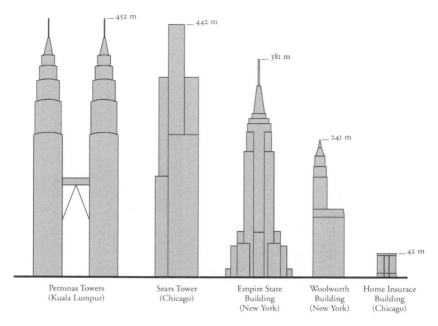

FIGURE 4.8. Height comparison of five notable steel-skeleton skyscrapers built between 1885 and 1998.

of diminishing size converging on the stunning unglazed central oculus (Lucchini 1966).

Better cement to produce better concrete has been available only since 1824, when Joseph Aspdin began firing limestone and clay at temperatures high enough to vitrify the alumina and silica materials and to produce a glassy clinker (Shaeffer 1992). Its grinding produced a stronger Portland cement (named after limestone whose appearance it resembled when set) that was then increasingly used in many compressive applications, mainly in foundations and walls. Hydration, a reaction between cement and water, produces tight bonds and material that is very strong in compression but has hardly any tensile strength, and hence it can be used in beams only when reinforced: such a composite material has monolithic qualities, and it can be fashioned into countless shapes.

Three things had to happen in order to have a widespread use of reinforced concrete: a good understanding of tension and compression forces in the material, the realization that hydraulic cement actually protects iron from rust, and understanding that concrete and iron form a solid bond (Peters 1996). But even before this knowledge was fully in place, there were isolated trials and projects that embedded cast or wrought iron into concrete beginning during the 1830s, and the first proprietary system of reinforcement was introduced in England and France during the 1850s (Newby 2001). In 1854 William Wilkinson patented wire rope reinforcement in coffered ceilings, and a year later François Coignet introduced a system that was used in 1868 to build a concrete lighthouse at Port Said. Fireproofing rather than the monolithic structure was seen as the material's most important early advantage.

During the 1860s, a Parisian gardener, Joseph Monier (1823–1906), patented a reinforced concrete beam (an outgrowth of his production of garden tubs and planters that he strengthened with simple metal netting in order to make it lighter and thinner), and in 1878 he got a general patent for reinforced structures. Many similar patents were granted during the two following decades, but there was no commensurate surge in using the material in construction. The pace picked up only after 1880, particularly after Adolf Gustav Wayss (1851–1917) bought the patent rights to Monier's system for Germany an Austria (in 1885) and after a Parisian contractor, François Hennebique (1842–1921), began franchising his patented system of reinforced construction, particularly for industrial buildings (Straub 1996; Delhumeau 1999). Inventors designed a variety of shaped reinforcing steel bars in order to increase the surfaces at which metal and concrete adhere (figure 4.9).

During the 1890s came the modern technique of cement production in rotary kilns that process the charge at temperatures of up to 1,500°C, and engineers developed standard tests of scores of various cement formulas and concrete aging. Thomas Edison got into the act with his cast-in-place concrete houses that he built in New Jersey (they still stand), but such utilitarian structures did not make concrete a material of choice for residential construction. Much more creatively, two architects, Auguste Perret and Robert Maillart (1872–1940), made the reinforced concrete esthetically acceptable. Perret did so with his elegant apartments, including the delicately facaded 25 Rue Franklin in Paris, and with public buildings, most notably the Théâtre des Champs-Élysées. Maillart designed and built (in Switzerland between 1901 and 1940) more than 40 elegant concrete bridges (Billington 1989). By the 1960s, new high-strength rein-

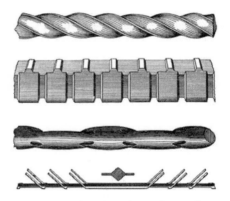

FIGURE 4.9. Shaped steel bars for reinforcing of concrete: spiral twist by E. L. Ransome, a square design with projections by A. L. Johnson, alternating flat and round sections by Edwin Thacher, and a bar with bent protrusions by Julius Kahn and the Hennebique Co. Reproduced from Iles (1906).

forced concrete began to compete with the steel frame in skyscraper construction (Shaeffer 1992).

By 1990 the world's tallest reinforced concrete structure, the 65-story 311 South Wacker Drive building in Chicago, reached 292 m, and even taller structures, with concrete able to withstand pressures of more than 100 MPa, were planned. After 1950 came widespread commercial applications of prestressing, another fundamental pre-WWI invention involving concrete and steel. Its origin can be dated exactly to 1886 when Carl Dochring came up with an ingenious idea to stretch the reinforcing bars while the concrete is wet and release the tension after the material had set (Abeles 1949). This puts the individual structural members, which are usually precast offsite, into compression and makes it possible to use much less steel and concrete (about 70% and 40% less) for the same load-bearing capacity and hence to build some amazingly slender structures. During the first decades of the 20th century, Eugène Freyssinet (1879–1962) developed much of the underlying technical understanding, and he also came up with the idea of poststressing by tensioning wires that are threaded through ducts formed in precast concrete (Grotte and Marrey 2000; Ordóñez 1979).

Reinforced concrete is simply everywhere—in buildings, bridges, highways, runways, and dams. The material assumes countless mundane and often plainly ugly shapes (massive gravity dams, blocky buildings) as well as daring and visually pleasing forms (thin shells). Its unmatched accretion is the world's largest dam—China's Sanxia on the Yangzi, 185 m tall and 2.3 km long—which contains nearly 28 million cubic meters of concrete reinforced with nearly half a million tons of steel bars. In terms of both global annual production and cumulatively emplaced mass, reinforced concrete is now the leading man-made material, seen as flat expanses and stunning arches, unimaginative boxes and elegant projections.

The material has been used for the shoddily built apartment blocks of Beijing's Maoist period (they looked dilapidated even before they were completed) and for the elegant sails of Sydney Opera House, which look permanently inspirational. Unadorned reinforced concrete forms the world's tallest free-standing structure, Toronto's oversize CN tower, as well as Frank Lloyd Wright's perfectly proportioned cantilevered slabs that carry Falling Water over a cascading Pennsylvanian stream. As Peters (1996:58) noted, "[R]einforced concrete has been variously attacked as the destroyer of our environment and praised as its savior," but whatever the reaction may be, his conclusion is indisputable: "Our world is unthinkable without it." Yet only a few history-minded civil engineers and metallurgists know its origins, and so reinforced concrete may be perhaps the least appreciated of all fundamental pre-WWI innovations, an even more obscure foundation of modern civilization than the electrical transformer described in chapter 2.

Aluminum

This light metal is much more common in nature than is iron: this most common of all metallic elements is, after oxygen and silicon, the third largest constituent of Earth's crust, amounting to about 8% by mass (Press and Siever 1986). But because of its strong affinity for oxygen, the metal occurs naturally only in compounds (oxides, hydroxides, fluorides, sulfates, and silicates) and never in a pure state. And unlike iron, which has been known and worked for more than 3,000 years, aluminum was identified only in 1808 by Humphry Davy (of the first electric arc fame), isolated in a slightly impure form as a new element only in 1825 by Hans Christian Ørsted (of the magnetic effect of electric currents fame), and two years later produced in powder form by Friedrich Wöhler (of the first ever urea synthesis fame). Despite many subsequent attempts to produce it in commercial quantities, aluminum remained a rare commodity until the late 1880s: Napoleon III's infant son got a rattle made of it, and 2.85 kg of pure aluminum were shaped into the cap set on the top of the Washington Monument in 1884 (Binczewski 1995).

The quest for producing the metal in quantity was spurred by its many desirable qualities. The most notable attribute is its low density: with merely 2.7 g/cm^3, compared to 8.9 g/cm^3 for copper and 7.9 g/cm^3 for iron, it is the lightest of all commonly used metals. Only silver, copper, and gold are better conductors of electricity and only a few metals are more malleable. Its tensile strength is exceeded only by best steels, and its alloys with magnesium, copper, and silicon further enhance its strength. Moreover, it can be combined with ceramic compounds to form composite materials that are stiffer than steel or titanium. Its high malleability and ductility mean that could be easily rolled, stamped, and extruded, and it offers attractive surface finishes that are highly resistant to corrosion. The metal is also generally regarded as safe for use with food products and for beverage packaging. Finally, it is easily recyclable, and its secondary production saves 95% of the energy needed for its production form bauxite.

Bauxite, named after the district of Les Baux in southern France where Pierre Berthier first discovered the hard reddish mineral in 1821, is the element's principal ore, and it contains between 35% and 65% of alumina (aluminum oxide, Al_2O_3). But the first attempts to recover the metal did not use this abundant ore. Henri Saint-Claire Deville (1818–1881) changed Wöhler's method by substituting sodium and potassium and began producing small amounts of aluminum by a cumbersome and costly process of reacting metallic sodium with the pure double chloride of aluminum and sodium at high temperature. This expensive procedure was used on a small scale for about 30 years in France. In 1854 Robert Bunsen separated aluminum electrolytically from its fused compounds, and shortly afterward Deville revealed his apparatus for

reducing aluminum from a fused aluminum-sodium chloride in a glazed porcelain crucible that was equipped with platinum cathode and with carbon
anode.

Although this process did not lead to commercial production, Borchers
(1904:113) summed up Deville's contribution by noting that he introduced the
two key principles that were then "repeatedly re-discovered and patented, viz:
1. The use of soluble anodes in fused electrolytes; and 2. The addition of
aluminium to fused compounds of the metal during electrolysis, by the agency
if alumina." This created a situation similar to pre-Edisonian experiments with
incandescing carbon filaments: the key ingredients of a desirable technique
(filament and vacuum for the light bulbs, soluble anodes, and addition of
alumina for the electrolysis of aluminum) did not have to be discovered, but
the techniques had to be made practical and rewarding. And, as with the
history of incandescent lights, many impractical solutions were put forward
between 1872 and 1885 to solve the challenge of large-scale smelting of aluminum. Worldwide production of the metal was just 15 t in 1885, and even
after Hamilton Y. Castner's new production method cut the cost of sodium
by 75%, aluminum still cost nearly US$20/kg in 1886 (Carr 1952).

A Duplicate Discovery

When the commercially viable solution finally came—during that extraordinarily eventful decade of the 1880s—it involved one of the most remarkable
instances of independent concurrent discovery. In 1886 Charles Martin Hall
and Paul Louis Toussaint Héroult (figure 4.10) were two young chemists of

FIGURE 4.10. Portraits of Charles Hall (left) and Paul Héroult from the late 1880s when
the two young inventors independently patented a nearly identical aluminum reduction
process. Photographs courtesy of Alcoa.

the same age (both born in 1863, and both died in 1914), working independently on two continents (the first one in a small town in Ohio, the second one near Paris) to come up with a practically identical solution of the problem during the late winter and early spring months of 1886. Just two years later, their discoveries were translated into first commercial enterprises producing electrolytic aluminum.

Hall's success was the result of an early determination and a critical chance encounter: his aspiration to produce aluminum dated to his high school years, but after he enrolled at the Oberlin College in Ohio in 1880, he had the great luck of meeting and working with Frank F. Jewett, an accomplished scientist who came to the college as a professor of chemistry and mineralogy after four years at the Imperial University in Tokyo and after studies and research at Yale, Sheffield, Göttingen, and Harvard (Edwards 1955; Craig 1986). Jewett provided Hall with laboratory space, materials, and expert guidance, and Hall began intensive research effort soon after his graduation in 1885.

Hall first excluded the electrolysis of aluminum fluoride in water because it produced only aluminum hydroxide. His next choice was to use water-free fused salts as solvents for Al_2O_3, and his first challenge was to build a furnace that would sustain temperatures high enough to melt the reactants, but he was not able to do so with his first externally heated gasoline-fired furnace. He then began to experiment with cryolite, the double fluoride of sodium and aluminum (Na_3AlF_6), and on February 9, 1866, he found that this compound with melting point of 1,000°C was a good solvent of Al_2O_3. As with so many early electrochemical experiments, Hall had to energize his first electrolysis of cryolite, which he did on February 16, 1886, with inefficient batteries. Graphite electrodes were dipped into a solution of aluminum oxide and molten cryolite in a clay crucible, but after several days of experiments he could not confirm any aluminum on his cathode.

Because he suspected that the cathode deposits originated in the crucible's silicon lining, he replaced it by graphite. On February 23, 1886, after several hours of electrolysis in a woodshed behind Hall's family house at 64 East College Street in Oberlin, he and his older sister Julia found deposits of aluminum on the cathode. Hall's quest is a near-perfect example of a fundamental technical advance that was achieved during the most remarkable decade of the Age of Synergy by the means so emblematic of the period: the combination of good scientific background, readiness to experiment, to persevere, and to make necessary adjustments (Hall built most of his experimental apparatus and also prepared many of his reactants), and all of this followed by prompt patenting, fighting off the usual interference claims, and no less persistent attention to scaling-up the process to industrial level and securing financial backing.

Hall's first American specification was received in the Patent Office on July 9, 1886—but it ran immediately into a problem as the office notified him of a previous application making a very similar claim: Héroult filed his first

French patent (Patent 175,711) on April 23, 1886, and by July 1887 he formally contested Hall's priority. But as Hall could prove the date of his successful experiment on February 23, 1886, he was eventually granted his U.S. rights, divided among two patents issued on April 2, 1889. U.S. Patents 400,664 and 400,766 specified two immersed electrodes, while U.S. Patents 400,667 specified the use of cryolite (figure 4.11).

While his patents were pending, Hall looked for financial backing, and after some setbacks and disappointments he finally found a committed support of a Pittsburgh metallurgist Alfred E. Hunt, and a new Pittsburgh Reduction Co. was formed in July 1888 and began reducing alumina before the end of the year (Carr 1952). This was a fairly small establishment powered by two steam-driven dynamos and producing initially just more than 20 kg of aluminum per day, and 215 kg/day after its enlargement in 1891 (Beck 2001). Its electrolytic cells (60 cm long, 40 cm wide, and 50 cm deep) were made of cast iron, held up to about 180 kg of cryolite, and had 6–10 carbon anodes (less than 40 cm long) suspended from a copper bus bar, with the tank itself acting as the cathode.

A new plant at New Kensington near Pittsburgh reached output of 900 kg/day by 1894, and in the same year an even larger facility was built at Niagara Falls to use the inexpensive electricity from what was at that time the world's largest hydro station. In 1907, the Pittsburgh Reduction Co. was renamed Aluminum Company of America, and Alcoa, now a multinational corporation

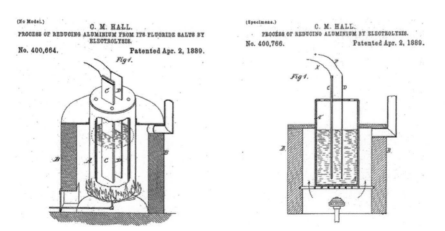

FIGURE 4.11. Illustrations of aluminum reduction process in Hall's two U.S. patents (400,664 on the left, 400,766 on the right) granted on April 2, 1889. In both images, A is a crucible, B a furnace, and C the positive and D the negative electrode. The first patent proposed $K_2Al_2F_8$ as the electrolyte; the second one, $Na_2Al_2F_8$.

with operation in 39 countries, remains the metal's largest global producer (Alcoa 2003). Canada's first plant, the precursor of today's large multinational Aluminum Company of Canada (Alcan), was built in 1901.

Similarities between Hall's and Héroult's independent specifications are striking. Héroult, who conducted his experiments on the premises of his father's small tannery and who did not rely on batteries but on a small steam-powered Gramme dynamo to produce a more powerful current, described his invention as a "process for the production of aluminium alloys by the heating and electrolytic action of an electric current on the oxide of aluminium, Al_2O_3, and the metal with which the aluminium shall be alloyed" (cited in Borchers 1904:127). Like Hall, he also made a provision for external heating of the electrolytic vessel, and his aluminum furnace also had carbon lining and a carbon anode. Héroult's technique was commercialized for the first time in 1888 by the Aluminium Industrie Aktiengesselschaft in Neuhausen in Switzerland and by the Société Electrométallurgique Française (Ristori 1911). Larger works, using inexpensive hydroelectricity, were soon established in France, Scotland, Italy, and Norway.

Mature Hall-Héroult Process

More than a century after its discovery, the Hall-Héroult process remains the foundation of the world's aluminum industry. Nothing can change the facts that the abundant Al_2O_3 is nonconducting and that molten cryolite, in which it gets dissolved, is an excellent conductor and hence a perfect medium for performing electrochemical separation. Heat required to keep the cryolite molten comes from the electrolyte's resistance and not from any external source; during the electrolysis, graphite is consumed at the anode with the release of CO_2 while liquid aluminum is formed at the cathode and collects conveniently at the bottom of the electrolytic vessel, where it is protected from reoxidation, because its density is greater than that of the supernatant molten cryolite.

Another important step toward cheaper aluminum that took place in 1888 and on which we continue to rely more than a century later was Karl Joseph Bayer's (1847–1904) patenting of an improved process for alumina production. After washed bauxite is ground and dissolved in NaOH, insoluble compounds of iron, silicon, and titanium are removed by settling, and fine particles of alumina are added to the solution to precipitate the compound, which is then calcined (converted to an oxide by heating) and shipped as a white powder of pure Al_2O_3 (IAI 2000). Typically, 4–5 t of bauxite will produce 2 t of alumina, which will yield 1 t of the metal.

As so many other late 19th-century innovations, the electrolytic process was greatly scaled up while its energy consumption fell considerably. During the

20th century, average cell size in Hall-Héroult plants had doubled every 18 years while energy use decreased by nearly half between 1888 and 1914 and then dropped by nearly as much by 1980 (Beck 2001; figure 4.12). By 1900 the worldwide output of aluminum reached 8,000 t, and its price fell by more than 95% compared to the cost in 1885 (Borchers 1904). By 1913 its global production was 65,000 t, and the demand rose to nearly 700,000 t by the end of WII. By the end of the 20th century, annual global output of aluminum from the primary electrolysis of alumina surpassed 20 Mt and was about 50% higher than the worldwide smelting of copper (IAI 2003). The metal is pro-

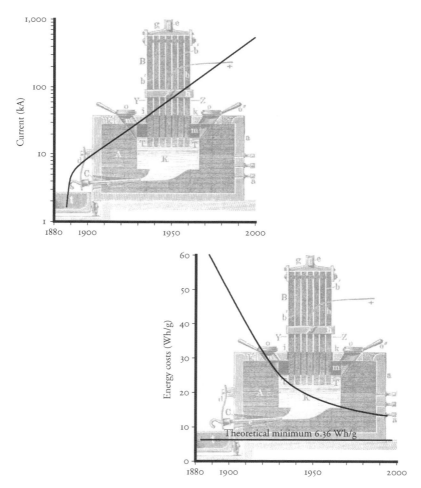

FIGURE 4.12. Increasing cell size (indicated by highter current) and declining energy cost of the Hall-Héroult aluminum production, 1888–2000. Based on Beck (2001).

duced both in pure (99.8%) form and as various alloys, and most of it is sent for final processing in rolls of specified thickness.

Inexpensive electrolytic aluminum was a novelty that had to take away market shares from established metals and create entirely new uses (Devine 1990a). During the early 1890s, the metal's main use was for deoxidation of steel, but a much larger market soon emerged with aluminum's use for cooking utensils, and then with parts for bicycles, cameras, cars, and many other manufactures. No machine is more unimaginable without aluminum and its alloys than the modern aircraft: steel alloys are the quintessential material in land and sea transport, but they are too heavy to be used for anything but some special parts in aircraft.

Aluminum bodies began displacing wood and cloth in aircraft construction during the late 1920s, after the mid-1930s demand for the metal was driven mainly by the need for large numbers of military planes (Hoff 1946). After WWII the market for aluminum alloys expanded rapidly with the large-scale adoption of military jet airplanes during the late 1940s and of commercial jets during the late 1950s. High-strength aluminum alloys (mainly alloy 7075 made with copper and zinc) make up 70–80% of the total airframe mass of modern commercial planes, which means that for the Boeing 747, this metal adds up to more than 100 t per plane (figure 4.13). The rest is accounted for by steel alloys, titanium alloys (in engines, landing gear) and, increasingly, composite materials (CETS 1993). Space flight and Earth-orbiting satellites have been another important market for such alloys.

The metal's light weight and resistance to corrosion also guaranteed its expanding uses in construction (from window and door frames to roofing and cladding), and aluminum extrusions are widely used in railroad rolling stock (large hopper cars that carry bulk loads), in automobiles and trucks (in 2000 a typical sedan contained twice as much Al, mostly in its engine, as it did in 1990), in both recreational boats and commercial shipping, and in food preparation (cooking and table utensils) and packaging (ubiquitous pop and beer cans, foils). And aluminum wires, supported by steel towers, are now the principal long-distance carriers of both HVAC and HVDC.

The Hall-Héroult process is not without its problems. Even after all those efficiency gains, it is still highly energy intensive: in 2002, primary aluminum production in the United States averaged about 15.2 MWh of electricity per ton of the metal (IAI 2003), or about 55 GJ/t. Adding the inevitable electricity losses in generation and transmission and energy costs of bauxite mining, production of alumina and electrodes, and the casting of the smelted metal raises the total rate to almost 200 GJ/t, and well above that for less efficient producers. This is an order of magnitude more than producing steel from blast-furnace smelted pig iron (EIA 2000). But as there are no obvious alternatives, it is safe to conclude that our reliance on those two great 1888 commercial innovations in metal industry—Bayer's production of alumina and Hall-

FIGURE 4.13. The Boeing 747, the largest passenger aircraft of the late 20th century whose construction requires more than 100 t of aluminum alloys. Photo from author's collection.

Héroult's electrolysis of the mixture of cryolite and alumina—will continue well into the 21st century.

New Syntheses

Even reasonably well-informed people do not usually mention explosives when asked to list remarkable innovations that helped to create, rather than to imperil, the modern world. While it is an indisputable historic fact that the immense increase in human destructive power attributable to more powerful explosives has been one of the key reasons behind inflicting greater losses in successive large-scale military conflicts, the constructive power of modern explosives has received a disproportionately small amount of attention. Yet their constant use is irreplaceable not only in construction projects but also in destruction of old buildings and concrete and steel objects and, above all, in ore and coal mining. How many Americans would even guess that during the late 1990s coal extraction (mostly in surface mines) accounted for 67% of all industrial explosives used annually in the United States (Kramer 1997)?

There was only a brief delay between the patenting of the first modern explosive and its widespread use. Dynamite patents were granted to Alfred Nobel (1833–1896; figure 4.14) in 1867 (G.B. Patent 1345) and in 1868 (U.S. Patent 78,317), and by 1873 the explosive was such a great commercial success that he could settle in Paris in a luxurious residence at 59 Avenue Malakoff, where a huge crystal chandelier in his large winter garden lit his prize orchids and where a pair of Russian horses was waiting to take his carriage for a ride in the nearby Bois de Boulogne (Fant 1993). In contrast, the first new synthetic materials—nitrocellulose-based Parkesine (1862) and Xylonite (1870) used for combs, shirt collars, billiard balls, and knife handles—made little commercial difference after their introduction. And the first commercial product made from the first successful thermoset

FIGURE 4.14. Alfred Nobel. His wealth and fame did little to ease his near-chronic depression and feelings of worthlessness (see chapter 6). Photograph © The Nobel Foundation.

plastic—a Bakelite gear-lever knob for a Rolls Royce—was introduced only in 1916, nearly a decade after Leo Baekeland filed his patent for producing the substance (Bijker 1995).

Numerous, and always only black, Bakelite moldings, sometimes elegant but more often either undistinguished or outright ugly-looking (and still too brittle)—ranging from the classic rotary telephone to cigarette holders, and from early radio dials to electric switch covers—began appearing only after WWI. So did products made of polyvinyl chloride, now one of the most commonly used plastics encountered in cable and wire insulation as well as in pipes and bottles, a polymer that was first produced in 1912. Many new plastics were introduced during the 1930s, but their widespread commercial use came only after 1950. Modern plastics are thus much less beholden to the pre-WWI period than is electricity or steel, and they fill the image of quintessential 20th-century materials. Consequently, their introduction, proliferation, and impact will be examined in this book's companion volume.

But there is another chemical synthesis, one involving a deceptively simple reaction of two gases, whose laboratory demonstration was followed by an unusually rapid commercialization that established an entirely new industry just before the beginning of WWI. Nearly a century later this accomplishment ranks as one of the most important scientific and engineering advances of all times. As such, this process deserves closer attention, which is why this chapter closes with a fairly detailed narrative and appraisal of the Haber-Bosch synthesis of ammonia from its elements.

Powerful Explosives

Introduction of new powerful explosives during the 1860s ended more than half a millennium of gunpowder's dominance. Clear directions for preparing different grades of gunpowder were published in China by 1040, and its European use dates to the early 14th century (Needham et al. 1986). The mixture of the three components capable of detonation typically included about 75% saltpeter (potassium nitrate, KNO_3), 15% charcoal, and 10% sulfur. Rather than drawing oxygen from the surrounding air, ignited gunpowder used the element present in the saltpeter to produce a very rapid (roughly 3,000-fold) expansion of its volume in gas. Historians have paid a great deal of attention to the consequences of gunpowder use in firearms and guns (Anderson 1988), particularly to its role in creating and expanding the European empires after 1450 (Cipolla 1966; McNeill 1989).

Gunpowder's destructive capacity was surpassed only in 1846, when Ascanio Sobrero (1812–1888), a former student of Justus von Liebig who also worked with Alfred Nobel when they both studied with Jules Pelouze in Paris in the early 1850s, created nitroglycerin. This pale yellow, oily liquid is highly explosive on rapid heating or on even a slight concussion, and it was produced by replacing all the $-OH$ groups in glycerol—$C_3H_5(OH)_3$—with $-NO_2$ groups (Escales 1908; Marshall 1917). Glycerol, a sweet syrupy liquid, is the constituent of triglycerides, and hence it is readily derived from lipids or from fermentation of sugar; among its many commercial uses are antifreeze compounds and plasticizers. Nitroglycerol is the most common of nitroglycerin's many synonyms, but the proper chemical designation is 1,2,3-propanetriol trinitrate (figure 4.15). The compound is best known for its use as an antianginal agent and coronary vasodilator.

After an explosion at the Torino armory, Sobrero stopped all experiments with the dangerous substance and assuaged his anxiety about discovering such a terrible compound with the thought that sooner or later some other chemist would have done it. But Sobrero did not become famous for his invention: who knows his name today besides the historians of chemistry? Enter Nobel, whose great fame does not rest on discovering a new explosive compound but, first, on a process to ignite nitroglycerine in a controllable and predictable fashion and then on the preparation and marketing of the substance in a form that is relatively safe to handle and practical to use for industrial blasting. Nobel's nitroglycerin experiments began in 1862 when he was detonating the

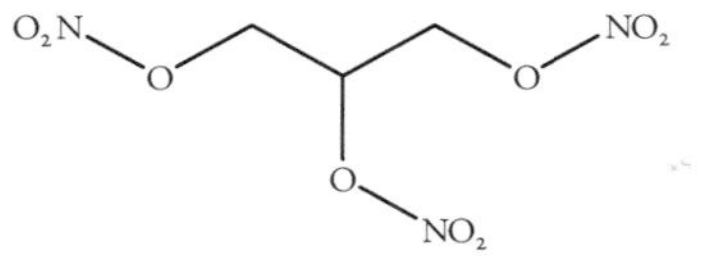

Nitroglycerine

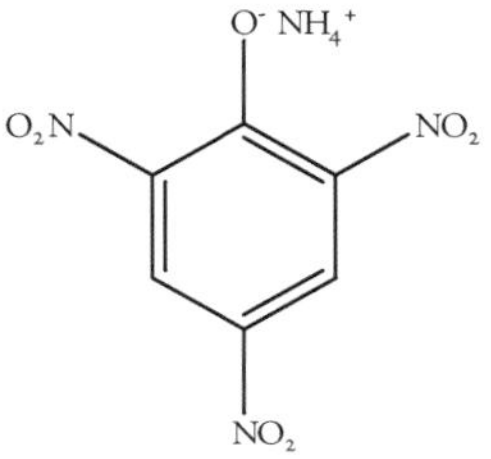

Ammonium picrate 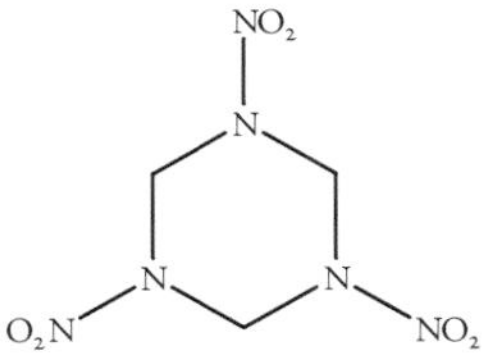   Cyclonite 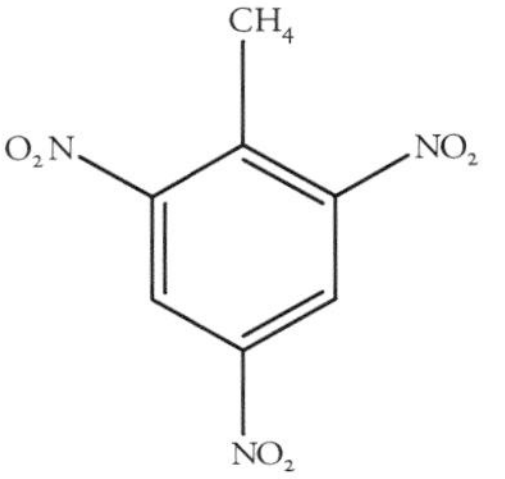   TNT

FIGURE 4.15. Structures of 1,2,3-propanetriol trinitrate (nitroglycerine), ammonium picrate (2,4,6-trinitrophenol ammonium salt, $C_6H_6N_4O_7$), cyclonite (hexahydro-1,3,5-trinitro-1,3,5-triazine, $C_6H_6N_6O_6$), and TNT (1-methyl-2,4,6-trinitrobenzene, $C_7H_5N_3O_6$).

compound by using small amounts of black powder attached to a slow fuse. In 1863 he got his first patent for using nitroglycerin in blasting and introduced an igniter (a wooden capsule filled with black powder) to trigger the compound's explosion.

In November of the same year he invented dynamite, a new explosive that, to quote his U.S. Patent 78,317 granted on May 26, 1868 (Nobel 1868:1),

> consists in forming out of two ingredients long known, viz. the explosive substance nitro-glycerine, and an inexplosive porous substance, hereafter specified, a composition which, without losing the great explosive power of nitro-glycerine, is very much altered as to its explosive and other properties, and which, at the same time, is free from any quality which will decompose, destroy, or injure the nitro-glycerine, or its explosiveness.

Nobel decided to give this combination a new name, dynamite, "not to hide its nature, but to emphasize its explosive traits in the new form; these are so different that a new name is truly called for" (cited in Fant 1993:94). Nobel's dynamite was made with highly porous *Kieselguhr*, a mineral that was readily available along the banks of the Elbe and the Alster (the Elbe's small northern tributary) near Hamburg, where Nobel moved from Sweden in 1865. The mineral is diatomaceous earth formed of siliceous shells of unicellular microscopic diatoms, aquatic protists found in both fresh and marine waters that often display nearly perfect radial symmetry. Diatomaceous earth—also used as an insulator, as abrasive in metal polishes, an ingredient in toothpaste, to clarify liquids, and as a filler for paper, paints, and detergents—can absorb about three times its mass of nitroglycerine.

A mixture of 25% diatomaceous earth and 75% nitroglycerine produces a reddish yellow, crumbly mass of plastic grains that is slightly compressible and safe from detonation by ordinary shocks (Shankster 1940). The explosive has been usually sold pressed into cartridges enclosed in waxed paper wrappers. Today's standard dynamite composition is basically the same as Nobel's ratio: 75% nitroglycerol, 24.5% diatomaceous earth, and 0.5% sodium carbonate, the last compound being added in order to neutralize any acidity that might be formed by the decomposition of nitroglycerin. Nobel returned to complete the invention in 1866, when he came up with a detonator, without which the new explosive would be nearly impossible to use. This igniter was a metal capsule filled with $Hg(CNO)_2$ (mercury fulminate): a strong shock produced by this primary explosive, rather than high heat, set off the nitroglycerin.

The Nobel igniter, patented in 1867, opened the way for widespread use of dynamite, making it one more critically important item to mark the beginning of the Age of Synergy, whose ethos Nobel embodied as much as did Edison, Hall, Parsons, or Daimler: he wanted to see his inventions commercialized and diffused as rapidly as possible. By 1870 he had five nitroglycerine factories in Europe, established the first dynamite factory in Paulilles in southern France in 1871, and watched the explosive conquer the blasting market worldwide. By 1873 there were 15 dynamite factories in Europe and North America, and between 1871 and 1874 the total output of the explosive had quadrupled to just more than 3,000 t. As the most powerful explosive yet available—its velocity of detonation is as much as 6,800 m/s, compared to less than 400 m/s for gunpowder (Johansson and Persson 1970)—the compound was more effective in breaking rocks apart rather than just displacing them. Consequently, blasting for mining operations, railroad construction, and other building projects became not only safer but also more efficient and less expensive.

Nobel had tried unsuccessfully to gelatinize nitroglycerin already in 1866, and he finally succeeded in 1875 by preparing a gelatinous dynamite, the combination of nitroglycerine (about 92%) and nitrocellulose (cotton treated with nitric acid, also known as nitrocotton or gun cotton). This tough yet plastic and water-resistant compound became the explosive of choice for packing boreholes in rock-blasting operations. Cheaper ammonium nitrate was later used to replace a relatively large share of nitroglycerin and to produce many kinds of plastic and semiplastic ammonia gelatin dynamites of different explosive intensity. These substances contain 50–70% explosive gelatin (a mixture of nitroglycerine and nitrocotton), 24–45% ammonium nitrate, and 2–5% wood pulp (Shankster 1940).

Ammonium nitrate—a compound that was first made by J. R. Glauber in 1659 by reacting ammonium carbonate with nitric acid—is difficult to detonate, but it is a powerful oxidizing agent, and its heating under confinement results in highly destructive explosions. Mixtures of ammonium nitrate and carbonaceous materials (charcoal, sawdust) were patented as explosives by C. J. Ohlsson and J. H. Norbin in 1867, but mixtures of the compound with fuel oil (generally referred to as ANFO) became commonly used only during the 1950s (Marshall 1917; Clark 1981). Since that time, explosives based on ammonium nitrate that are inexpensive and safe to use have captured most of the blasting market by replacing dynamite; some 2.5 Mt of these mixtures accounted for 99% of U.S. industrial explosives sales during the late 1990s (Kramer 1997).

Unfortunately, ANFO has been used not only as a common industrial explosive but also in powerful car and truck bombs by terrorists. The two worst terrorist attacks that took place in the United States before September 11, 2001—the bombing of the World Trade Center in New York City on February 26, 1993 (when, fortunately, only six people were killed), and the destruction of the Alfred Murrah Federal Office Building in Oklahoma City on April 19, 1996 (169 adults and children killed)—involved truck bombs using ANFO mixtures. The bombing also resulted in a renewed interest in additives able to desensitize ammonium nitrate, and new tests showed that this option does not work very well with larger volumes of explosives that are used in terrorist acts: the Oklahoma bomb contained about 1.8 t of nitrate (Hands 1996).

Although Nobel was frequently reproached for inventing such a destructive substance, dynamite did not take place of gunpowder for military uses. While it made it easier to commit terrorist attacks—in 1882 Czar Alexander II was among the first victims of a dynamite bomb—slower acting, and preferably smokeless propellants were needed for guns. Such explosives were produced during the 1880s: *poudre B* (nitrocellulose is first gelatinized in ether and alcohol and then extruded and hardened to produce a propellant that could be handled safely) by Paul Vieille (1854–1934) in 1884, Nobel's own Ballistite (using nitroglycerine instead of ether and alcohol) in 1887, and cordite, patented in England by Frederick Abel and James Dewar in 1889. The combination of these new powerful propellants and better guns made from new alloys resulted in considerably longer firing ranges.

Destructiveness of these weapons was further increased by introducing new explosive fillings for shells. Ammonium picrate was prepared in 1886; trinitrotoluene (TNT), which was synthesized by Joseph Wilbrand in 1863 and which must be detonated by a high-velocity initiator, was first manufactured in Germany in 1891 and by 1914 became a standard military compound; and the most powerful of all prenuclear explosives, cyclotrimethylenetrinitramine, known either as cyclonite or RDX (for royal demolition explosive), was first made by Hans Henning in 1899 for medicinal uses by treating a formaldehyde derivative with nitric acid, and its widespread use came only during WWII.

Structural formulas of these three powerful explosives clearly show that they are variations on the common theme, namely, a benzene ring with three nitro groups (figure 4.15). Cyclonite production took off only after cheap supplies of formaldehyde became available, and its combination of high velocity of detonation (8,380 m/s) and large volume of liberated gas make it an unsurpassed choice for bursting charges. Consequently, this substance was produced and used on a massive scale during WWII, as was TNT, which was deployed in more than 300 million antitank mines.

Ammonia from Its Elements

Inclusion of this item may seem puzzling, as the synthesis of the simplest triatomic nitrogen compound does not figure on commonly circulating lists of the greatest invention in modern history. Whether they were assembled by the canvassing of experts or of the general public, these compilations (a few of them were noted in chapter 1) almost invariably contain automobiles, telephones, airplanes, nuclear energy, space flight, television, and computers as the most notable technical inventions of the modern era. And yet, when measured by the most fundamental of all yardsticks, that of basic human physical survival, none of these innovations has had such an impact on our civilization as did the synthesis of ammonia from its elements (Smil 2001). The two keys needed to unlock this puzzle are one of the peculiar attributes of human metabolism and immutable realities of nitrogen's presence and cycling in the biosphere.

In common with other heterotrophs, humans cannot synthesize amino acids, the building blocks of proteins, and have to ingest them (eight essential amino acids for adults, and nine for children) in plant or animal foods in order to grow all metabolizing tissues during childhood and adolescence and to maintain and replace them afterward. Depending on the quality of their diets, most adults need between 40 and 80 g of food proteins (which contain roughly 6–13 g of nitrogen) each day. Crops, or animal foods derived from crop feeding, provide nearly 90% of the world's food protein, with the rest of the nutrient coming from grasslands, fresh waters, and the ocean. In order to produce this protein, plants must have an adequate supply of nitrogen—and this macronutrient is almost always the most important factor that limits the yields in intensive cropping. Unfortunately, the natural supply of nitrogen is restricted. Nitrogen makes up nearly 80% of Earth's atmosphere, but the molecules of dinitrogen gas (N_2) are nonreactive, and only two natural processes can split them and incorporate the atomic nitrogen into reactive compounds: lightning and biofixation.

Lightning, producing nitrogen oxides, is a much smaller source of reactive nitrogen than is biofixation, whereby free-living cyanobacteria and bacteria and, more important, bacteria symbiotic with legumes (above all, those belonging to genus *Rhizobium*), use their unique enzyme, nitrogenase, to cleave the atmospheric dinitrogen's strong bond and to incorporate the element first into ammonia and eventually into amino acids. In addition, reactive nitrogen is easily lost from soils through leaching, volatilization, and denitrification. As a result, shortages of nitrogen limit the crop yields more than shortages of any other nutrient. Traditional agricultures could improve the nutrient's supply only by cultivating leguminous crops or by recycling crop residues and animal and human wastes. But this organic matter has only very low concentrations of nitrogen and even its (utterly impracticable) complete recycling would be able to support only a finite number of people even if they were to subsist on a largely vegetarian diet.

In contrast, it is much easier to cover any shortages of phosphorus and potassium. The latter nutrient is not commonly deficient in soils, and it is easily supplied just by mining and crushing potash. And the treatment of phosphate rocks by diluted sulfuric acid, the process pioneered by John Bennett Lawes (1814–1900) in England during the 1860s, yields a fairly concentrated source of soluble phosphorus in the form of ordinary superphosphate. But at the close of the 19th century, there was still no concentrated, cheap, and readily available source of fertilizer nitrogen needed to meet higher food requirements of expanding populations. And not for the lack of effort.

Two new sources were commercially introduced during the 1840s—guano (bird droppings that accumulated on some arid islands, mainly in the Pacific) and Chilean nitrate ($NaNO_3$) extracted from near-surface deposits in the country's northern desert regions—but neither had a high nitrogen content (15% in nitrate, up to 14% in the best and less than 5% in typical guanos), and both sources offered only a limited supply of the nutrient. So did the recovery of ammonium sulfate from coking ovens, which was introduced for the first time in Western Europe starting in the 1860s: the compound contained only 21% N. As the 19th century was coming to its close, there was a growing concern about the security of future global food supply, and no other scientist summarized these worries better than William Crookes (1832–1919), a chemist (as well as physicist) whom we will again encounter in chapter 5 in relation to his comments on wireless transmission.

Crookes based his presidential address to the British Association for the Advancement of Science that he delivered in September 1898 (Crookes 1899) on "stubborn facts"—above all, the limited amount of cultivable land, growing populations, and relatively small number of food-surplus countries. He concluded that the continuation of yields that prevailed during the late 1890s would lead to global wheat deficiency as early as in 1930. Higher yields, and hence higher nitrogen inputs, were thus imperative, and Crookes saw only one possible solution: to tap the atmosphere's practically unlimited supply of non-reactive nitrogen in order to produce fixed nitrogen that could be assimilated by plants.

He made the existential nature of this challenge quite clear (Crookes 1899: 45–46):

> The fixation of nitrogen is vital to the progress of civilised humanity. Other discoveries minister to our increased intellectual comfort, luxury, or convenience; they serve to make life easier, to hasten the acquisition of wealth, or to save time, health, or worry. The fixation of nitrogen is a question of the no far-distant future.

And he had no doubt as to how this will happen: "It is the chemist who must come to rescue . . . It is through the laboratory that starvation may ultimately be turned into plenty" (Crookes 1899:3).

But in 1898 there was no technical breakthrough in sight. The German cyanamide process—in which coke was reacted with lime and the resulting calcium carbide was combined with pure nitrogen to produce calcium cyan-

FIGURE 4.16. Fritz Haber's portrait (left) taken during the early 1920s. Photograph courtesy of Bibliothek und Archiv zur Geschichte der Max-Planck-Gessselschaft, Berlin. Carl Bosch's photograph (right) from the early 1930s. Photograph courtesy of BASF Unternehmensarchiv, Ludwigshafen.

amide ($CaCN_2$)—was commercialized in that year, but its energy requirements were too high to become a major producer of nitrogen fertilizer. That the sparking of nitrogen and oxygen can produce nitrogen oxides (readily convertible to nitric acid) was known for more than a century, but very high temperatures needed for this process (up to 3,000°C) could be produced economically only with very cheap hydroelectricity.

The first commercial installation of this kind (the Birkeland-Eyde process) was set up in Norway in 1903, and it was enlarged before 1910, but the total output remained small. The breakthrough came 11 years after Crookes's memorable speech, when Fritz Haber (1868–1934; figure 4.16) demonstrated his process of ammonia synthesis from its elements. And just four years after that, his small bench-top apparatus was transformed into a large-scale commercial process by a dedicated engineering team led by another outstanding German chemist, Carl Bosch (1874–1940; figure 4.16).

Genesis of the Haber-Bosch Process

Fritz Haber was born to a well-off merchant family (dyes, paints, chemicals) in Breslau, and as a young man took advantage of the advancing integration of well-educated Jews into privileged German society. He studied in Berlin, where Carl Liebermann (1842–1914), best known for the synthesis of alizarin (a red dye), was one of his supervisors, and in Heidelberg under Robert Wilhelm Bunsen. Afterward, he went through a number of jobs before Hans Bunte offered him an assistantship at the Technische Hochschule in Karslruhe, where Haber stayed for the next 17 years, working initially in electrochemistry and then in thermodynamics of gases.

His first experiments with ammonia took place in 1904 at the request from the Österreichische Chemische Werke in Vienna, and as noted in his 1919 Nobel lecture, the challenge was only seemingly simple (Haber 1920:326):

> We are concerned with a chemical phenomenon of the simplest possible kind. Gaseous nitrogen combines with gaseous hydrogen in simple quantitative proportions to produce ammonia. The three substances involved have been well known to the chemist for over a hundred years. During the second half of the last century each of them has been studied hundreds of times in its behaviour under various conditions during a period in which a flood of new chemical knowledge became available.

But none of the many illustrious chemists who attempted to perform that simple reaction—$N_2 + 3H_2 \leftrightarrow 2NH_3$—during the 19th century had succeeded in the task, and by the beginning of the 20th century it appeared that such a synthesis may not be possible, a perception that deterred many new attempts.

Haber's first experiments to synthesize ammonia at 1,000°C and at atmospheric pressure were soon abandoned due to very low yields obtained with iron catalyst, which he chose after testing more than 1,000 materials. He returned to the task only three years later, in a large part in order to answer an unjustified criticism that Walther Nernst, one of the leading chemists of the early 20th century, leveled against his previous experiments. But Nernst was right in pointing out that increased pressure should lower the reaction temperature, and almost immediately after Haber, with his English assistant Robert Le Rossignol, began new experiments under a pressure of 3 MPa, he obtained much higher, although still clearly noncommercial, yields (Smil 2001).

His new calculations showed that a pressure of 20 MPa, at that time the limit he could obtain in his laboratory, and a temperature of 600°C would yield about 8% ammonia—but he had no catalyst that would work at such a relatively low temperature. By 1908, his search for ammonia synthesis received assistance from BASF, at that time the world's largest chemical company, which was interested in advancing research on nitrogen fixation by the Birkeland-Eyde process. As already noted, this technique used a high-voltage electric arc to produce nitrogen oxide that could be later converted to nitric acid. In 1907, Haber and his pupil A. König published a paper on that topic, and their work attracted the attention of BASF. On March 6, 1908, the company concluded two agreements with Haber: the first one provided fairly generous financial assistance to do more work on the electric arc process; the second one, a smaller sum of money to continue his previous work on high-pressure ammonia synthesis.

Nothing pathbreaking ever came out of arc experiments, while the high-pressure catalytic synthesis, thanks to Le Rossignol's skills in constructing the necessary apparatus (including a small double-acting steel pump) and Haber's perseverance in searching for better catalysts and for the best combination of pressure and temperature, began soon yielding promising results. Haber's first German patent for ammonia synthesis (Patent 235,421 valid since October 13, 1908) is generally known as the circulation patent, and its basic principle is still at the core of every ammonia plant today (BASF 1911). Commercialization of the process could not be done without catalysts that could support rapid conversion at temperatures of 500–600°C.

Initially Haber believed that osmium would be the best choice, and the finely divided metal supporting ammonia yields of about 6% and higher was clearly of economic interest; hence, he advised BASF to buy up all the osmium that was at that time in the possession of a company that used it to produce incandescent gas mantles. Haber filed his osmium catalyst patent on March 31, 1909, and continued to search for other catalysts; uranium nitride looked particularly promising. But above all, he concentrated on perfecting his laboratory apparatus built to demonstrate the potential of high-pressure synthesis with gas recirculation, and by the end of June the setup was finally ready.

Convincing demonstration of ammonia synthesis is one of those rare great technical breakthroughs that can be dated with great accuracy. On July 3, 1909, Haber sent a detailed letter to the BASF directors (now in the company's Ludwigshafen archives) that described the events of the previous day when Alwin Mittasch and Julius Kranz witnessed the demonstration of Haber's synthesis in his Karslruhe laboratory (Haber 1909):

> Yesterday we began operating the large ammonia apparatus with gas circulation in the presence of Dr. Mittasch, and we were able to keep its uninterrupted production for about five hours. During this whole time it had functioned correctly and it produced liquid ammonia continuously . . . The steady yield was 2 cm³/minute and it was possible to raise it to 2.5 cm³. This yield remains considerably behind the capacity for which the apparatus has been constructed because we have used the catalyst space very insufficiently.

The apparatus used in the experiment is shown in figure 4.17. Inside an iron tube kept under the pressure of 20.3 MPa was a nickel heating coil used to raise the gas temperature to a desired level; the synthesized ammonia was separated from the flowing gas at a constant high pressure, and the heat from the exothermic reaction could be removed from exhaust gases and used to preheat the freshly charged gas replacing the synthesis gas removed by the conversion. Rossignol designed a small double-acting steel pump to circulate

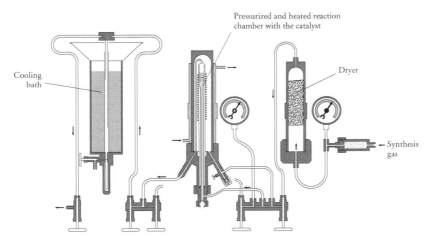

FIGURE 4.17. Laboratory apparatus that was used by Haber and Le Rossignol in their successful experiments to synthesize ammonia from its elements. Reproduced from Smil (2001).

the gases, and their recycling over the catalyst made it possible to sustain relatively high production rates even though the thermodynamic equilibrium was not particularly favorable. Haber's third key German patent application (Patent 238,450 submitted on September 14, 1909, and issued on September 28, 1911) detailed the synthesis under very high pressure (Haber 1911).

Once there was no doubt about the feasibility of high-pressure catalytic synthesis of ammonia from its elements, BASF moved swiftly to commit its resources to the commercialization of the process. This challenge called for scaling up Haber's bench-top apparatus—a pressurized tube that was just 75 cm tall and 13 cm in diameter—into a large-throughput assembly operating under pressures that were at that time unprecedented in industrial synthesis and filling it up with a relatively inexpensive yet highly effective catalyst. And it also called for, first and foremost, producing economically large volume of the two feedstock gases. Rapid success could not have happened without Carl Bosch's decisive managerial leadership and his technical ingenuity.

Bosch came to BASF in 1899 directly after graduating from Leipzig University and eventually was put in charge of BASF's nitrogen fixation research. That goal eluded him as it did so many other outstanding synthetic chemists of the time. But when Haber demonstrated his bench-top process to BASF, it was Bosch who was in charge of turning it into a profitable commercial operation. I have already related in chapter 1 how in March 1909 Bosch's understanding of steel metallurgy persuaded the BASF leadership to proceed with the commercialization of Haber's process. But before any large-scale production of ammonia could begin, Bosch and his team had to solve a number of unprecedented engineering challenges, including the failure of the first scaled-up converters. These 2.5-m-long experimental tubes (heated electrically from the outside and filled with the catalyst) exploded after about 80 hours of operation under high pressure. Fortunately, Bosch anticipated such a mishap and placed them in reinforced concrete chambers.

Examination of failed converters revealed the total loss of elasticity as they became hard and brittle after being subjected concurrently to high temperature and high pressure. Bosch's first approach was to fit the pressure tube with a soft inner lining, but his eventual, and better, solution was to contain high temperature within an inner wall of soft (low-carbon) steel and to force a mixture of pressurized cold hydrogen and nitrogen into the space between the inner and outer walls. Consequently, the inner shell was subject to equal pressure from both sides, and the strong outer steel shell was under high pressure but it remained much cooler. With this solution, converter sizes were increased to 4 m and 1 t in the first pilot plant in 1912, and units 8 m long weighing 3.5 t were installed in the first commercial process in 1913.

Perhaps the most notable challenge among other engineering problems that had to be overcome before setting up the entire operation was the production of hydrogen (nitrogen was derived from Linde's air liquefaction). In 1912 Bosch and Wilhelm Wild patented their *Wasserstoffkontaktverfahren*, a catalytic shift reaction that transforms CO and steam into CO_2 and hydrogen. The second major set of challenges involved the search for the best and inexpensive catalyst. Alwin Mittasch (1869–1953) undertook a systematic testing of all metals known to have catalytical abilities in pure form or as binary (e.g., Al-Mg, Ba-Cr, Ca-Ni) or ternary catalysts, as well as more complex mixtures (Mittasch 1951). Magnetite (Fe_3O_4) from the Gällivare mines in northern Sweden supported the highest yield, and Mittasch then sought the best possible combination with catalytic promoters. The first patent for such mixed catalysts was filed on January 9, 1910.

Nothing illustrates the thoroughness and efficacy of Mittasch's exhaustive search better than the fact that most commercial catalysts used in ammonia synthesis during the 20th century were just slight variations on his basic combination that relied on magnetite with additions of Al_2O_3, K_2O, CaO, and Mg. These promoted iron catalysts are also extraordinarily stable, able to serve for up to 20 years without deactivation. Only recently did Kellogg, Brown & Root introduce their advanced ammonia process (KAAP), which is not based on the classical magnetite catalyst but instead a precious metal, ruthenium, with co-promoters (Smil 2001).

Construction of the first ammonia plant at Oppau near Ludwigshafen began on May 7, 1912, and the production started on September 9, 1913, with the gas used as a feedstock for the synthesis of ammonium sulfate fertilizer. But less than a year after the plant's completion, Oppau's ammonia was diverted from making fertilizer to replacing Chilean nitrate, whose imports were cut off by the British naval blockade. There is no doubt that the Haber-Bosch synthesis was one of the factors that helped to prolong WWI. Blockaded Germany could do without Chilean nitrates that were previously used for producing explosives, and its troops had enough ammunition to keep launching new offensives for nearly four years, until the spring of 1918. In order to satisfy this large new wartime demand, the second, much larger, ammonia plant was completed at Leuna near Halle in 1917, and when the war ended Germany was the only country with a considerable capacity to produce inexpensive inorganic nitrogen fertilizer.

But the Haber-Bosch process, and its variants introduced during the 1920s, had a limited impact on the world's food production before the 1950s as only a few countries were relatively intensive users of inorganic fertilizers and as the economic downturn of the 1930s and WWII set back the industry. Global synthesis of ammonia remained below 5 Mt until the late 1940s, and while several European countries had fairly high average rates of fertilizer applications already before WWI, in the United States more than one-third of all farmers did not use any fertilizer nitrogen even by the late 1950s. But then the situation changed rapidly, particularly as nitrogen fertilizers became the key nutrient that unlocked the yield potential of new cultivars everywhere except in the conflict-torn and mismanaged economies of the sub-Saharan Africa.

Importance of Synthetic Ammonia

While the importance of electricity or internal combustion engines for the modern society is obvious, even reasonably educated urbanites do not appreciate the essential role played by nitrogen fertilizers, and even most scientists are not aware of the extent to which the global civilization depends on the Haber-Bosch synthesis of ammonia. Global production of ammonia nearly doubled during the 1950s, further quadrupled by 1975, and then, after a brief period of stagnation that began during the late 1980s, it rose to nearly 130 Mt by the end of the 20th century. During the late 1990s, the global output of ammonia and sulfuric acid was virtually identical, but because of ammonia's much lower molecular weight (17 vs. 98 for H_2SO_4), the gas is the world's leading chemical in terms of synthesized moles.

Both the scale and the efficiency of the process have become far superior to the initial performance, above all thanks to the combination of innovations introduced during the 1960s. Replacement of reciprocating compressors by centrifugal machines cut the consumption of electricity by more than 90% and led to larger, and more economical, plants. Almost complete displacement of coal by natural gas as both the fuel and the feedstock (source of hydrogen) further lowered the energy cost of ammonia synthesis. During the late 1990s, about two-thirds of ammonia, or an equivalent of about 85 Mt of nitrogen a year, were used by fertilizer industry (figure 4.18), but in contrast to the pre-1960 period, when the gas was a feedstock for making a variety of compounds, most of it now is used for the synthesis of urea.

Unlike the gaseous ammonia, urea is a solid compound, containing 45% nitrogen (more than any other nitrogen fertilizer) and is produced in small granules that are easily stored, shipped, and applied to fields. Only small shares of nitrogenous fertilizers are applied to tree plantations, pastures, and lawns, and the bulk of all applications go to cereals and to oil and tuber crops. A detailed balance of nitrogen flows in the global agroecosystems shows that synthetic nitrogenous fertilizers supplied about half of the nutrient available to the world's crops, with the other half coming from leguminous crops, organic recycling, and atmospheric deposition (Smil 1999a). Does this also mean—with nitrogen being the leading limiting input in food production—that roughly one-half of humanity is now alive thanks to the Haber-Bosch synthesis? This question cannot be answered without referring to a specific food supply: how many people would not be alive without ammonia synthesis depends on the prevailing diets.

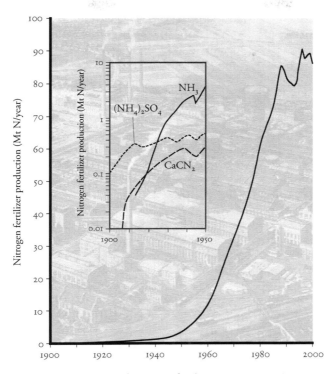

FIGURE 4.18. Global production of nitrogen fertilizers, 1900–2000 (in megatons of nitrogen per year). Pre-1913 output includes ammonium sulfate from coking and cyanamide and nitrate from the electric arc process. Based on data in Smil (2001).

Detailed accounts of the global nitrogen cycle indicate that about three-quarters of all nitrogen in food proteins available for human consumption comes from arable land (with the rest coming from pastures and aquatic species). If synthetic fertilizers provide about half of all nitrogen in harvested crops, then at least every third person, and more likely two people out of five, get the protein in their diets from the Haber-Bosch synthesis. As with most global averages, such a revelation both overestimates and underestimates the importance of this great invention. In affluent nations, fertilization helps to produce an excess of food in general and assures high animal food intakes in particular (most of those countries' crops are fed to animals rather than being eaten directly by people), and they also help to produce more food for export. Consequently, even significant cuts in the amount of applied fertilizers would result in less meaty diets and in lower exports but would not at all imperil the overall adequacy of the food supply. Depending on the meatiness of their diets, no more than 1 in 10 of the world's 1 billion affluent consumers would then derive their protein from synthetic ammonia.

In contrast, rising fertilizer applications were essential for lifting populous low-income countries from the conditions of bare subsistence and widespread malnutrition. Norman Borlaug, one of the leaders in the development and diffusion of new high-yielding varieties of crops, captured the importance of nitrogen fertilizers in his speech accepting the Nobel Prize for Peace in 1970 (Borlaug 1970): "If the high yielding dwarf wheat and rice varieties are the catalysts that have ignited the Green Revolution, then chemical fertilizer is the fuel that has powered its forward thrust."

As a result, both China and India have become basically self-sufficient in staple cereals, and (one of the insufficiently appreciated achievements of the 1990s) average per capita food availability in China is now at least as high as in Japan (FAO 2003). This means that the remaining, and not insignificant, extent of India's and China's malnutrition (effecting, respectively, about 200 and nearly 140 million people during the late 1990s) is a matter of unequal access to food and not of food shortages (Smil 2000b). Perhaps the best illustration of different degrees of dependence on nitrogen fertilizers is a comparison of the United States and Chinese situation.

In the late 1960s Americans applied just less than 30 kg of nitrogen per hectare of their farmland, while the Chinese rate was less than 5 kg/ha. Then came a stunning reversal of China's international politics: the country's return to the United Nations and Kissinger's secret trip in 1971 were followed by Nixon's state visit in 1972. The consensus opinion was that this historic shift was motivated almost solely by the need for a strategic partner against the mortal threat of the expansionist USSR: *Realpolitik* made yesterday's despised aggressors, raining bombs on Vietnam, today's valued allies as Mao chatted with Nixon in Zhongnanhai. But such major policy shifts have rarely a single cause, and, as far as I am aware, at the time of China's opening nobody pointed out a reason ultimately more powerful than the fear of the Soviet hegemony: the need to avert another massive famine.

The world's greatest and largely man-made famine claimed at least 30 million lives between 1958 and 1961 (Smil 1999b). In its wake, China's population expanded at an unprecedented rate: with no population controls in place, it grew from 660 million people in 1961 to 870 million by 1972. This addition of more than 200 million people in a single decade represented the fastest population growth in China's long history and the highest ever national increment in global terms. At the same time, the slowly rising yields could not keep up even with basic food needs: by 1972 China's average per capita food supply was below the levels of the early 1950s! The only effective solution was to increase rapidly the synthesis of nitrogenous fertilizers—and soon after Nixon's visit China placed orders for 13 of the world's largest and most modern ammonia-urea complexes.

Such an order could not be filled without turning to M. W. Kellogg, America's and the world's leader in ammonia synthesis: eight of the 13 ammonia plants came from Kellogg, and they fed their product to urea plants delivered by Kellogg's Dutch subsidiary. Was China's opening to the world a matter of grand politics and strategic alliances? Undoubtedly—but it was also a matter of basic survival. More purchases of ammonia-urea complexes followed, as did more fertilizer imports: by the early 1980s China became the world's largest consumer, and a decade later also the world's largest producer, of fertilizer nitrogen. During the late 1990s, American applications averaged about 50 kg of nitrogen per hectare—but the Chinese rate surpassed 200 kg/ha and, in five of the most intensively cultivated provinces with the total population equal to that of the United States, topped 300 kg/ha! Fertilizer nitrogen already provides about 60% of the nutrient in China's crops, and as more than 80% of the country's protein is derived from crops, roughly half of all nitrogen in China's food comes from inorganic fertilizers.

Similarly high degrees of high existential dependence have been evolving, nationally or regionally, in other land-scarce countries. When these countries reach the limit of their cultivated area, then they must turn, even if they remain largely vegetarian, to higher nitrogen inputs. During the last two generations of the 20th century, India increased its total applications of nitrogenous fertilizers by roughly the same rate as did China (more than 40-fold rise), while Indonesia applied during the late 1990s nearly 25 times as much synthetic fertilizer as it did in 1960 (FAO 2003). With most of the population growth during the next two generations taking place in Asia and Africa, this dependence will have to grow as there is still no practical alternative to supply the needed mass of the most important macronutrient.

While it is not difficult to make plausible scenarios of a world that will eventually function without the two great mainstays of modern civilization— internal combustion engines and electricity generated by the combustion of fossil fuels (the ubiquitous prime mover could be replaced by fuel cells, electricity can be generated by converting direct and indirect solar flows)—there is no substitute for dietary proteins whose production requires adequate nitrogen supplied to crops. During the late 1970s, it appeared that it might be possible to confer the fixation capability possessed by legumes on wheat and rice but, much as with commercial nuclear fusion or with mass-produced electric cars, this goal has remained elusive, and the dependence on the Haber-Bosch synthesis of ammonia will likely extend far into the 21st century, a lasting legacy of one the most important, yet so inexplicably little appreciated, pre-WWI technical advances.

BOSTON

SALEM

THE TELEPHONE

5

Communication and Information

The apparatus . . . is all contained in an oblong box about 7 inches high and wide, and 12 inches long. This is all there is visible of the instrument, which during the lecture is placed on a desk at the front of the stage, with its mouthpiece toward audience. Not only was the conversation and singing of the people at the Boston end distinctly audible in the Salem Hall, 14 miles away, but Professor Bell's lecture was plainly heard and applause sent over the wires by the listeners in Boston.

> *Scientific American* of March 31, 1877, reporting on A. G. Bell's latest telephone demonstration

Less than a year after Alexander Graham Bell exhibited the first version of his telephone at the Centennial Exposition in Philadelphia in June 1876, he was demonstrating a better design during public lectures. Although much admired, the first device had a limited capacity as it could transmit only over short distances and did so with a much diminished signal; in contrast, the improved device made the first intercity audio communication possible. By the time of the Salem lecture, pictured in the frontispiece to this chapter, the record dis-

FRONTISPIECE 5. Alexander Graham Bell lecturing on his telephone to an audience in Salem, Massachusetts, on February 12, 1877, and a group in the inventor's study in Boston listening to his explanations. The early apparatus was housed in an oblong box about 18 cm wide and tall, and about 32 cm long (inset on the right). Reproduced from the cover of *Scientific American*, March 31, 1877.

tance was about 230 km, from Boston to North Conway in New Hampshire. Bell's telephone was only the first of several fundamental inventions that eventually revolutionized every aspect of modern communication. Almost exactly 10 years after Bell's first intercity telephone calls came a discovery with even farther reaching consequences.

What Heinrich Hertz succeeded in doing between 1886 and 1888 was to generate, send, and for the first time to receive electromagnetic waves whose frequencies 'range themselves in a position intermediate between the acoustic oscillations of ponderable bodies and the light-oscillations of the ether' (Hertz 1887:421). He generated sparks by an induction coil, sent electromagnetic waves across a room where they were reflected by a metal sheet, and then measured their frequency (distance between their crests) by observing, under microscope, tiny sparks that could be seen on the other end of the room across a small gap of a receiving wire loop. Expressed in units that honor his name, these broadcasts had frequencies of 50–500 megahertz (MHz), or 6 m to 60 cm (Aitken 1976). Lower frequencies of this range are now used for FM radio and broadcast TV; the higher ones for garage door openers, medical implants, and walkie-talkies.

Has there been any other physical experiment that was so simple—a few batteries, coils, and wires positioned in a 14-m-long lecture room—yet eventually brought such a wealth of amazing spin-offs? Hertz pursued this research only in order to confirm Maxwell's theoretical conclusions about the nature of electromagnetic waves and could not foresee any use for these very rapid oscillations. That perception changed soon after his premature death in 1894, and his fundamental discovery was first transformed into wireless telegraphy and soon afterward into broadcasts of voice and music. And more wonders based on those rapid oscillations were to follow during the 20th century: television, radar, satellite telecommunication, cellular telephony and, most recently, wireless WWW.

But the story does not end here as the Age of Synergy revolutionized *every* kind of communication, a term that I use in the broadest sense in order to embrace all forms of printed, visual, and spoken information as well as the means and nontechnical requirements of its large-scale diffusion. Consequently, in this chapter I examine first not only the progress in flexible and rapid printing of large editions of books and periodicals but also the new techniques that were introduced to meet the rising demand for printing paper. When describing the invention of the telephone, I first explain its telegraph pedigree, and before outlining the early history of wireless communications, I look closer at Hertz's experimental breakthrough.

Although all of these techniques have been instrumental in creating new economies and new social realities of the 20th century, only the American origins of telephone—or at least Bell's famous imploration, 'Mr. Watson—

come here—I want to see you' (Bell 1876a)—are more widely known. Public knowledge regarding the genesis of other communication techniques is either largely absent or perpetuates various misunderstandings: neither Thomas Edison nor the Lumière brothers invented moving pictures, nor did Marconi make the world's first radio broadcasts (Garratt 1994). Moreover, even many informed people would not rank sulfate pulp or linotypes along such key innovations of the two pre-WWI generations as internal combustion engines or electricity. Indeed, many people may not be even aware of the existence of the former two innovations and of their importance for modern societies.

Because of its topical sweep, this is inevitably the most heterogeneous chapter of this book. And yet a closer look reveals a number of critical commonalties that qualify all of these achievements as quintessential innovations of the Age of Synergy. All of the techniques examined in this chapter had pre-1860 pedigrees, but these early attempts were not good enough to be translated into commercial realities. Once introduced in their basic functional form during the Age of Synergy, all of them traversed the often formidable distance between a promising patent and commercial success in a matter of years. And immediately after their introduction, nearly all of them became subject to intensive refinement and redesign that boosted their performance and brought them close to technical maturity often in less than a generation after their initial patenting. And all of them deserve attention because of their combined impact on civilization.

Sulfite and sulfate pulp keeps furnishing most of our paper needs, and there are no radical innovations on the papermaking horizon to change that. Before they were displaced by photo typesetting, linotype machines were used to set all books, newspapers, and merchandise catalogs during the first two-thirds of the 20th century. Cellular phones are now doing away with wired touch-tone telephones that were introduced in 1963—but until that time the classical rotary telephone, whose diffusion began in the early 1920s, was the standard equipment. Development and commercialization of radio and television, as well as of sonar and radar, were only a matter of time after Hertz demonstrated the generation and reception of high-frequency waves. The age of inexpensive and simple-to-operate cameras using roll film started with the first of George Eastman's boxy Kodaks in 1888, exactly 75 years before Kodak Instamatics were introduced in 1963.

But before I start the topical surveys of major communication techniques, I must mention one of their forms that has been inexplicably neglected in standard accounts of modern logistics: parcel deliveries. There is no justification for this omission either on commercial or on emotional grounds. Few people are aware of the relatively late origins of postal parcel deliveries. British parcel post began operating in August 1883 (Samuelson 1896). In the United States, where international parcel post has been available since 1887, private companies did very good business by delivering mail orders to millions of rural customers. Not surprisingly, their opposition to postal parcel service delayed its enactment until 1912. The service began on January 1, 1913, and the enthusiastic response translated into impressive economic benefits; for example, during the first five years of parcel post delivery Sears, Roebuck & Co., the second of the two giant mail-order retailers (established in 1893, 21 years after Montgomery Ward), tripled its earnings.

Physical evidence of parcel deliveries (by trucks, vans, and airplanes) is now ubiquitous in all affluent societies. United Parcel Service, the world's largest parcel delivery company, was established in 1907 in Seattle as the American Messenger Co. and got its present name in 1919 (UPS 2003). In 2003 its capitalization exceeded $70 billion and its annual revenues surpassed $30 billion, the total higher than the gross domestic product of about 120 countries in 2002 (UNDP 2003). With diffusion of the Internet, millions of consumers became new converts to the convenience of receiving goods at home, and their repeat orders have been driving the expansion of on-line shopping.

Printed Word

I will start with the printed word simply because of the longevity of the art. Gutenberg's movable type (first used in 1452) led to the well-documented explosion of printing. By 1500, more than 1,700 presses were at work across Europe, and they produced more than 40,000 separate titles or editions amounting to more than 15 million incunabula copies (first printed editions) (Johnson 1973). Subsequent centuries brought slow but cumulatively important improvements that resulted in such remarkable typographic achievements as the first comprehensive encyclopedia edited by Denis Diderot and Jean le Rond D'Alembert (1751–1777). Mechanization of printing advanced during the first half of the 19th century with the introduction of high-volume presses.

A new press built in 1812 by Friedrich König still had a flat type bed, but the paper was pressed onto it by a revolving cylinder: when powered by a steam engine, this machine could print more than 1,000 sheets per hour. Introduction of durable and flexible *papier-mâché* matrix made it possible to cast curved stereos once the cylindrical rollers replaced the flat printing bed and the type itself was locked onto a circular surface. This took some time to accomplish, and Richard Hoe (in 1844) and William Bullock (in 1865) were the two principal designers of practical rotary machines (Sterne 2001). Hoe's revolving press was fed sheets of paper by hand, and Bullock's roll-fed rotary could produce 12,000 newspapers per hour. Mechanization of folding, stitching, and binding helped to made progressively larger runs of newspapers and books cheaper.

But there were still obvious and irksome complications and self-evident limits. Setting the text with movable type was a highly time-consuming process that cried for mechanization. And the collecting of rags—to be converted, after sorting and cleaning, by boiling in spherical or cylindrical vessels in alkaline liquor into new paper—was imposing obvious limits on the total mass of printed matter. Societies where used textiles were the only feedstock for papermaking could never hope to become both extensively and deeply literate and well informed. All of these challenges were resolved by new inventions and by their rapid improvement.

Typesetting and Typewriting

Setting texts with movable type was a highly repetitive, monotonous job but also one that required a high degree of alertness, combining persistence and patience. Every character had to be individually selected from a distribution box and lined up backward (as if seen in a mirror) with appropriate spacing (achieved by inserting faceless pieces of type) in a composing stick. Once a stick was filled, the type was lifted onto a shallow tray (galley); finished galleys were used to produce proofs, marked proofs were used to correct any typesetting errors, and pages were then imposed (arranged in numerical sequence) and fastened into a *forme* that could be easily moved around, inked, and printed. When the printing was done, the type had to be disassembled, laboriously cleaned, and returned to distribution boxes.

There were many attempts to mechanize this task, and many typesetting machines were invented and patented during the course of the 19th century. John Southward, writing in the late 1880s for the 9th edition of *Encyclopaedia Britannica*, noted that less than half a dozen of these machines had stood the test of practical experience, that a few relatively successful designs were confined to special classes of work, and that

> it is open to doubt whether the nimble fingers of a good compositor, aided by the brains which no machinery can supply, do not favourably compare on the ground of economy with any possible mechanical arrangement (Southward 1890:700–701).

But even as Southward was expressing his doubts about the future of practical typesetting machines, their tentative and limited success was turning into a commercial triumph. Indeed, his very words (and more than 10,000 pages of the 9th edition of the world's most famous encyclopedia) were actually typeset by an improved version of a relatively simple machine introduced in 1872 by Charles Kastenbein. Interestingly, Southward's encyclopedic entry does not contain any mention of the machine whose basic design was completed by 1884, that had been used to set newspapers since 1886, and that eventually

FIGURE 5.1. Ottmar Mergenthaler, the designer of the most successful commercial linotype. Reproduced from Coraglia (2003).

proved the superiority of mechanical arrangement over nimble fingers. Fingers, and the brains, were still needed—not to pick, place, and space the letters but simply to type: the machine was Mergenthaler's linotype.

Ottmar Mergenthaler (1854–1899; figure 5.1) belonged to that large group of German craftsmen who made up a notable part of the multimillion emigration from Western Europe's largest nation to the New World and whose mechanical skills were behind many innovations introduced in the United States during the two pre-WWI generations. He was apprenticed to a watchmaker, and after his arrival in the United States he began working in his cousin's engineering workshop in Washington, D.C. Among other things, the shop produced models to accompany patent applications, and that is how Mergenthaler met in 1876 Charles T. Moore, who brought in plans of a machine designed to transfer a page for printing by lithography. Building its model led Mergenthaler to the idea of designing a composing machine, a task that preoccupied him for the next 15 years (Mengel 1954; Kahan 2000).

Technical requirements for a successful composing machine were very demanding. The tasks of composing, justifying, casting, and distributing the type had to be automated; the machine had to be operated by a single person and perform at a rate sufficiently faster than hand composing in order to justify its cost. Mergenthaler's solution was first demonstrated in a prototype in 1884, and it was patented in 1885 (figure 5.2). The operation began when a keystroke let a small thin bar (matrix) containing the female mold of a character descend from a vertical storage bin (magazine). Matrices, and appropriate space bands, were then placed in alignment in a line, a task that required accuracy of 1/100 of a millimeter and uniform depth in order to print every letter and number clearly (Scott 1951).

Mergenthaler had to solve first an unprecedented problem of precision manufacturing of the matrices: the process had to be relatively inexpensive in order to produce 1,200 matrices needed for every machine, and they had to be strong in order to withstand repeated gripping, lifting, pushing, pressing, and spacing. This required building a matrix factory and designing complex machines to

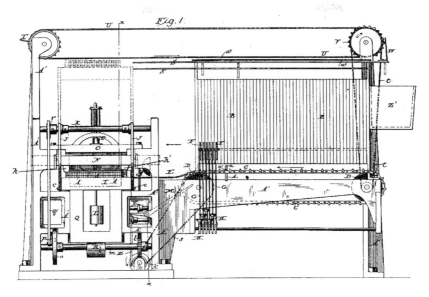

FIGURE 5.2. Basic design of Mergenthaler's linotype patented in 1885 (U.S. Patent 317,828). A few notable parts include matrix magazines (*B*), matrix distributor (*U*), separable mold (*N*), and melting pot (*O*).

mill, grind, emboss, cut, stamp, and finish the blanks punched from sheets of brass. At least one critical task could use an already commercialized process: cutting of characters was done by a punch machine invented by Linn Boyd Benton (1844–1932). And Mergenthaler used another already available invention, by J. D. Schuckers, for automated spacing of matrices.

Completed matrices were released from the assembler, so the operator could proceed to set the next line, and were transferred by the first elevator for justification and casting with hot (285°C) metal (an alloy of 85% lead and 12% antimony). After casting, the matrix line was carried upward by the second elevator, space bands were separated first and returned to the appropriate box for reuse, and then the matrix was lifted to the level of the distributor bar that was suspended over the magazine and passed along it until their unique combinations of keys caused them to fall into their original compartments to be ready for forming new matrix lines. Cast slugs were pushed into trays or galleys to be used for direct printing or for making a stereotype for a rotary press; afterward they could be melted and the metal could be reused. The first production version of linotype was used to set *New York Tribune* on July 3, 1886. More than 50 machines were at work the very next year, and in a fashion emblematic of the period's innovation, the invention became almost immediately subject of numerous improvements; eventually there were more than 1,500 patents related to its design and operation.

Mergenthaler put his Simplex linotype in production by 1890 (figure 5.3), and two years later came Model 1, the first truly successful design. But for some time it appeared that Mergenthaler's linotype would have a superior competitor in a typesetting machine designed by James W. Paige, who began to work on his design in 1873. After years of designing setbacks and delays in raising the necessary capital—Mark Twain was among the investors who lost large sums in this venture—the first machine was finally built in 1894. The entire assembly was made of 18,000 parts; it stood nearly 2 m tall and weighed about 2.3 t (Legros and Grant 1916). *Chicago Herald* agreed to test it; its verdict was favorable, but by that time Mergenthaler linotypes dominated the typesetting market and there was no enthusiasm to manufacture a machine whose chronically delayed development had already cost close to $1 million.

Eventually Philip T. Dodge, the president of the Linotype Co., purchased the patent and Cornell and Columbia universities were presented with the only two machines ever built. By 1895 there were already more than 1,000 Mergenthaler linotypes around the world, and about 6,000 copies of Model 1 were made before more reliable and more complex designs took over. The machine, whose invention has been called the second greatest event in the history of printing, and which admiring Edison saw as the eighth wonder of

FIGURE 5.3. Mergenthaler's 1890 Simplex linotype design (reproduced from *Scientific American*, August 9, 1890) and composing room of the *New York Times* full of linotype machines (this photograph was taken in 1942 by Marjory Collins).

the world, was used to set the news of defeats and victories of the two world wars, of economic expansion and crisis of 1920s and 1930s, and of the first decade of the Cold War (figure 5.3).

During the 1950s, photo typesetting and photo-offset printing began a new trend, and the last of the Linotype company's nearly 90,000 machines was produced in 1971. But thousands of linotypes were still used during the 1980s, and many survived into the 1990s. As linotypes began disappearing, the type-setters, while appreciative of the abilities of new computerized equipment, missed the sound, smell, and feel of those complex, clanging, heat-radiating machines. I also remember them fondly, from Prague of the 1960s, and I agree with a retired Italian linotype operator that there is something magic about *l'arte di fondere i pensieri in piombo* (Coraglia 2003).

The letters of the two leftmost vertical rows of keys on the Linotype console that were used to test the machine spell a magical "etaoin shrdlu." Those on the top row of typewriters give another enigmatic line: "qwertyuiop"—but the first successful machine could type it only in capitals. This machine had an even more extensive pedigree of failed designs than did practical incandescent light. The first real breakthrough came in 1867, when Christopher Sholes, Carlos Glidden, and Samuel Soule filed a patent for a type writing machine, which was granted in June of the following year (figure 5.4). The trio sold the rights for $12,000 to James Densmore and George W. N. Yost, who in turn arranged for production of the machine by E. Remington & Sons, a renowned maker of guns and sewing machines (Zellers 1948).

Design of the first commercial typewriter betrays its origins: it was done by William Jenne from Remington's sewing machine division, and this treadle typewriter of 1874 clearly resembled a sewing machine. Its sturdy upright body decorated with gold and bold flower decals sat on a four-legged table with a foot treadle used to advance the paper. The platen was made of vulcanized rubber; there was a wooden space bar and a four-line keyboard of capital letters starting with the QWERTY sequence. The type bars with raised-relief letters and numbers hit the inked ribbon against the underside of the platen (up-strike) so the typist could not see the text. About 5,000 treadle machines were sold, and changes, small as well as major ones, followed rapidly. A design that made it possible to print both capital and small letters by depressing the same key was patented (U.S. Patent 202,923) by Byron A. Brooks in 1878.

But the so-called visible machines, where the typist could see the text, became common only by the late 1890s, and portable typewriters (led by 1906 Corona) appeared before WWI, when standard machines acquired many noise-reducing features and other mechanical improvements (Typewriter Topics 1924; figure 5.5). On such machines were most of the 20th century great novels and stories written. Of the three great American Nobel Prize winners in literature, both Hemingway and Faulkner used Underwoods (Hemingway also had Roy-als, both desk and portable models); Steinbeck wrote a clear longhand on

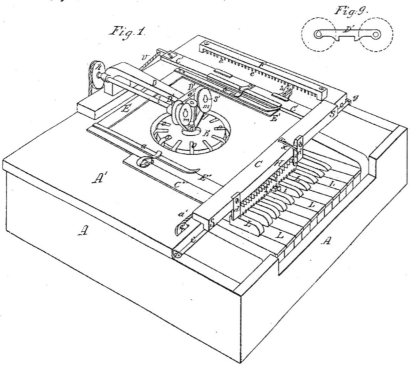

FIGURE 5.4. Patent illustration (U.S. Patent 79,265) of a "type-writing" machine designed in 1867 by Sholes, Glidden, and Soule..

yellow legal pads. But the typewriter became more than an indispensable tool of trade for novelists and journalists. During WWII, typewriters came under the control of the War Production Board as tens of thousands of them were taken by armies to battle (a ship sunk off Normandy during the D-Day carried 20,000 Royals and Underwoods) together with tripod stands, and as thousands of them were fitted with non-Latin alphabets and special characters to be used in different theaters of war (Frazier 1997).

FIGURE 5.5. Four notable early typewriters: Remington's Standard No. 6, an "invisible" design of 1894 (upper left); a Remington-Sholes Visible from 1908 (upper right); a portable Senta, which came out first in 1912 (lower left); and a standard Royal from the early 1920s (lower right). Reproduced from Typewriter Topics (1924).

Alternatives to frequently jamming type bars appeared even before the first visible typing: James Hammond's revolving cylinder in 1881, and George Blickensfelder's revolving wheel (the precursor of daisywheel design) in 1893. The first electric typewriter, with a typewheel printhead, was sold in 1902, but IBM's Selectric became the best-selling model only during the 1970s, just before the typewriters gave way to PCs. And the QWERTY arrangement? Sholes chose this irrational order deliberately in order to slow down the speed of typing and hence to minimize the jamming of type bars. Rational keyboard designs—for English language typewriters August Dvorak's famous statistics-based rearrangement that speeded up typing by more than a third (Dvorak et al. 1936)—have been available for decades, but the universal design was not only maintained by virtually all 20th-century typewriters but also has been transferred, without any second thoughts, to computer keyboards.

Papermaking

By the beginning of the 19th century, steadily rising demand for paper was putting greater strain on the supply of rags. In different places and at different times, this feedstock included cotton, linen, flax, hemp, and even silk fibers. Small runs of artisanal handmade sheets are still made from rags, while cotton gin trash and flax waste are used to make paper for the U.S. currency (Brennan 2000). Demand for larger volumes of fibrous feedstock further increased after the introduction of a new papermaking technique that produced long strips rather than single sheets of paper. The machine for this semicontinuous production was invented in 1798 by Nicolas Louis Robert, a clerk at the Essonne Paper Mills in France, and the English rights for its use were bought by Henry Fourdrinier (1766–1854), the owner of a mill in Kent.

By 1803 the Fourdrinier brothers and Bryan Donkin had constructed the first practical machine at a mill in Frognore in Hertfordshire (Wyatt 1911). This machine produced strips of paper up to 3.6 m wide in lengths of 15 m, and they had to be taken off wet and hung to dry. The basic principle remains two centuries later: pulp is laid on a continuous wire mesh at the wet end of the machine, most of water is expelled in the felt press section, and the process is finished by passing paper over a series of heated cylinders. Performance is a different matter: by 1910 the largest fourdriniers were 5 m wide and could produce rolls of paper nearly 13 km long at the speed of 240 m/min. Today's large fourdriniers can produce rolls of up to 18 km long at speeds surpassing 400 m/min.

During the first half of the 19th century, paper recycling became more common, and the search for new feedstocks led to the use of jute sacking, cereal straws, Manila hemp, and esparto grass, but wood was obviously the most abundantly available source of fiber. Mechanical pulping was developed for practical use by Heinrich Völter in Bautzen by 1845, but it has been used commercially on a larger scale only since the early 1870s. The process had high yield (up to 90% of wood becomes pulp), but this also means that it removed hardly any lignin. This polymer normally constitutes between 30% and 40% of wood and imparts to it strength, but its presence in the pulp makes for an inferior paper. Furthermore, wood resins in mechanical pulp contribute to paper's rapid deterioration, betrayed by distinct yellowing with age. These drawbacks make little difference for ephemeral paper uses, and that is why mechanical wood pulp still accounts for about 20% of all pulp produced worldwide (FAO 2003). Its most important use is for newsprint, and it also goes into toilet paper, towels, cardboard, and building board.

Experiments to remove lignin, resins, and other intercellular matter and to produce nearly pure cellulose began during the 1850s when the soda (or alkaline) process, patented by Charles Watt and Hugh Burgess in 1854, was introduced in the United States for the treatment of deciduous woods, mainly for poplars, aspen, beech, birch, maple, and basswood. The first American mill using the process was in Pennsylvania in 1863. Chipped wood was boiled with sodium hydroxide (NaOH) in large boilers for up to nine hours at pressures of up to 1 MPa, and the bleached pulp (total yield about 50%) was used for book and magazine papers. The soda process was gradually displaced by the sulfite (or acid) process that eventually produced most of the world's paper during the first third of the 20th century.

Benjamin Chew Tilghman (1821–1901), better known for his patenting of sandblasting, began his work on this treatment during the 1850s. His U.S. Patent 70,485 for treating wood with acid, boiling it under pressure, and adding a suitable base substance for easier bleaching (Tilghman 1867) was followed by many improvements contributed mostly by engineers in Europe (Steenberg 1995; Edge 1949). An indirect boiling process, with steam at 125–135°C circulating through copper or lead coils to boil the mixture for at least 20–30 hours, was patented in Sweden in 1871 by Carl Daniel Ekman (1845–1904) and in Germany in 1874 by Alexander Mitscherlich (1836–1918). With its longer boiling at lower temperature, this process yields stronger pulps than does the direct boiling method that uses hot steam (140–160° C) for 8–10 hours. In Europe this new method became generally known as Ritter-Kellner process. Carl Kellner (1851–1905) developed it in 1873 when he worked for Baron Hector von Ritter-Zahony's paper factory in Görz, and had it patented in 1882 (Dvorak 1998). America's sulfite pulp production began only in 1883 at the Richmond Paper Co.'s plant in Rumford, Rhode Island (Haynes 1954). Figure 5.6 shows an early design of American pressurized tanks used for pulp production.

Feedstock for both methods is ground coniferous wood that is boiled under pressure of 480–550 kPa in a solution of bisulfites of calcium or magnesium until all nonfibrous constituents of wood (resins, gums) are dissolved and cellulose fibers can be recovered, washed, and bleached to remove all traces of the liquor (Edge 1949). Just before WWI, sulfite pulp accounted for about half of all wood pulp produced in the United States (Keenan 1913). The sulfite process produces fairly strong pulp, but its acidity eventually embrittles the paper, whose disintegration is further accelerated by its susceptibility to air pollution and high humidity. Most matter that was printed between 1875 and 1914 has already entered the stage of fairly advanced disintegration: even when carefully handled, brittle pages of many 100–125-year-old publications that I read in preparation for writing this book tear easily, and some literally crumble. And the same fate awaits most of the pre-1990 publications as the first official paper standard for permanent papers was adopted in the United States only in 1984, and the international norm, ISO 9706, came a decade later.

The second pulpmaking innovation, and the one that has dominated global papermaking since the late 1930s, came shortly afterward with the introduction of the sulfate process. In 1879 Swedish chemist Carl F. Dahl invented a pulping process that used sulfates rather than sulfites to boil coniferous wood in large

FIGURE 5.6. Apparatus for treating wood with bisulfite of lime. Reproduced from *Scientific American*, March 29, 1890.

upright boilers at pressures between 690 and 890 kPa for about four hours (Biermann 1996). The sulfate process produces much stronger paper, hence the use of the Swedish (or identical German) word for strength to label the process kraft pulping. The residual lignin colors paper brown, and so bleaching is required to produce white paper, and additives (acids, bases, sizing agents, adhesives, resins, fillers) are used to control pH, strength, smoothness, and brightness.

While the kraft process yields superior pulp at a lower price, it also produces highly offensive hydrogen sulfide that used to escape uncontrolled from the mills, and sulfate pulps are also more difficult to bleach. The first American kraft mill was built only in 1911 in Pensacola, Florida, but eventually the advantages of sulfate pulping—the only chemical is inexpensive sodium sulfate, a large amount of energy is produced during the recovery process, and pine trees, abundant worldwide, are well suited for it—made it the leading form of papermaking in the United States and around the world. By the year 2000, the sulfate process accounted for nearly 70% of the world's wood pulp production; mechanical pulping produced about 20% of the total, and the rest comes from the sulfite and semimechanical treatment (FAO 2003).

Importance of these relatively simple chemical processes—both involving nothing else but pressurized boiling of ground wood in acid solutions—that were commercialized before WWI is not generally appreciated. Yet sulfite and sulfate pulping made it possible to convert the world's largest stores of cellulose sequestered in tree trunks into high-quality paper and ushered in the age of mass paper consumption. In 1850, near the end of the long epoch of rag paper, there were fewer than 1,000 papermaking machines in Europe and about 480 in the United States, and the annual per capita output of paper in Europe and North America amounted to less than 5 kg. By 1900 Germany alone had 1,300 machines and the worldwide total surpassed 5,000. U.S. paper production doubled during the 1890s to 2.5 Mt, and by 1900 it surpassed 30 kg per capita. Paper became one of the cheapest mass-produced commodities of modern industrial civilization.

Consequences of this change could be seen everywhere. Picture postcards, invented in Austria in 1869 and available in France and the United States by 1873, became very cheap and as popular as today's e-mail. In 1867 Margaret Knight (1838–1914) designed the first flat-bottom paper bag and soon afterward also a machine to make such bags; better machines, by Luther Childs Crowell in 1872 and by Charles Stillwell in 1883, made their production even cheaper. This seemingly simple innovation endured throughout the 20th century, and its dominance began to recede only during the 1980s with the rise of plastic grocery bags. By the mid-1890s the largest mail order business in the United States, Montgomery Ward & Co., whose first catalog in 1872 was a single page, was distributing seasonal catalogs containing more than 600 pages (Montgomery Ward 1895). Writing slates disappeared even from the poorest schools: the just cited catalog was selling 160 cream-colored pages of its School Spelling Tablet for three cents.

The combination of cheaper paper and halftone reproduction (see the next section) led to an enormous expansion of the publishing industry, particularly in the United States with its new 10-cent illustrated magazines (Phillips 1996). In addition to the rise of mass journalism, the printing of large editions of books, ranging from bibles and historical novels to how-to-do manuals, became affordable. As books became cheaper, the total number of public libraries also increased rapidly; in the United Kingdom the Local Government Act of 1894 made it possible to have them even in remote rural parishes (Fleck 1958).

Consumption of paper is still going up even in countries where its per capita use already amounts to more than 200 kg a year (FAO 2003). By the year 2000 the U.S. annual paper consumption surpassed 300 kg/capita, nearly 15% above the 1990 mean, and similar increases were recorded in Japan (now at 250 kg/capita) and in the European Union (now averaging about 220 kg/capita). The link between the use of printing and writing paper and affluence is even stronger than between the consumption of primary energy and gross domestic product: the correlation coefficient for the world's 35 most populous countries is 0.98! The world's affluent economies, whose population accounts for less than 15% of global population, now consume nearly 80% of all printing and writing paper. These realities expose the myth of the paperless office that should have become, according to the forecasts of electronic technoenthusiasts, the norm by now.

As Sellen and Harper (2002) demonstrate, most people prefer reading, evaluating, and summarizing information printed on paper spread around on their desks rather than displayed on multiple overlapping windows on their computer screens. I belong to this majority: although I have used a succession of IBM machines since the mid-1980s, I maintain only a small number of electronic files but I treasure and intensively use my archive of printed materials that now amounts to hundreds of thousand of pages. Of course, tree-saving electronic paper would be a truly revolutionary departure, and several research teams are at work to come with a practical, thin, and erasable display (Chen et al. 2003).

Accessible Images

There was no shortage of excellent images in the pre-industrial world, and some of them were reproduced in relatively large number of copies; they ranged from exquisite copper engravings of Renaissance and Baroque masters to fascinating multicolored *ukiyo-e* prints that depicted everyday life of Tokugawa Japan. But all of these images had two things in common: they were produced in artisanal fashion, one at a time, and at a relatively high cost. Invention of lithography—by Alois Senefelder (1771–1834)—made the reproduction process much cheaper, but prints still had to be pulled off one by one in single sheets. Introduction of steam-driven flat bed and, later, rotary presses speeded up the printing process, and electric motors made it even more flexible—but none of these innovations changed the painstaking process of cutting, engraving, or lithographing the originals.

And, of course, both engravings and woodcuts were inherently labor-intensive interior techniques entirely unsuited for any rapid capture of images in the open or for recording moving people, animals, or objects. Only the development of photography changed that but, despite many impressive improvements, the scope of this new art was limited for decades by cumbersome and inconvenient procedures as well as by the absence of simple techniques to reproduce these images in newspapers and books and to make inexpensive copies for personal use and distribution.

In 1826 Joseph Nicéphore Niepce (1765–1833) produced the world's first photograph, a very crude image of nearby buildings seen from an upper window of his home that was captured on light-sensitive bitumen and fixed on a pewter plate. During the late 1830s the first *nature morts* and photographs of buildings and people appeared on polished silver-plated copper plates made by Louis Jacques Mandé Daguerre (1789–1851). And William Henry Fox Talbot (1800–1877) and John Herschel (1792–1871) invented independently the process of photography on sensitized paper (Frizot 1998; Schaaf 1992). In 1839 Herschel used for the first time the terms "negative" and "positive," and in 1847 Claude Niepce, Joseph's cousin, invented the glass plate. Four years later Frederick Scott Archer introduced a cumbersome wet collodion process that became the standard procedure and has left behind a remarkable record of portraits, documents of daily life, and landscape images.

Wet plates were dominant for about three decades. The first dry, silver-bromide—sensitized, gelatin-emulsion plates were produced in 1864 by W. B. Bolton, and B. J. Seyce and many inventors, including Joseph Wilson Swan of the incandescent light fame, improved them during the 1870s (Abbott 1934). An easily produced and highly sensitive kind, prepared by Charles Bennett, reached the market only in 1878. Still, taking pictures required patience, skill, and considerable expense due to relatively long exposures and heavy, bulky, and fragile plates. Moreover, darkrooms were needed to develop the images. The three kinds of innovations that revolutionized photography were (1) the reduction of exposure time that allowed for capturing an expanding range of natural settings and movements; (2) development of convenient photographic film to replace heavy glass plates; and (3) availability of affordable handheld cameras that could take fairly good snapshots without any special manipulation, and access to inexpensive custom film development that obviated the need for an in-house darkroom and the knowledge of photographic chemistry.

Cameras, Films, Photographs

The earliest photographs required very long exposure times. Niepce's faint view from his window took eight hours to capture; two decades later typical daguerrotypes required 30 seconds when the plate was exposed in a studio. Talbot took the first high-speed flash image in 1851 by using a spark from the discharge of a Leyden jar, but that was clearly an impractical arrangement for casual use. Progressive improvements that began during the 1850s reduced the typical exposure times, and by 1870 Talbot was able to expose his plates for just 1/100 of a second. The most revealing breakthrough in high-speed photography came when Eadweard Muybridge (1830–1904), an Englishman who started his American photographic career with the images of Yosemite in 1867, developed a technique to capture motion in rapid sequences with a camera capable of shutter speed of as much 1/1,000 of a second (Hendricks 2001).

This work was prompted by Leland Stanford—at that time a former governor of California, the president of the Central Pacific, the future founder of the eponymous university, and an avid horse breeder—who wanted to confirm his belief that a galloping horse will, at one point, have all of its legs off the ground simultaneously. In 1878 Muybridge proved this to be correct: perhaps his most famous photo sequence confirmed that all four hooves of a trotting horse are momentarily off the ground (figure 5.7). After his return from a European tour, where he showed his Zoopraxiscope, a viewbox that displayed a rapid series of stills that conveyed motion (the principle used later in animated cartoons), Muybridge made an agreement with the University of Pennsylvania to prepare a comprehensive series of sequences of animal and human locomotion.

This work began in 1884, and it used a fixed battery of 24 cameras positioned parallel to the 36-m-long track with a marked background, and two portable batteries of 12 cameras each that were sited at both ends of the track. These cameras captured the motion of animals and people as they broke strings stretched across the track and activated the shutters in turn. The complete set of 781 sequences was published in 1887 (Muybridge 1887), and it includes such predictable images as running at full speed and throwing a baseball as well as such oddities as a woman pouring a bucket of water over another woman. But more widely practicable and truly instantaneous photography was made possible only when highly sensitive dry plates, which reduced exposures to fractions of a second in well-lit environments, became readily available. Even amateur photographers could now capture movement, and by 1888 their task was made easier by new devices that measured the time of required exposure.

When George Eastman (1854–1932), the man who removed the other two

FIGURE 5.7. Four of the 12 silhouetted frames of Muybridge's famous sequence of "automatic electro-photographs" showing a trotting horse (Abe Edgington, owned by Leland Stanford) at the Palo Alto track on June 15, 1878.

obstacles that precluded the popular diffusion of photography, developed his interest in taking pictures, he was a poorly paid clerk at Rochester Savings Bank. He bought a heavy camera and the associated wet-plate outfit for a planned trip to Santo Domingo in 1877. The trip fell through, but Eastman found a new calling: he experimented with coatings, began making Bennett's dry plates, and in 1879 he devised, and patented, a new coating machine to do the job (Kodak 2001; Brayer 1996). Searching for a lighter and more flexible support than glass, he introduced a three-layered negative stripping film (paper, soluble gelatin, gelatin emulsion) in 1884, and a year later, in cooperation with camera maker William H. Walker, he invented a roll holder that could be attached to standard plate apparatus. All that was needed was to put a smaller size of his film into a handheld, easy-to-use camera.

His solution was a simple and heavily advertised camera whose principal selling slogan was "You press the button, we do the rest." Eastman created its name, Kodak, in order to use K, his favorite letter, to start and end a word that would resonate in his advertisements. The U.S. patent for the camera was issued in September 1888, and Kodak's first version was in production between June and December of that year (Coe 1988; GEH 2001). This design did not include any technical innovations as it used only already tested components, but it produced the first handheld camera suitable for amateur photography; its mode of operation (not its size) made it an obvious forerunner of all modern point-and-shoot models that became particularly popular during the last third of the 20th century. Kodak's operation was not quite as simple as the slogan had it: the film had to be advanced and rewound, and the first 13 Kodak models had a string connected to the shutter mechanism whose protruding top had to be pulled two or three times to wind up the barrel shutter spring to its full tension (figure 5.8).

The camera's wooden box (a rectangular prism, 16.3 cm long) was covered with smooth black leather; the 57 mm lens was centered on one narrow face, the rewinding lever and the string pull were on the top, and the release button for a cylindrical shutter was just above the center on the left side. As the lens had a great depth of field, it produced focused images from as close as 1.2 meters, and its angle (60°) made it possible to dispense with a viewfinder. Kodak spools were available with 50 or 100 exposures, and the entire camera was returned to Rochester, New York, factory, where the film was removed, developed, and printed. The camera took circular photographs (so the users would not have to worry about keeping it level) with a diameter of 6.25 cm, and as it had no exposure counter, photographers had to keep the track of pictures taken. The original list price was $25, a considerable amount in 1888, and about 5,200 units were produced before new models were introduced in 1889.

That was also the year when the stripping film, whose development required time-and labor-intensive operations, was replaced by the precursor of modern

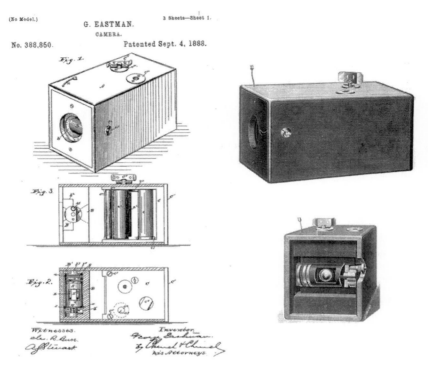

FIGURE 5.8. Patent drawings for Eastman's Kodak camera, and its actual appearance and a detail of its shutter mechanism (reproduced from *Scientific American*, September 15, 1888).

films made of modified celluloid. This nitrated cellulose was first made by Alexander Parkes (1813–1890) in 1861, and under the name Parkesine it was used to make a variety of small items such as combs, knife handles, pens, and boxes. Henry M. Reichenbach, a chemist working for Eastman, discovered that by dissolving the cellulose nitrate in alcohol and by using several additives, he could produce thin, flexible, and perfectly clear films by casting the mixture on glass plates (Coe 1977). Once set, this celluloid was coated with gelatin emulsion to make the world's first transparent roll film, whose production was patented in April 1889 and which was used in the Number 1 Kodak camera released in October 1889.

Several technical advances of the 1890s made amateur photography even easier. They included the easy-to-load cartridge film that was patented by Samuel N. Turner in 1895. This film was backed by black paper marked with numbers that could be read through a small window and thus allowed for advancing a precise length of a film for each exposure, a technique that was displaced only by the advent of electronic cameras. The Folding Pocket Kodak, the obvious ancestor of all roll-film cameras marketed during the next six decades, was introduced in 1897. Kodak's subsequent most notable pre-WWI innovation was the world's first practical safety film that used cellulose acetate instead of the highly flammable cellulose nitrate (Kodak 2001).

By the beginning of the 20th century, amateur photography was thus one of the most accessible, if often disparaged, arts. Alexander Lamson's reaction in *The Beacon* of July 1890 expressed the feeling of "real" amateurs: "Photographers who merely 'press the button' and leave someone else to 'do the rest' are not really entitled to the honorable name" (Lamson 1890:154). This was a misplaced judgment, as the overwhelming majority of picture takers had no special intents and aspirations beyond producing small mementos of their lives, families, friends, and places. And that is what they still do today: they snap, without any focusing, billions of images every year. And while too many of these pictures are fairly useless or merely frivolous, millions of them carry a unique emotional message as they help us to make links between generations, to recall special occasions, to reknit the social fabric—and to leave behind a visual record of changing times.

The 1890s were also the first years when just about everybody could also enjoy inexpensively reproduced photographs thanks to the invention and perfecting of the halftone printing process. Previous reproduction techniques could either produce limited numbers of excellent (continuous tone) but costly copies, while several processes aimed at larger print runs (most notably photogravure) were cumbersome as well as very expensive. That is why, for decades after the invention of photography, pictures were used only to replace earlier field sketches as the basis for preparing reproducible images and why most publications contained no, or only a few, illustrations. Because the most common relief printing processes could render only solid blacks and whites, large numbers of skilled artists were engaged during the second half of the 19th century in converting photographs into reproducible engravings.

An alternative to this laborious process became available during the 1880s with a photomechanical reproduction that could convert black line (or dotted) drawings into relief matrices by producing negatives on sensitized zinc plates and then etching them. But, obviously, this technique was also unable to reproduce photographs or wash drawings. Only halftone reproduction could do that by breaking up the continuous tones of the photographic image into a pattern of tiny dots of variable size that could be printed by the inexpensive relief technique (Phillips 1996). Because of the limited resolution of our vision, these variably sized dots blend and produce the illusion of shaded grays from an image that is composed entirely of black and white. Moritz and Max Jaffé used a gauze screen as early as 1877, and Charles-Guillaume Petit and Georg Meisenbach were other inventors credited with commercialization of the process whose first versions were quite complicated and relatively slow.

By 1885 Frederic Eugene Ives (1856–1937) perfected the process by inventing a screen that was made from two sheets of glass with a fine series of etched lines (six per millimeter are now standard for fine resolution) forming a cross-line grid that was filled with an opaque substance (Ives 1928). Rephotographing a full-tone photograph through the screen produces the halftone effect as a sensitized copper plate is marked by a dotted pattern. Ives never patented his cross-line design, and halftone printing diffused rapidly once the Levy Co. began producing cross-line screens in 1892. The new technique cut the cost newspaper illustrations by about 95% and made illustrated printing common-place. During the late 1880s, full-page engraving in such upscale publications as *The Illustrated London News* or *Harper's Magazine* cost about $300; a few years later it was an inexpensive option (less than $20 for a full page) for every printer (Mott 1957).

Halftones introduced an unprecedented realism to printed illustration and merged the market for photographs with the market for printed word; this combination brought an enormous expansion of the publishing industry after 1890. And because it was cheaper to print halftones than the text, the new technique made illustrated newspapers and magazines not only less expensive to produce but also more attractive, further stimulating their sales (Phillips 1996). And, obviously, photographs quickly became an indispensable tool for advertising. Nothing has changed in this respect: the halftone technique re-mains as essential today to convert photographs into inane advertising images as when it was introduced during the 1890s.

Ives's other great printing invention also dates to the early 1880s: producing primary-color separations (Ives used blue, red, and green filters), making their halftones and then overprinting precisely aligned layers of yellow, cyan, and magenta ink to get a full-spectrum color reproduction of the original image. We still use three colors of ink (cyan, magenta, yellow) and black and you can see their overlaps by examining newspaper or magazine illustrations with a magnifying glass or notice their alignment markers that are often left near the edges of color-printed pages.

Movies

After Edison refused Muybridge's offer to transform Zoopraxiscope into a more sophisticated machine, he became determined to develop a device that, in his own words, would do for the eye what his phonograph did for the ear. But, unlike in the case of incandescent light, most of the work on Edison's concept to combine a camera and a peephole viewer was done by his assistant, William Kennedy Laurie Dickson (1860–1935), who worked previously on Edison's ore milling. Development of Edison's Kinetograph and Kinetoscope is a perfect example of complexity of technical inventions and of the inappropriateness of narrow attributions. Edison's initial idea was to follow his phonograph design and have tiny photographic images affixed to a rotating cylinder. But after his visit to Paris in 1889—where he met Étienne-Jules Marey (1830–1904), professor of physiology, whose Chronophotographe used a continuous roll of film to produce a sequence of still images for research in cardiology—he abandoned that impractical approach and directed Dickson to pursue the roll design.

Edison, irrepressible as always, told a reporter of the *New York Sun*: "Now I've got it. That is I've got the germ or basic principle. When you get your base principle right, then it's only a question of time and a matter of details about completing the machine" (cited in Musser 1990:71). The U.S. patent for the Kinetograph (a camera) and Kinetoscope (the viewer) was filed in August 1891 (figure 5.9). Dickson ordered 35-mm film from Eastman (whose cameras used 70-mm film) and fed it vertically through the Kinetoscope. Edison's first film studio was completed in May 1893, and on May 9 was the first public showing of an electrically powered Kinetoscope at the Brooklyn Institute of Arts and Science. The reward for lining up and peering into the machine was to see three Edison employees hammering on an anvil and passing around a bottle of beer. Such was the sublime beginning of American motion pictures.

Soon, with Dickson and William Heise in charge of the motion-picture production, Edison's studio began releasing an increasing variety of full-length (in 1894 that meant less than a minute) films. The first copyrighted product, a would-be comical clip that showed Fred Ott, an Edison employee, sneezing into a handkerchief, was followed by pictures of an amateur gymnast, of a lightning shave in a barbershop, and of Eugene Sandow, an Austrian strong man, flexing his upper torso. Nearly 80 pictures were made in 1894, and they included wrestlers, cock fights, terriers attacking rats, Buffalo Bill and Indian war council, and the gyrations of skimpily dressed dancers. Comparisons with some leading subjects of mainstream U.S. filmmaking a century later are unavoidable: wrestling, violence, Westerns, sex, even another muscle-displaying Austrian (Arnold Schwarzenegger, of course).

The pattern of moviemaking was set right at the beginning; all that remained was to embellish it and make it gradually more provocative and, technical advances permitting, much more elaborate. Although the resulting images could be seen only by one person at a time, Kinetoscope parlors (the first one opened in New York in April 1894) showing short clips became fairly popular in many large U.S. cities. While Edison's derivative invention of individual peephole viewing was the first commercial display of filmed moving pictures, it was not clearly the beginning of movies, the collective experience of rapidly photographed images projected on screen in darkened rooms.

The real movies were born and greatly improved between 1895 and 1910 (Nowell-Smith 1996; Rittaud-Hutinet 1995; Musser 1990; Mitry 1967). Precise attributions of particular beginnings (first camera, first projection) are more exercises in selective definitions than in describing the complex reality. Unlike in the case of such decisive solo inventions as Parsons's steam turbine or such accurately datable performance breakthroughs as Edison's durable incandescent light, moving pictures have no precise moment of origin, nor can any country or any individual claim a clear-cut priority for their earliest commercialization and artistic development. But the cinematographic protagonists clearly fall into two groups: the largely forgotten ones, and a few iconic names.

On the technical side, the first group includes about a dozen Frenchmen, Germans, and Britishers. Louis Aimé Augustin Le Prince (1842–1890) patented his 16-lens "apparatus for producing animated pictures of natural scenery and life" in January 1888 (U.S. Patent 376,247), and the images of his in-laws and his son he took with it in October of that year in the garden of their house in Leeds were the first successfully photographed and projected motion pictures. Léon Bouly designed his Cinématographe in 1892 and patented an improved version a year later. Max and Emil Skladanowsky held the first public show of their 15-min movie program produced with their Bioskop on November 1, 1895, in Berlin.

And the first portable apparatus to make films was Kineopticon made by Bert Acres, who began his work on a kinetic camera in 1893 (de Vries 2002). In April 1895 Woodwille Latham demonstrated his Eidoloscope, a kind of projecting Kinetoscope. Dickson, who left Edison's company, joined with three partners to form the American Mutoscope Co. to market his projector (subsequently widely know as the Biograph, the name that later became synonymous with cinema in a number of countries), which used wider film (70 mm) and advanced it by friction, without sprockets.

The best-known names are, of course, those of Edison and the Lumière brothers, but the invention attributed to Edison was a case of outright deception and the Lumières' principal contribution was skillful marketing. As the Kinetoscope was enjoying increasing popularity, Edison was in no rush to develop a competing device. But when Francis Jenkins and Thomas Armat introduced their Phantoscope (figure 5.9), Edison's company bought the rights to the machine and agreed to manufacture the projector (slightly altered by Armat) on the condition that it would be advertised as Edison's new invention, the Vitascope. The promotional brochure claimed that Edison invented "a new machine . . . which is probably the most remarkable and startling in its results of any that the world has ever seen" (quoted in Musser 1990:113). Edison's role in all of this was indisputably unethical: not only did he agree to this deception, but he also appeared at the first private screening of the Vitascope, on March 27, 1896, and acted for the benefits of assembled reporters as the inventor of yet another magical machine, assuring widespread press coverage of the event.

America's first Vitascope theater opened its doors on April 23, 1896, in a New York City music hall, and a year later there were several hundred of these

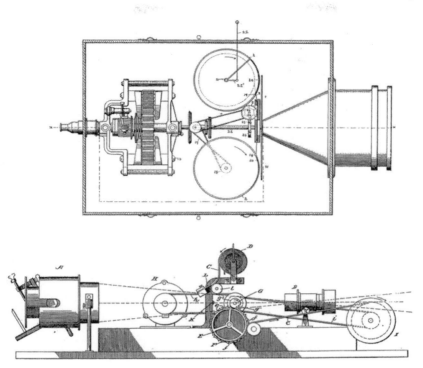

FIGURE 5.9. Patent drawings for Dickson's (Edison's) kinetographic camera (U.S. Patent 589,168) filed in August 1891 and granted six years later (top), and for Jenkins and Armat's Phantoscope (U.S. Patent 586,953), whose slightly altered version was produced as Edison's Vitascope (bottom).

projectors across the country and even in Honolulu. But this rapid diffusion was accompanied by many recurrent problems: only a few experts could initially set up the system, which ran on DC (choice dictated by Edison's bias) and hence could not be accommodated without converters in an increasing number of AC settings. Poor quality of early films that caused repeated perforation tears, and fire was the greatest hazard. Film tearing was solved in 1896 by doubling the thickness of the nitrate base, but the first nonflammable acetate film was introduced only in 1908.

Standard film histories concur that all those inventive Americans pioneering their competing scopes were not the creators of cinema, and that that primacy belongs to Louis Lumière (1864–1948), the younger son of Lyon photographer Antoine Lumière (Conreur 1995; Rittaud-Hutinet 1995; Mitry 1967). At the age of 17, Louis improved the preparation of a gelatin-bromide plate used for snapshots, and the family's company began producing these plates in 1882. His brother Auguste (1862–1954) is usually named as a co-inventor of motion

pictures, although he left a clear account that credits Louis with the key revelation. But Auguste also conceded that Louis did not construct an entirely
new machine from fundamental principles: "The only problem that remained
to be solved consisted of finding a drive system that would ensure uniform
exposure and substitution time for the images" (quoted in Mitry 1967:70).

Origins of the Lumières' machine (French Patent 245,032) thus go directly
to Edison's Kinetoscope, Marey's Chronophotographe, and Bouly's Cinématographe. After Antoine Lumière saw the display of Edison's new device on a
visit to Paris in 1894, he bought it and had one of his workers to disassemble
it. And not only was he familiar with Marey's work, but he also picked up
Bouly's expired patent and with it the name of the device, because Bouly,
much like Beau de Rochas, did not pay his annual patent fees. Perhaps the
most important advantage of *Cinématographe Lumière* was that it combined
camera and projector in a small box that weighed less than 5.5 kg and hence
could be easily used for filming in *plein air*.

But if the Lumières' apparatus was highly derivative, their promotion effort
was very effective. Indeed, Sauvage (1985) saw it as the biggest stunt in the
film history. In any case, by the mid-1890s the public interest was piqued by the
Kinetoscope, and it was clear that the market was ready for a new commercial
display of moving images. The first demonstration of the Cinématographe took
place on March 22, 1895, in a lecture at the Societé d'Encouragement
l'Industrie Nationale in Paris (CineFrance 2003). Three days before that, the
brothers shot their first, 52-second movie, *La sortie des usines*, which showed
workers passing through the front gate of the Lumière et fils factory in Lyon.
During the subsequent months, they were preparing for the commercialization
of their invention by making more short films and lecturing to select groups to
generate widespread interest. The first public screening of 10 short clips they
filmed between March and December 1895 was in the Salon Indien in the basement of the Grand Café on the Boulevard des Capucines on December 28,
1895, and it was attended by 33 people.

The show lasted only 20 minutes, and it consisted mainly of scenes of
everyday Parisian life. By January 1896, up to 2,500 people a day were paying
for the show, whose most impressive clip was *L'arrivée d'un train en gare de
La Ciotat*, filmed in June 1895. This consisted of nothing else but a steam
locomotive pulling carriages and moving diagonally across the screen as it
approached the platform in a town on the Côte d'Azur, where the Lumières
had a summer residence. Many spectators felt that the machine will leave the
screen, and O'Brien (1995:34) rightly argued that this was

> an incomparable and unrepeatable shock effect that filmmakers have
> perennially, and with ever more difficulty, tried to emulate: the roller-
> coaster dynamics of *Die Hard* or *Speed* deploy a century's worth of
> technological advances toward achieving the same visceral surprise that

the Lumières accomplished less strenuously by planting a camera on the platform of La Ciotat Station as a train was arriving.

London saw the first Lumière movies on February 20, 1896, New York on June 29, and by the end of summer shows took place in cities from Madrid to Helsinki. Cinématographe and Vitascope movies were incompatible, as the former used film with a single pair of round perforations for each frame, while the latter had two perforations. After a New York showing, some critics found that the Cinématographe produced clearer images with less vibration. The Lumières eventually made more than 1,500 films, each lasting about 50 seconds. A representative selection of 85 of Lumière vignettes is now available from Kino on Video (Kino International 2003), while the Library of Congress now has 341 early Edison motion pictures on its *Inventing Entertainment* website (LOC 1999).

Movie shows eventually ceased to be dependent on saloons, theaters, and arcades as specialized movie theaters proliferated by the turn of the century (LOC 1999). By 1908 there were about 8,000 of these establishments in the United States (nickelodeons, usually for fewer than 200 people), and this created such a huge demand for new films that it could be satisfied only by imports, mainly from France, where by the end of 1906 at least one new film was finished every day, with Pathé-Frères studio in Vincennes being the leading producer. Georges Méliès (1861–1938), a magician and theater impresario, became the first great innovator of cinematographic art (Conreur 1995). He pioneered narrative cinema, fantastic subjects, and science fiction, and he also developed such special effects as double exposure (in *La caverne maudite* in 1898), actors performing with themselves on a split screen (in *Un homme de tête* in the same year), and the dissolve (in *Cendrillon* in 1899).

Among the technical innovations tried before WWI were wide screens (70–75 mm film), which became popular only during the 1950s, and the first talking movies (Edison's Kinetophone in 1913). The earliest short films included most of the subjects that became staples of 20th-century cinematography: mundane events, trivia, beauty, conflicts. And the Spanish-American War of 1898 was the first conflict whose preparations were shown on screen: actual armed engagements were reenacted in New Jersey using the National Guard troops. As the movies lengthened (to 5–10 minutes by 1905, and to about 15 minutes by 1910) storytelling became common. In 1906 the world's first feature film (*The Story of the Kelly Gang*) that lasted more than one hour was not made either in France or in New York but in Australia by J. & N. Tait theatrical company (Vasey 1996).

By the end of 1900s, a number of film companies opened their studios in the western suburb of Los Angeles, and within a decade these Hollywood newcomers dominated not only America's but the world's moviemaking (Gomery 1996). Their success broke up the power of the Motion Pictures Patents

Co., a combination of leading U.S. and European equipment and moviemakers that was formed in order to charge inflated prices for cameras, projectors, and movies. But that short-lived monopoly group had at least one salutary effect: in 1909 it adopted 35-mm film as its standard-size stock (*pélicule format standard*), and a century later, after every aspect of moviemaking had changed drastically, the 35-mm film is still with us (Rogge 2003).

Only historians of cinema are now aware of the trust's brief existence, but every movie fan recognizes the name Paramount, the company that was established by Adolph Zukor in July 1912. And so the pattern of 20th-century Hollywood moviemaking was essentially set before WWI: use newspapers and pulp magazines, classical novels whose copyright had expired, and successful theatrical plays for suitable plots, integrate vertically all the activities from production to promotion, and go after foreign markets. Nearly a century later, they are still playing the same show, except that the potential for success and failure has been greatly magnified. And just one last notable pre-WWI movie event before I turn to sound: during his tour of the United States, Charlie Chaplin was eventually persuaded to join Mack Sennett's independent Keystone Co. late in 1913.

Reproducing Sound

Invention of the first of the two distinct ways of reproducing sound—its conversion into electromagnetic waves, virtually instantaneous transmission across increasingly long distances, and reconversion to audible frequencies by the means of telephony—led in a matter of months to the success of the second method whereby words or music were stored ingeniously for later replay by the first of many possible recording techniques. The first invention, as already noted in chapter 1, is certainly the best-known instance of nearly simultaneous independent filing in the history of patenting: Alexander Graham Bell (1847–1922) of Salem, Massachusetts, filed his application for "transmitting vocal sounds telegraphically" on February 15, 1876, just two hours ahead of patent caveat submitted by Elisha Gray (1835–1901) of Chicago. The second invention, finalized just before the end of 1877, further increased Edison's fame, built on his improved stock ticker and quadruplex telegraph, and only his demonstration of a complete system of practical incandescent light received a greater acclaim when it was unveiled two years later.

And much as in the case of incandescent lights, or internal combustion engines, the appearance of practical telephony was preceded by several decades of interesting experiments and proposals. These developments failed to reach commercial stage but provided essential stepping stones toward an invention whose many socioeconomic impacts have not been dulled even more than 125 years later. The latest stage in the evolution of telephones—the worldwide diffusion of wireless devices whose latest models can also receive and transmit Internet messages and take and send digital images—has rejuvenated and reconfigured this now classic technique. As a result, telephones have become, once again, one of the most desirable personal possessions.

The phonograph, despite Edison's determined effort, was never a great commercial success, and it was gramophone, Emile Berliner's most famous invention, which became the standard means of music and voice reproduction for most of the 20th century. Its dominance began to decline only during the 1960s with the introduction of compact audiocassettes (by Philips in 1963), and its nearly complete demise came with the introduction of compact discs in the early 1980s. Many gramophones are still around, but the entire field of recorded sound is one of the best examples of how some key innovations of the pre-WWI period were transformed during the 20th century into new, and qualitatively superior, systems.

Wireless transmission of sound and faithful reproduction of words and music at receiving points that were increasingly distant from transmitters rank among the most significant innovations of the Age of Synergy. Although regular radio broadcasts began to take place only during the early 1920s, all of the key components and procedures for widespread diffusion of wireless transmission were put in place before 1914. And it is a much less appreciated fact that the same was true as far as television was concerned: nearly all critical components for wireless transmission of images were either proposed or invented by 1912.

Inventing Telephones

The gestation period of devices that could successfully transmit human voice was less than half a century, a span shorter than the time that elapsed between the first failed trials of incandescent lights and Edison's 1879 success. The first known rudimentary demonstration of telephony was a magic lyre that Charles Wheatstone (1802–1875) showed in 1831: when he connected two instruments by a rod of pine wood, a tune played on one was faithfully reproduced by the other. Wheatstone did not pursue the implications of this short-distance non-electric telephony, but in 1837 he became famous as one of the co-inventors of the telegraph. The first possibility of electric telephony was noted in 1837 by Charles G. Page, who heard the sound produced by an electromagnet at the instant when the electric circuit is closed and described the phenomenon as Galvanic music (Gray 1890).

The first explicit formulation of the idea of electric telephony came in 1854, when Charles Bourseul concluded that a flexible plate that would vibrate in response to the fluctuating air pressure generated by voice could be used to open and close an electric circuit and that the transmitted signals could be reproduced electromagnetically by a similar plate at the receiving end. Philip Reis was the first experimenter who tried to make the idea work (and who coined the term telephony). In an 1861 lecture to the Frankfurt Physical Society, he described an apparatus whose receiving electromagnet produced a sound of a pitch that corresponded to the frequency of on and off switchings.

The quest was renewed by two Americans during the 1870s, by Alexander Graham Bell (figure 5.10), an amateur inventor, and Elisha Gray, a partner in the Western Electric Manufacturing Co. in Chicago. Bell and Gray eventually found about each other, and the Bell's triumph was not a result of superior design, and fundamentally not even of luck in filing the patent application just hours ahead of Gray's caveat, but of a very different perception of the importance of telephony (Hounshell 1981; Garcke 1911b; Gray 1890). Both inventors were initially motivated by coming up with a multiplex system for telegraphic transmissions: after decades of sending just one message in one direction at a time, a duplex system, introduced in 1872, effectively doubled the capacity of telegraphic networks, and multiplexing could multiply it even more dramatically and become one of the most lucrative inventions.

Gray began by investigating the possibility of a telegraph wire carrying several frequencies; this musical telegraph could become a multiple transmitter, given a receiver that could segregate the individual tones. And if sending several

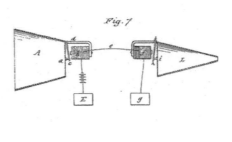

FIGURE 5.10. Alexander Graham Bell and the key figure of his patent application for "improvement in telegraphy." The photograph was taken in 1904 when Bell was 57 years old. The patent drawing is reproduced from Bell (1876b).

tones at once, why not human voice? By 1874 reports on Gray's experiments anticipated that once perfected the invention will do away with telegraph keys as the operators will simply talk over the wire. At the same time, there was a widespread skepticism about the usefulness of voice communication: after all, Reis's experiments led nowhere, and Gray's own patent attorney believed that telephony would be just a toy. In the summer of 1874 Gray demonstrated one practical result of his work, a keyboard device with single-tone transmitters that covered an octave range, but subsequently he concentrated on developing the phenomenon into a multiplex telegraph.

Meanwhile, Bell was pursuing the same goal, approaching the task from an acoustical point of view: his appreciation of speech properties came from his father, a professor of elocution and a creator of a system for teaching the deaf. After he learned about Gray's work, in the spring of 1874, he was able to get financial support from Gardiner G. Hubbard, whose deaf daughter he was tutoring, and by July of 1875 he was able to transmit sounds by shouting into the diaphragm of the transmitter (Rhodes 1929; Bruce 1973). This caused the attached reed to vibrate and generate weak currents in the transmitting electromagnet, which were then reproduced by the receiving diaphragm. Gray, still pursuing his multiplex goal, finally decided to file at least a caveat, the formal notice of telephone's basic concept: he did so on February 14, 1876, and was preceded by just two hours by Bell's full patent application. Comparison of the two concepts, both using a liquid transmitter, shows how nearly identical they were (figure 5.11).

But there was no immediate legal contest: as soon as Gray learned that Bell anticipated him by just hours, he agreed with his attorney and with his financial sponsor against going ahead with an immediate filing of a patent and contesting Bell's claim. Less than a month after the patent filing, and on the evening of the day when the first sentences were exchanged telephonically between Bell and Watson, exultant Bell wrote to his father: "I feel that I have at last struck the solution of a great problem—and the day is coming when telegraph wires will be laid on to houses just like water or gas—and friends converse with each other without leaving home" (Bell 1876a:3).

Bell's specification for "improvement in telegraphy" described "undulatory currents" that are produced "by gradually increasing and diminishing the resistance of the circuit" and that, unlike the intermittent ones, can transmit simultaneously a much larger number of signals and hence can be used to construct an apparatus for "transmitting vocal or other sounds telegraphically" (Bell 1876b). The last of seven figures of Bell's patent application sketched the simple operating principle of such a device (figure 5.10):

The armature *c* is fastened loosely by an extremity to the uncovered leg *d* of the electro-magnet *b*, and its other extremity is attached to the center of a stretched membrane, *a*. A cone, *A*, is used to converge sound-

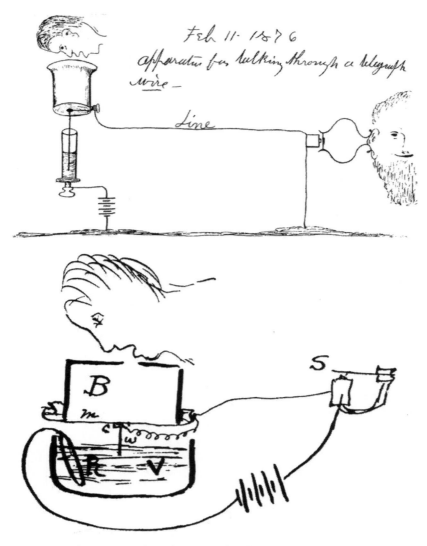

Feb 11· 1876
apparatus for talking through a telegraph
wire—

Line

FIGURE 5.11. Comparison of Gray's (top) and Bell's concepts of the telephone shows two near-identical designs with liquid-bath transmitters. Gray's sketch was done on February 11, 1876; Bell's on March 9, 1876, more than three weeks after the two men filed their patent applications. Reproduced from Hounshell (1981).

vibrations upon the membrane. When a sound is uttered in the cone the membrane *a* is set in vibration, the armature *c* is forced to partake of the motion, and thus electrical undulations are created upon the circuit *A b e f g* . . . The undulatory current passing through the electromagnet *f* influences its armature *h* to copy the motion of the armature *c*. A similar sound to that uttered into *A* is then heard to proceed from *L*. (Bell 1876b:3)

You could build your own basic telephone following these simple instructions—but it would be a very primitive electromechanical device. You would have to shout into it, and a single apparatus at each end would have the double duty of both transmitter and receiver. Still, this great novelty got enthusiastic reception at the 1876 Centennial Exposition held in Philadelphia, and a few months later Bell was demonstrating an improved design (see the frontispiece to this chapter). But it was Edison, rather than Bell, who came up with the first practical transmitter. Edison's involvement began after the Western Union had second thoughts about entering into the telephone business once it saw the attention Bell's device got in Philadelphia. Initially, this leader of telegraph industry declined to invest in a new form of telecommunication and rejected Bell's offer to sell his patents to the company and thus to get the telephone monopoly. Once Western Union reconsidered its telephone policy, it turned to Edison to come up with a commercial device (Josephson 1959).

Belatedly, Edison, who built himself a replica of Reis's device in 1875, reentered the field of telephony. After several months of intensive experimentation (in which he was greatly disadvantaged due to his poor hearing), he was able to replace Bell's liquid transmitter by a superior variable-resistance carbon device, essentially a microphone controlled by the received sound as the vibration of a diaphragm modulated an externally supplied electric current. This was achieved by changing the resistance of carbon granules enclosed in a small button due to the changing pressure applied by a plunger attached to the diaphragm (figure 5.12). This design produced much higher volume of sound and with sharper articulation than did Bell's primitive device; a caller could thus speak into what was essentially a microphone while listening at the same time through the receiver.

Edison filed the basic patent in April 1877, and soon afterward came other improved microphone designs by David Hughes and Emile Berliner. The next step was to insert carbon granules between the diaphragm and a metal backplate, and this design, patented by Henry Hunnings in 1879, was improved first by Edison in 1886 by packing the granules in a small button container and then, in 1892, by Anthony White, who interposed the granules between polished carbon diaphragms, with the front one of these placed against a diaphragm made of mica (Garcke 1911b). Known as the solid-back transmitter,

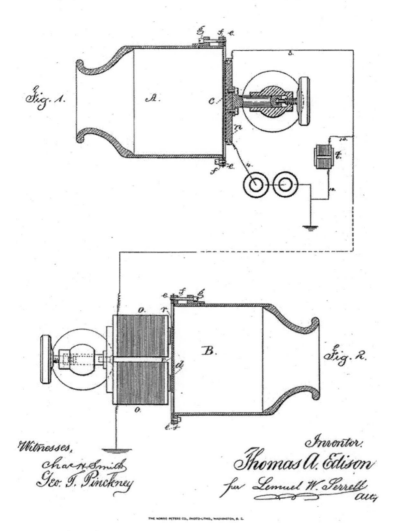

FIGURE 5.12. Edison filed his application for speaking telegraph on April 27, 1877, but it was granted only on May 3, 1892. The illustration shows transmitter (top) and receiver. Resonators are marked *A* and *B*, sheet metal diaphragms *c* and *d*; clamping rings (*e* and *f*) and tightening screws (*g*) keep the diaphragms tensioned; the disk *n*, made of conducting material, was placed in front of the transmitter diagram, and the electromagnet *o* in front of the receiver diaphragm.

this device was used basically unchanged until the mid-1920s, and its improved versions were installed in all telephones until the 1970s, when they were displaced by dynamic transducers.

The Bell Telephone Co. was formed in July 1877, and a year later, with 10,000 phones in service, Theodore Vail (1845–1920) became its new general manager; he eventually guided the corporation to become a giant monopoly (Jewett 1944). Edison's carbon transmitter became a centerpiece of a collection of other telephone patents (including Gray's receiver) that Western Union used in November 1877 to set up its own telephone business. But it left the field entirely to Bell's company by October 1879, when it acknowledged the priority of Bell's patent and when Gray received a consolation payment of $100,000. Meanwhile, Bell continued to elaborate and adjust his basic design. Almost exactly a year after filing the patent, he was able to make a call between Boston and North Conway in New Hampshire over a distance of nearly 230 km (Anonymous 1877), and by October 1877 he had a simplified device in a more compact portable form, with an elongated cylinder forming a handle with a flaring mouthpiece (figure 5.13).

In a society saturated with telephones, it is hard to imagine that the invention was not embraced as rapidly as possible—but there were many technical, financial, and organizational challenges to overcome (Casson 1910). Getting and staying connected were major hurdles. Only when the new service borrowed the established telegraph concept of a central switching office could the numbers of subscribers grow: the first manual switchboard that connected many phones through a single exchange was opened in January 1878 in New Haven, Connecticut, and a decade later more than 150,000 subscribers were served by nearly 750 main exchanges. In 1888 came the introduction of central battery supply to energize all telephones in an exchange instead of relying on subscriber's own batteries or hand cranking. This important principle assures that modern telephone systems—still powered from a large central battery system backed up by emergency generators—remain open even during electricity outages.

Although better transmitters and receivers soon obviated the need for shouting, and signal amplifiers (first in 1892) extended the length of useful telephone lines, technical problems and poor connections were common. Calling distances continued to lengthen: in 1877, the first regular line from Boston to Somerville was less than 10 km; by 1881 the link between Boston and Providence spanned 60 km, and by 1892 a call could be made from New York to Chicago. But the first very expensive transcontinental call (which took more than 20 minutes to arrange) could not be made until 1915, when New York was connected with San Francisco. Total number of American telephones rose from 48,000 in 1880 (or one phone for about every 10,000 people) to 134,000 by 1885 and to 1.35 million (or one phone for every 56 people) by 1900 (Garcke 1911b).

FIGURE 5.13. Bell's improved 1877 telephone and some of his demonstrations of the new device. Reproduced from *Scientific American*, October 6, 1877.

Trials with direct dialing began as early as 1891, when an unlikely pioneer was a Kansas City undertaker Almon Strowger (U.S. Patent 447,918) who suspected that operators were switching his business to competitors. A much better automatic system introduced in 1901 produced "a peculiar humming sound" when the station called up was busy, and it also had an early version of rotary dial "provided at its circumference, opposite each figure, with an aperture into which the finger may be inserted" (Anonymous 1901:85). But none of these early automatic switching devices diffused widely before WWI, and it took even longer for the universal adoption of dial tone that was pioneered by Siemens in 1908. The first four decades of telephone service thus required large numbers of switchboard operators to run local, and later long-distance, service, and an effective solution of this major precondition of telephone's diffusion relied on employing women.

Growth of manual telephone exchanges provided many women from middle-class families with their first entry into the labor force. But even modest wages of telephone operators could not make the service cheap. By 1900, more than 20 years after the technique was invented, basic monthly telephone charge in major U.S. cities was between a quarter and a third of the average monthly wage. This meant that the overwhelming majority of customers were businesses and that, for most Americans, telegraph remained the dominant mode of intercity communication well into the first decades of the 20th century. And while Europe of the last two decades of the 19th century pioneered such critical inventions as steam turbines and internal combustion engines, its adoption of telephones lagged far behind the U.S. pace.

The United States and Europe diverged early in their approach to the problems of capital investment, affordable rates, and reliable service. In the United States, Bell's monopoly lasted until 1894 and allowed the company to charge high prices and make high profits; the subsequent increase in the number of independent telephone companies was only temporary, and virtually all of them were connected to Bell's long-distance service, which was monopolized by the company's new parent organization, American Telephone and Telegraph. Bell's strategy of high long-distance charges used to subsidize low-cost local services lasted for most of the 20th century until the deregulation and breakup of the company on January 1, 1984.

In the United Kingdom, small companies initially operated independently under the supervision of the British Post Office, and as a result the already noted contrast between the speed of diffusion of electric lighting in the United Kingdom and the United States applied to telephones as well. In Samuelson's (1896:128) words, "the telephone has been coldly received in England" and compared to the United States the country "has lagged behind . . . to an extent which is almost ludicrous." After the Post Office, which had the long-distance monopoly from 1896, took over all services in 1912, the United Kingdom had one phone for about 60 people, as did Germany, while the U.S. rate fell to just below 10 people/telephone; French ratio was still more than 150, and the Russian one just dipped below 1,000.

The large investments of the monopolistic companies in installed equipment, which they expected to last for decades, was by far the most important reason for the slow progress of telephonic innovation during most of the 20th century. The first one-piece black set incorporating transmitter and receiver in the same unit was introduced only in the late 1920s. Subsequently, the telephone device stagnated—with the exception of pushbutton dialing introduced in 1963—until the beginning of a rapid conversion to electronic telephony begun in the late 1970s (Luff 1978). Only then was the carbon microphone replaced by a dynamic transducer (identical device acting as a receiver), and small integrated circuits eliminated bulky bell ringer, transformer, and oscillator. Phones got smaller, lighter, and cheaper, and soon the entire innovation cycle was repeated with wireless cellular devices.

Recorded Sound

The pedigree of Edison's phonograph can be traced to both telegraph and telephone. While trying to improve his telegraph transmitter, Edison discovered that when the recorded telegraph tape was played at a high speed, the machine produced a noise that resembled spoken words (Josephson 1959). Could he then record and replay a telephone message by attaching a needle to a diaphragm of a receiver and producing a pricked paper tape? Soon he tried a tinfoil cylinder and was surprised as he heard himself reciting "Mary had a little lamb." As this was the first device able to record and to reproduce sound, it attracted a great deal of public attention. In 1878 Edison took the phonograph on the tour that included a demonstration to U.S. President Rutherford Hayes, and set up a new company to sell the machine that he wanted to see in every American home (figure 5.14).

During the early 1880s, when most of his attention was given to the development of new electric system, Edison looked for a substance to replace the foil as the recording medium—and so were other inventors. In 1885, Chichester Bell and Charles Tainter (1854–1940) got the patent for the Graphophone, a phonograph with wax-coated cylinders incised with vertical grooves, and by 1887 Edison patented his improved phonograph, which used wax cylinders as well as a battery-powered electric motor (figure 5.14). Both devices gave rise to a new recording industry, but neither was a great commercial success. Phonograph was marketed as a multifunctional device: as a recorder of family's vocal mementos, a music box, a dictation machine for businesses, an audio text for the blind, and later also in a miniature form inside Edison's talking dolls.

The two main problems with the phonograph were the recording mode

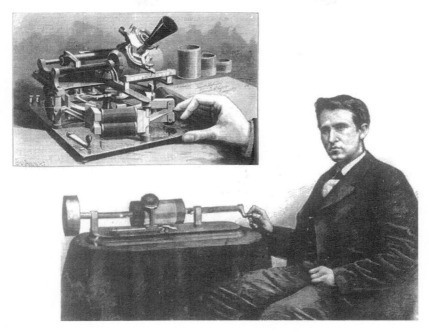

FIGURE 5.14. Edison with his first perfected hand-cranked tinfoil phonograph (reproduced from *Scientific American*, July 25, 1896) and his improved wax-cylinder, electric motor-powered phonograph (inset; reproduced from *Scientific American*, December 24, 1887).

and the recording medium. A vertical (hill-and-dale) cut needed a mechanism that would prevent the stylus from jumping out of a groove, and the wax cylinder was obviously too soft and too fragile to make permanent recording; it broke easily with repeated handling, it needed a storage box, and it could not be reproduced cheaply in large quantities. But Edison—displaying his penchant for lost causes that led him later to invest years, and large sums of money, in fruitless efforts to beneficiate iron ore and mass produce concrete houses—kept improving the home phonograph for decades after its introduction, and the Edison company continued to make recorded cylinders until 1919. By that time the phonograph was a true antique as the era of gramophone was in full swing (literally, given the popularity of new jazz recordings).

Origins of the gramophone can be traced directly to the failings of the phonograph: vertical-cut grooves, soft recording surface, and difficult mass production of recorded sound. All of these were successfully resolved in a relatively short time by Emile Berliner (1851–1929), a German immigrant who after a succession of jobs in New York became a cleanup man in the laboratory of Constantine Fahlberg (1850–1910), the discoverer of saccharine. There he became interested in experimenting with electricity, and in 1876 he invented a simple, loose-contact telephone transmitter (a type of microphone), which he patented in 1877 and which earned him a job with the Bell Telephone Co. of Boston (LOC 2002). Six years later Berliner returned set up a small research laboratory in his house in Washington, D.C., and just four years later he patented (U.S. Patent 372,786) a new system for recording sound with a lateral stylus and playing the records on a hand-cranked machine. Its first public presentation took place at the Franklin Institute on May 16, 1888 (Berliner 1888).

Berliner's solution was clearly anticipated by ideas about the "process of recording and reproducing audible phenomena" that Charles Cros (1842–1888) submitted to the French Academy of Sciences in 1877—but of whose existence Berliner, and even the U.S. patent examiners, were ignorant (Berliner 1888). Cros's key idea, based in turn on Scott's phonautograph, was to attach a light stylus to a vibrating membrane and use it to produce undulating tracings on a rotating disk—and then, by means of a photoengraving process, to convert the undulatory spiral into relief or intaglio lines and to produce a playable record. Nothing practical came out of Cros's suggestions, but when Berliner eventually learned about them he very generously acknowledged their priority.

Berliner's first recordings were done by a stylus tracing a fine undulatory line in a thin layer of fatty ink, which he prepared by mixing one part of paraffin oil with 20 parts of gasoline and drying the deposit after the gasoline evaporated (Berliner 1888). A stylus attached to a mica diaphragm recorded the vibrations by moving laterally; the grooves were etched by acid, and a 15-cm zinc record, with 2-minute capacity, was placed on a hand-cranked turntable to be played by a steel stylus at 30 rpm. Multiple reproduction was done by electroplating the master zinc record and using the metal negative to stamp out the desired number of positive copies. After trying many different substances, Berliner chose celluloid, and soon after he switched to hardened rubber, which could withstand better the pressure exerted on steel needles by the cumbersome combination of the tone arm and horn.

Berliner (1888:18) correctly anticipated that the records (he called them phonautograms) would provide both substantial royalties for performers and a great deal of enjoyment to their collectors:

> Prominent singers, speakers or performers may derive an income from royalties on the sale of their phonautograms, and valuable plates may be printed and registered to protect against unauthorized publication. Collections of phonautograms may become very valuable, and whole evenings will be spent at home going through a long list of interesting performances.

Berliner's U.S. Gramophone Co. sold 1,000 machines in 1894 and 25,000 15-cm hard rubber records. In 1895 Berliner patented (U.S. Patent 534,543) the recording on a horizontal disk (figure 5.15); in 1896 he discovered that shellac (an organic compound prepared from a gummy secretion of Asian scale insect *Coccus lacca*) from the Duranoid Co. was superior to hard rubber and switched to it in 1897, and the material remained dominant until the late 1940s, when it was replaced by vinyl.

Berliner's key recording feature, whereby the groove both vibrates laterally and propels the stylus, remained the standard audio technique until stereo LPs, introduced during the late 1950s, combined the lateral and vertical cut. Soon after gramophone's introduction, Berliner began improving the machine's basic design, by replacing hand cranking with either windup spring or electric motors and by recording an increasing variety of performances. Rising sales

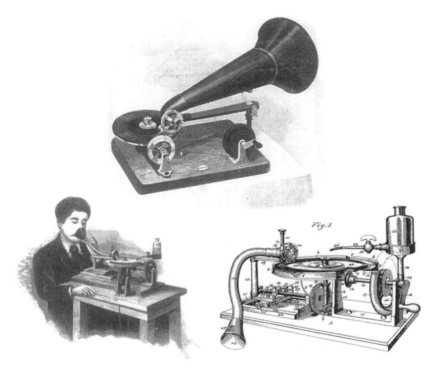

FIGURE 5.15. Emile Berliner's drawing of the apparatus he invented (U.S. Patent 534,543 filed in 1892 and granted in February 1895) to produce gramophone records (lower right) and use of the instrument for recording speech (lower left) and an early design of a hand-cranked reproducing gramophone (top; reproduced from the cover of *Scientific American*, May 16, 1896). With Berliner's design, it was necessary that "the record tablet be covered with a thin film of alcohol, and for this purpose a thin stream of alcohol (stored in the vessel on the right) is directed upon the center of the tablet . . . from which the alcohol spreads in all directions by centrifugal force . . ."

brought a number of illegal competitors, and in 1900 Berliner turned over his U.S. patents to Eldridge R. Johnson of the National Gramophone Co. in Camden, New Jersey, at that time the main supplier of playback machines (LOC 2002). Johnson formed a new business that eventually changed its name to the Victor Talking Machine Co. and, after its 1929 merger with RCA, became the world's largest and best-known recording enterprise, RCA Victor. Berliner relocated his company to Montreal in 1900.

During the second year of his Montreal operation, Berliner sold 2 million single-sided records: the other side featured what was to become one of the world's most famous trademarks, Francis Barraud's painting of his dog, Nipper, listening to a gramophone (*His Master's Voice*, source of the company abbreviation HMV), which Berliner registered on July 10, 1900. Victor's two major milestones were the first recording that sold a million copies (the first Red Seal record with Enrico Caruso, on an exclusive contract, singing *Vesti la giubba* from Leoncavallo's *I Pagliacci*) in 1903, and the introduction of the Victrola (in 1906), the first entirely enclosed cabinet phonograph, which despite its high cost ($200) became a big seller for the next two decades.

The first double-sided disks were produced in 1902, but even so the playtime remained limited, with both sides containing no more than 10 minutes of music. Recording complete symphonies or operas was thus still impractical, but it was done anyway: HMV's first complete opera, Verdi's *Ernani*, filled 40 single-sided disks in 1904. But segments of classical compositions and key parts of famous operas became common, opening up entirely new audio experiences. Prior to that, even musical aficionados might have heard a favorite composition only once or twice in a live performance or had to be content just studying its score. More important, recordings introduced great music to millions of listeners who would could not make it to a concert hall. On the pop side, the earliest hits included folk songs, band music, and beginning in 1912, the craze for ballroom dancing led to a proliferation of dance band records (but the first Dixieland Jazz record, which sold 1.5 million copies, came in 1917).

Before leaving the subject of recorded sound, I must note the genesis and delayed commercialization of magnetic recordings. Valdemar Poulsen (1869–1942), a Danish telephone engineer, patented his system of sound recording and reproduction on steel piano wire in 1898, and a working model of his Telegraphone was one of the electrical highlights at the Paris Exposition in 1900 (Daniel, Mee, and Clark 1999). An improved design, able to record 30 minutes of sound, became available in 1903, but better established and cheaper Edisonian phonographs and dictaphones delayed commercial development of the device, and the first Magnetophon recordings were made only in 1935.

Hertzian Waves

There were no fundamental obstacles to prevent the discovery of high-frequency electromagnetic waves years, even decades, before Hertz did so. Already in 1864, James Clerk Maxwell (1831–1879; figure 5.16) had presented to the Royal Society of London a dynamical theory of the electromagnetic field (Maxwell 1865), and his ideas generated great deal of interest, and disbelief, particularly after they were published in a two-volume *Treatise on Electricity and Magnetism* (Maxwell 1873). Implications of Maxwell's theory were clear: the existence of waves of varying length that would propagate at the speed of light, curl around sharp edges, and be absorbed and reflected by conductors; moreover, it was known how to calculate their length and how to produce them (Lodge 1894). But Maxwell himself did not make any attempts to test the wavelike nature of electromagnetic radiation, and it took more than two decades after the publication of his paper before somebody investigated their existence.

Or more precisely, before somebody did so and published the results, because here I must make an important detour in order to describe one of the most remarkable cases of lost priority and misinterpreted invention. Sometime in December 1879, seven years before Hertz began his experiments, David Edward Hughes (1831–1900), a London-born physicist who returned to the city from Kentucky, where he was a professor of music as well as physics, began sending and receiving electromagnetic pulses over distances of as much

FIGURE 5.16. James Clerk Maxwell (left) postulated the existence of electromagnetic waves longer than light but shorter than sound more than two decades before Hertz's experiments confirmed the theory. Photo from author's collection. This portrait of David Edward Hughes (right) was taken during the mid-1880s, just a few years after his pioneering experiments with "aërial electric waves." Photo courtesy of Ivor Hughes.

as several hundred meters (Hughes 1899; figure 5.16). As already noted, Hughes was, in 1878, one of the inventors of a loose contact microphone, and experiments with this device led him to suspect for the first time the existence of what he called "extra current" produced from a small induction coil.

He found that microphones made sensitive receivers of these waves, and he tested the transmission first indoors over distances of up to 20 m; then, walking up and down Great Portland Street with the receiver in his hand and the telephone to the ear, he could receive signals up to 450 m away. Between December 1879 and February 1880 eight people, most of them members of the Royal Society, witnessed these experiments, but George Gabriel Stokes (1819–1903), famous Cambridge mathematician and a future president of the society, concluded that everything could be explained by well-known electromagnetic induction effects. Although Hughes continued his experiments, he was so discouraged by his failure to persuade the Royal Society experts "of the truth of these aërial electric waves" that he refused to write any paper on the subject until he had a clear explanation of their nature, a demonstration that came from Hertz's experiments.

Crookes, who witnessed the December 1879 experiment, wrote two decades later to Fahie that "it is a pity that a man who was so far ahead of all other workers in the field of wireless telegraphy should lose all credit due to his great ingenuity and prevision" (Fahie 1899:305). How far ahead? A writer in *The Globe* of May 12, 1899 (cited in Fahie 1899:316), summed up this best when he said that the 1879 experiments "were virtually a discovery of Hertzian waves before Hertz, of the coherer before Branly, and of wireless telegraphy before Marconi and others." Incredibly, a few years after Hughes's first broadcasting experiments, there was another near-discovery by none other than Thomas Edison, and it actually went public. One of the world's great missed fundamental discoveries was a small item displayed in 1884 in a corner of the Philadelphia Exhibition's largest electrical exhibit, which belonged, predictably, to Thomas Edison: it was an "apparatus showing conductivity of continuous currents through high vacuo." In its patent application, filed on November 15, 1883, Edison (1884:1) noted that

> if a conducting substance is interposed anywhere in the vacuous space within the globe of an incandescent electric lamp, and said conducting substance is connected outside of the lamp with one terminal, preferably the positive one, of the incandescent conductor, a portion of the current will, when the lamp is in operation, pass through the shunt circuit thus formed, which shunt includes a portion of the vacuous space within the lamp.

This device became known as a tripolar incandescent lamp, but the observed phenomenon, called by Edison etheric force and commonly known as the

Edison effect, remained just a curiosity without any practical applications. Another 12 years had to pass after Edison secured this useless patent before the current passing through the vacuous space was recognized as electromagnetic oscillations at wavelength far longer than light but far shorter than audible sound. Interestingly, neither Edison, who through his long career experienced a full measure of triumphs and failures stemming from his obsessive inventive drive, nor Hughes, who ended his life as one of the most honored inventors of his generation, regretted their near misses.

Hughes's 1899 correspondence with Fahie reveals a man who looked back without bitterness, crediting Hertz with "a series of original and masterly experiments." And they, done between 1886 and 1889 without any knowledge of Hughes's work, were exactly that. Heinrich Rudolf Hertz (1847–1894; figure 5.17) was steered in the direction of these fundamental experiments by his famous teacher, Hermann von Helmholtz (1821–1894). In 1879 Helmholtz made an experimental validation of Maxwell's hypotheses the subject of a prize by the Berlin Academy of Sciences, and he believed that Hertz would be the best candidate to solve the problem. Although Hertz soon abandoned this line of inquiry, his "interest in everything connected with electric oscillations had become keener" (Hertz 1893:1).

Hertz finally began investigating what he termed *sehr schnell elektrische*

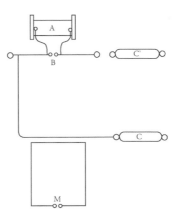

FIGURE 5.17. Heinrich Hertz's experiments opened the way for the entire universe of wireless communication and broadcast information. Simplicity of his epochal discovery is best illustrated by one of his early instrumental arrangements (right). An inducing circuit (top) contained the induction coil (*A*) and straight copper wire conductors (*C* and *C'*) with the discharger (*B*) at the center; the induced circuit was a rectangular wire with the gap adjustable by a micrometer (*M*). Reproduced from Hertz (1887).

Schwingungen (very rapid electric oscillations) in 1886 at the Technische Hochschule in Karlsruhe (where two decades later Fritz Haber made another epochal discovery, that of ammonia synthesis from its elements). His experimental setup was very simple, and all of his work is well documented in his diaries, letters, and papers (Hertz 1893; Hertz and Süsskind 1977). He produced electromagnetic waves by using a large induction coil, basically a transformer that received pulsed voltage from batteries to the primary winding and produced a much higher voltage in the secondary winding. In his first set of experiments, he used a straight copper wire with a small discharge gap for the inducing circuit and a rectangle of insulated wire for the induced circuit (figure 5.17). Subsequently, his discharger consisted of a straight copper wire 5 mm in diameter with a 7.5 mm spark gap in the middle that was attached to spheres 30 cm in diameter made of sheet zinc and placed 1 m apart. Once this simple dipole antenna was connected to an induction coil, its alternate charging and discharging produced sparks across the gap. His secondary circuit was even simpler, what we came later to call a loop receiving antenna, without any rectifier or amplifier, just a coil of wire 2 mm thick formed in a circle of 35 cm in radius left with just a small spark gap that could be regulated by a micrometer screw (Hertz 1887). Hertz set up all the experiments in such a way that the spark of the induction coil was visible from the place where the spark in the micrometer took place.

The most important reason for Hertz's success was his choice of frequencies. All of his experiments were done in a lecture room 15 m long and 14 m wide, but rows of iron pillars reduced the effective area to about 12 × 8 m. This meant that in order to radiate the waves from one end of the room, bounce them off a metal sheet at the other end, and measure the crests or nodes of the standing waves (by the strength of sparks) at least one half wavelength (and preferably several) had to fit into the length of the room. Hertz understood this requirement, and the best reconstruction of his experiments indicates that he was using wavelengths between 6 m and 60 cm, that is, frequencies of 50–500 MHz. His ingenious experiments proved that these invisible electromagnetic waves behave much as light does as they are reflected and refracted by surfaces and that they travel through the air with finite, lightlike velocity. In retrospect, the discovery of the waves that "range themselves in a position intermediate between the acoustic oscillations of ponderable bodies and the light-oscillations of the ether" (Hertz 1887:421) was obviously one of the most momentous events in history.

Hertz's feat was akin to identifying a new, enormous continent that is superimposed on a previously well-explored planet, accessing an invisible but exceedingly bountiful realm of the universe, the existence of an entirely new phenomenon of oscillations. Less than a decade after Hertz's experiments came the first wireless telegraph transmissions, and less than two decades later the first radio broadcasts. By the late 1930s, radio was the leading means of mass communication and entertainment, the first scheduled television broadcast-swere taking place, and radar was poised to become a new, powerful tool in warfare and, after WWII, also in commercial aviation. By the century's end, Hertzian waves had changed the world, and there is yet no end to their impact: their subdivision, modulation, transmission, and reception for the cellular telephony and wireless WWW are just the latest installment in the still unfolding story.

But Hertz did not foresee any practical use of his invention, and because of his premature death, he never got a chance to revise this conclusion once research into the communication with high-frequency waves got underway. The process that transformed Hertz's discovery into a new inexpensive communication technique that eventually reached all but a very small fraction of the world's population is an even better example of a collectively created innovation than is the creation of first electricity-generating and transmission systems. Although there can be little doubt that such systems would have emerged sometime during the 1880s or the 1890s, even without Edison, his drive, determination, and holistic approach had significantly speeded up the process. In contrast, none of the principal contributors to the development of radio occupies a similarly indisputable pivotal position, because no individual carried the invention from the basic investigation of newly discovered waves to commercial broadcasts.

Wireless Communication

Guglielmo Marconi's (1874–1937; figure 5.18) fame rests on filing the first wireless telegraph patent in 1896 (G.B. Patent 7,777; U.S. Patent 586,193), and being the first sender of wireless signals over increasingly longer distances and finally across the Atlantic in 1901 (Boselli 1999; Jacot and Collier 1935; Marconi 1909). But, David Hughes aside, there are at least four other pioneers of wireless telegraphy whose work made Marconi's achievements possible, or preceded his first broadcasting demonstrations. In 1890 Nikola Tesla's invention of his eponymous coil, the device that could step up ordinary currents to extremely high frequencies, opened the way for transmission of radio signals (Martin 1894). In 1890 Edouard Branly (1844–1940) noticed that an ebonite tube containing metal filings that would not normally conduct when placed in a battery circuit became conductive when subjected to oscillatory current from a spark generator. This device made a much better detector of electromagnetic waves than did Hertz's spark gap.

In June 1894, at the Royal Institution, and again in August in Oxford, Oliver Joseph Lodge (1851–1940; see figure 6.3) used an improved version of

FIGURE 5.18. Guglielmo Marconi, who, without a formal scientific or engineering education, put into practice what several much more experienced men only contemplated doing: sending and receiving wireless signals by using a simple patented apparatus. This photograph was taken in 1909 when Marconi's work was rewarded by a Nobel Prize in physics. Photograph © The Nobel Foundation.

Branly's tube, which he named coherer in order to demonstrate what might have been the world's first short-distance wireless Morse broadcasts (Jolly 1974). *Nature* reported on the first event, and in 1897 Lodge recalled that on the second occasion, when signals were sent some 60 m (and through two stone walls) from the Clarendon Laboratory to the Oxford Museum, he used Morse signals. On balance, Aitken (1976) believes that Lodge, not Marconi, was the inventor of wireless telegraphy. But Lodge himself acknowledged that "stupidly enough no attempt was made to apply any but the feeblest power so as to test how far the disturbance could really be detected" (Lodge 1908:84). He, much like Hertz, had a purely scientific interest in the matter and took steps to the first commercial steps only after Marconi's 1896 patent.

And yet already two years before Lodge's London and Oxford demonstrations, William Crookes— who also witnessed Hughes's 1879 experiment, and whose concerns about the world food production that he expressed in 1898 were noted in chapter 4—spelled out (albeit still timidly, as his vision was limited to Morse signals) commercial potential of Hertzian waves. He noted that

> an almost infinite range of ethereal vibrations or electrical rays . . . unfolded to us a new and astonishing world—one which it is hard to conceive should contain no possibilities of transmitting and receiving intelligence . . . Here, then, is revealed the bewildering possibility of telegraphy without wires, posts, cables, or any of our present costly appliances . . . This is no mere dream of a visionary philosopher. All the requisites needed to bring it within the grasp of daily life are well within the possibilities of discovery . . . (Crookes 1892:174, 176)

And Russian historians routinely claim that the inventor of wireless telegraphy is Alexander Stepanovich Popov (1859–1906), who began his work with electromagnetic waves as he tried to detect approaching thunderstorms (Radovsky 1957). He designed an improved version of Lodge's coherer and used a vertical antenna to pick up the discharges of atmospheric electricity. In his lecture to the Russian Physicist Society on May 7, 1895, Popov reported that he transmitted and received wireless signals over the distance of 600 m (Constable 1995). Popov's first public demonstration of wireless transmission took place in March 1896, and a year later he installed the first ship-to-shore link between the cruiser *Africa* and the Russian Navy headquarters in Kronstadt.

While none of these four inventors pushed for commercialization of their discoveries, 22-year-old Marconi came to England from Italy in 1896 with determination to patent the achievements of his Italian work and to commercialize his system of wireless telegraphy (Boselli 1999; Jacot and Collier 1935). He arrived in February 1896, and on March 30 (helped by his cousin Henry Jameson-Davis: Marconi's mother was Annie Jameson, a daughter of the famous Irish whiskey maker who married a well-off Bolognese merchant and landowner) he got a letter of introduction to William Preece, chief engineer of the British Post Office (Constable 1995). What Marconi brought to England was nothing fundamentally new: his transmitter was a version of high-frequency oscillator originally developed by his mentor Augusto Righi (1850–1920) in Bologna, his receiver was an improved version of Branly's coherer, and his antenna was of grounded vertical type.

What was decisive was his confidence that this system will be, with improvements, eventually able to send signals across long distances and his determination to achieve this goal in the shortest possible time. As Preece correctly observed in an 1896 lecture, many others could have done it, but none of them did (Aitken 1976). Marconi's preliminary British patent application was filed in London on June 2, 1896, and the first field tests sponsored by the Post Office were done on the Salisbury Plain, where in September 1896 he transmitted 150 MHz signals over the distance of up to 2.8 km. His U.S. patent application was filed on December 7, 1896 (figure 5.19). In the same month, after his signals crossed 14 km of the Bristol Channel, Marconi incorporated the Wireless Telegraph and Signal Co. and began to look for the best possible markets for his system. In September 1899 his signals spanned 137 km across the English Channel between Wimereux and Chelmsford. This achievement convinced Marconi that the signals follow Earth's curvature, and by July 1900 the directors of his company approved the ultimate trial of trans-Atlantic transmission.

John Ambrose Fleming (1849–1945) became the scientific adviser of the project, and he built a new, more powerful transmitter while Marconi himself

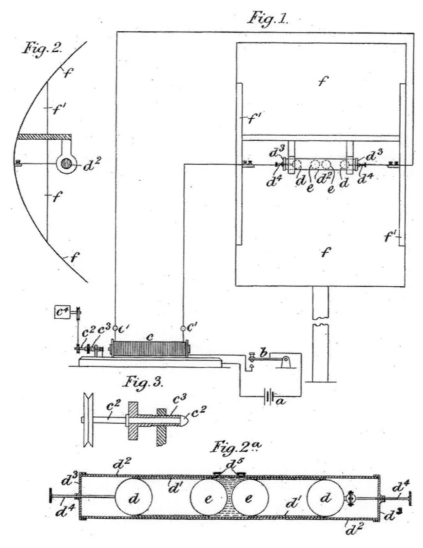

FIGURE 5.19. Illustration of the transmitting component of Marconi's pioneering patent (U.S. Patent 586,193) for sending and receiving electrical signals. Fig. 1 is a front elevation and Fig. 2 a vertical section of the transmitter with parabolic reflector (*a* is a battery, *b* a Morse key, *c* an induction coil, and *d* metallic balls). Fig. 2a is a longitudinal section of the oscillator.

designed the spectacular inverted wire cone antenna with 60 m diameter to be suspended in a ring of 20 60-m-tall wooden masts. The site selected for the transmitter, an alternator driven by an 18.7-kW oil engine whose output was transformed to 20 kV, was at Poldhu in Cornwall, where construction began in October 1900, and Marconi's mast at Cape Cod was to be the receiver. Setbacks delayed the first attempt: a nearly completed Poldhu aerial collapsed on September 17, 1901, and on November 26, 1901, a gale took down the Cape Cod aerial as well. But Marconi quickly redesigned the Poldhu transmitter, an inverted wire pyramid anchored by four 60-m wooden towers, and decided to shorten the distance and to build a new receiving station on Signal Hill that overlooks St. John's Newfoundland, 2,880 km from Poldhu.

The first trans-Atlantic signal, sent by the most powerful spark transmitter of that time, was picked up on December 12, 1901, on the simplest untuned receiver (figure 5.20): Marconi heard faint triple dots of Morse S at 12:30 and then again at 1.10 and 2.20 P.M., but nothing afterward. He hesitated before releasing the information to the press four days later. The news was greeted with enthusiasm in both Ottawa and New York, but there was also skepticism about the claim, and some of it remains even today. Major uncertainty surrounds the actual wavelength of the first transmission. Immediately after the event, Marconi claimed it was about 800 kHz; he did not quote any figure in his Nobel lecture, but in his later recollections, in the early 1930s, he put it at about 170 kHz (Bondyopadhyay 1993).

MacKeand and Cross tried to settle this uncertainty by using the best possible information in order to reconstruct Marconi's complete system and then to model its most likely performance. They concluded that the transmission centered on 475 and 540 kHz, that its total power at Poldhu reached about 1 kW and at St. John's was about 50 pW, and that it is likely that Marconi received high-frequency wide band signals, "spurious components of the spark transmitter output, propagated across the Atlantic by sky waves near the maximum usable frequency" (MacKeand and Cross 1995:29). In contrast, Belrose (1995, 2001) found it difficult to believe that Poldhu signals could have been heard on Signal Hill because the broadcast took place during the day (hence their heavy attenuation) and during a sunspot minimum period, and as the untuned receiver used by Marconi had no means of amplification.

Definitive demonstration of long-range wireless transmission took place in February 1902, when Marconi and a group of engineers sailed from Southampton to New York on the *Philadelphia* fitted with a four-part aerial: coherer-tape reception was up to 2,500 km from Poldhu, and audio signal reached as far as 3,360 km (Marconi 1909; McClure 1902). Although Marconi was completely surprised by the difference between the maximum daylight reception (1,125 km) and the nighttime maxima (he was not aware of the ionospheric reflection of radio waves during the night), he was now confident that as soon as he set up an American station similar to that at Poldhu, he would be able

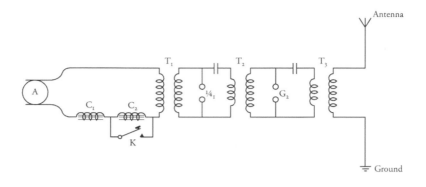

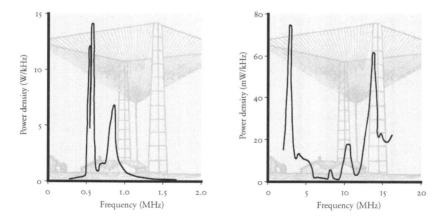

FIGURE 5.20. The basic circuit of Marconi's Poldhu transmitter contained an alternator (*A*), chokes (*C₁* and *C₂*), low-frequency (*T₁*) and high-frequency (*T₂* and *T₃*) transformers, spark gaps (*G₁* and *G₂*), capacitors (*P₁* and *P₂*), and telegraph key (*K*). Transmitted power at low frequencies had the highest density at about 0.6 MHz (bottom left); at high frequencies, at about 3 MHz (bottom right). Based on Fleming's original drawing in Bondyopadhyay (1993) and on Aitken (1976) and Belrose (1995).

to transmit and receive easily across the Atlantic. The Canadian government provided financial assistance to set up a high-power station at Glace Bay, Nova Scotia, and the exchange of messages with Poldhu began in December 1902. In 1907 an enlarged Glace Bay station and a new European station at Clifden in Ireland were used for the first commercial signaling across the Atlantic (Marconi 1909; Marconi Corporation 2003).

Marconi's first major customer was the British Royal Navy, which made a substantial purchase of radio sets in 1900, and wireless communication was used (unsuccessfully) for the first time during the Boer War (1899–1902) and shortly after that routinely during the Russo-Japanese War of 1904–1905. Commercially much more important was Marconi's contract with the Lloyds of London, the world's premier shipping insurer, which committed itself in 1901 to 14 years of exclusive use of Marconi's wireless telegraph system. But the diffusion of wireless telegraphy in merchant shipping was not a rapid affair. Only in 1914—two years after the signals sent on the night on April 14, 1912, by the senior radio operator Jack Phillips from the *Titanic* (he drowned, and several nearby ships did not have their receivers on) called for help for the sinking passenger liner—did an international conference mandate the presence of wireless on all ships carrying more than 50 people (Pocock 1995).

Marconi's aggressive advances had other, and much more experienced, engineers playing catch-up. Lodge filed his syntony (two circuits tuned to the same frequency) patent in 1897 (Lodge 1908). Adolf Slaby (1849–1913), the first Professor of Electro-technology at Berlin's Technische Hochschule, was present at Marconi's May 1897 experiments, and after his return to Germany he and Georg Wilhelm Alexander von Arco (1869–1940), working for Allgemeine Elektrizitätsgesselschaft (AEG), began to patent various improvements to receivers and antennas. So did, independently, Karl Ferdinand Braun (1850–1918), working for Siemens & Halske in Strassburg. After the two rival companies joined to set up Gesselschaft für drahtlose Telegraphie (much better known as Telefunken), those inventions provided the foundation for Germany's leading role in the early development of wireless broadcasting (Telefunken 2003).

In the United States, Tesla filed his two key radio patent applications (U.S. Patents 645,576 and 649,621) in 1897. He specified the apparatus that will produce "a current of excessively-high potential" to be transmitted through the air, and he also recognized that the receiving coils may be moveable, "as, for instance, when they are carried by a vessel floating in the air or by a ship at sea" (Tesla 1900:2; figure 5.21). As these filings preceded by more three years Marconi's U.S. patent application for an apparatus for wireless telegraphy (U.S. Patent 773,772 filed on November 10, 1900), Tesla's priority seemed secure: his reaction to the news of Marconi's trans-Atlantic broadcast was to note that that achievement was based on using 17 of his patents. But in 1904 the U.S. Patent Office reversed itself and recognized Marconi as the inventor. This only aggravated Tesla's already precarious financial position, but worse was to come in 1909 when Marconi but not Tesla got the Nobel Prize in physics.

This bitter story had its unexpected ending in 1943 when, a few months after Tesla's death, the U.S. Supreme Court finally ruled in his favor (Cheney 1981). But this was no righting of intellectual wrongs, merely a way for the court to avoid a decision regarding Marconi Co. suit against the U.S. government for using its patents. Marconi's success is a perfect illustration of the fact that no fundamental innovations are needed to become a much respected

N. TESLA.
APPARATUS FOR TRANSMISSION OF ELECTRICAL ENERGY.

(Application filed Feb. 19, 1900.)

(No Model.)

Witnesses:
Benjamin Miller.
G. W. Martling.

Nikola Tesla, Inventor

by Ken, Page & Cooper
Attys

innovator. Being first to package, and slightly improve, what is readily available, being aggressive in subsequent dealings, and making alliances with powerful users can take an entrepreneur and his company a lot further than coming up with a brilliant new idea. The preceding sentence, describing Marconi's moves, also sums up what William Gates and his Microsoft Corporation did with Windows for the IBM personal computer. The fact that Marconi was not a great technical innovator is best illustrated by the fact that he considered Morse signals quite adequate for the shipping business and that he, unlike Tesla and Fessenden, did not foresee the multifaceted development of radio and broadcasting industry (Belrose 1995; Cheney 1981).

Spark generators, favored by Marconi, could only transmit the Morse code. The first continuous wave signals that could be modulated by audio frequencies for the transmission of voice or music were produced by arc transmitters and high-frequency (HF) alternators. Valdemar Poulsen designed the first effective arc transmitter in 1902, and the largest devices were patterned on Kristian Birkeland and Samuel Eyde's process, which was commercialized in the same year in order to produce nitrogen oxides as feedstocks for the synthesis of inorganic fertilizers. These transmitters became eventually truly giant: Telefunken's Malabar in the Dutch East Indies, commissioned in 1923, had input of 3.6 MW and a 20-t electromagnet to quench the arc. Europe's largest arc transmitter, completed in 1918 in Bordeaux, had eight 250-m masts and covered an area of nearly 50 ha.

Tesla, already famous because of his electric motor and AC inventions, built the first low-radio-frequency alternator (operating at 30 kHz) in 1899, but two men whose work made the first radio broadcasts possible—a Canadian, Reginald Aubrey Fessenden (1866–1932), and a Swede, Ernst Frederik Werner Alexanderson (1878–1975)—are well known only to students of radio history. Fessenden (figure 5.22) was first to conceive and Alexanderson first to build, and then to perfect, HF alternators that could produce continuous wave signals. Fessenden's early accomplishments included the work in Thomas Edison laboratory and professorships of electrical engineering at Purdue and Pittsburgh University. In 1900 he began to work for the U.S. Weather Bureau and pursued

FIGURE 5.21. Tesla filed his application for an apparatus for transmission of electrical energy on September 2, 1897, and received U.S. Patent 649,621 in May 1900. The transmitter consists of a suitable source of electric current (G), transformer (A and C being, respectively, high-tension secondary and lower voltage primary coils), conductor (B), and terminal (D), "preferably of large surface, formed or maintained by such means as balloon at an elevation suitable for the purposes of transmission." In the receiving station, the signal is led from the elevated terminal (D') via a conductor (B') to the transformer, whose coils are reversed, with A being the primary, and the secondary circuit contains "lamps (L), motors (M), or other devices for utilizing the current."

FIGURE 5.22. In contrast to Tesla and Marconi, only radio experts and historians of invention now know the name of Reginald Fessenden, a Canadian working in the United States, who made the world's first radio broadcast in December 1906. Photograph courtesy of the North Carolina State Archives.

his idea that HF well above the voice band should make wireless telephony possible.

His first success came on December 23, 1900, when he transmitted a couple of sentences— "One-two-three-four, is it snowing where you are Mr. Thiessen? If it is, would you telegraph back to me?"—over a distance of 1.6 km between two 15-m masts built on Cobb Island in the Potomac River in Maryland (Raby 1970). After he left the bureau, he set up his broadcasting hut and antenna at Blackmans Point on Brant Rock in Massachusetts and its trans-Atlantic counterpart at Machrihanish in Scotland, so his challenge was to span a distance longer than did Marconi's wireless telegraphy experiments. His eventual success was made possible by a new 50 kHz alternator that was built by General Electric according to the designs of Ernst Alexanderson, a young Swedish engineer who was hired by GE in 1902 (Nilsson 2000). Alexanderson's machine had the periphery of its tapered disk rotating at 1,100 km/h (i.e., the speed of sound) and yet wobbling less than 0.75 mm.

This was an inherently wasteful, and hence costly, way to broadcast: due to high heat losses in windings and armature, and to high frictional losses arising from up to 20,000 rpm, efficiency of smaller units was less than 30%. Moreover, the technique was limited to relatively low frequencies of no more than 100 kHz, compared to 1.5 MHz for spark generators. But it was the first means of generating true continuous sine waves, and it could compete well with spark generators at longer wavelengths and lower frequencies that were favored at that time. After WWI, vacuum-tube transmitters began replacing all of the pioneering transmitting devices, but some spark generators remained in operation until after WWII, long after arc transmitters and most of the alternators were gone.

The first long-distance broadcast using the Alexanderson's alternator took place accidentally in November 1906, when an operator at Machrihanish clearly overheard the instructions relayed from Fessenden's Brant Rock head-quarters to a nearby test station at Plymouth. Encouraged, Fessenden prepared the world's first programmed radio broadcast: on Christmas Eve of 1906 he gave a short speech and then played a phonographic recording of Handel's *Largo* followed by his own short violin solo of Gounod's *O holy night*, a reading from the Bible, and wishes of merry Christmas (Fessenden 1940). This broadcast was heard by radio operators on ships of the U.S. Navy along the Atlantic coast and on the vessels of the U.S. Fruit Co. as far away as the Caribbean, and its variant was repeated on New Year's Eve. For Fessenden, this was a personal triumph because a widespread belief, attested by Fleming's 1906 book, held that an abrupt impulse is a necessary for wireless transmissions and that HF generators could not to the job.

Fessenden also developed heterodyning—a way to transfer a broadcast signal from its carrier to an intermediate frequency in the receiver in order to avoid retuning the receiver when changing channels—and amplitude modulation (AM). AM frequencies, now used in the range of 540 to 1,600 kHz, remained the best way to broadcast voice and music until the diffusion of frequency modulation (FM, now in the band between 88 and 108 MHz) that was invented by Edwin Armstrong (1890–1954). But these admirable achievements required further development and modification in order to make broadcasting a commercial reality (White 2003). Contributions by John Fleming, Edwin Armstrong, Lee De Forest, and others were essential in that respect. The year of the first broadcast brought another milestone in the history of radio because of Lee De Forest's (1873–1961) invention of triode or, as he called it, the Audion tube.

In 1904 Fleming invented the diode (a device based on Edison's unexploited effect), which was essentially an incandescent light bulb with an added electrode and could be used as a sensitive detector of Hertzian waves as well as a converter of AC to DC (figure 5.23). But, as he later recalled in his biography,

> Sad to say, it did not occur to me to place the metal plate and the zig-zag wire in the same bulb and use an electron charge, positive or negative, on the wire to control the electron current to the plate. Lee de Forest, who had been following my work very closely, appreciated the advantage to be so gained, and made a valve in which a metal plate was fixed near a carbon filament in an exhausted bulb and placed a zig-zag wire, called a grid, between the plate and the filament. (Fleming 1934:144)

The added grid electrode, interposed between the incandescing filament and the cold plate, acted as a modulator of the flowing current and made the Audion the first highly sensitive practical amplifier (figure 5.23). In 1912, when Armstrong introduced the first acceptable AM receiver, de Forest began using a series of his Audions in order to amplify HF signals, and he also discovered

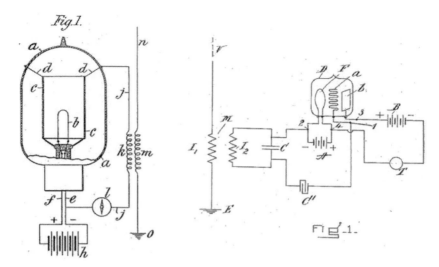

FIGURE 5.23. Detail of John A. Fleming's patent drawing (U.S. Patent 803,684 filed on April 10, 1905) for the diode (left), an "instrument for converting alternating electric currents into continuous currents." A glass bulb (*a*) contains a carbon filament (*b*) operating at 6–8 V and 2–4 A and connected to leads (*e* and *f*) by platinum wires. An aluminum cylinder (*c*), suspended by platinum wires (*d*), is open at the top and bottom, and it surrounds the filament without touching it. Lee De Forest's diagram (right) of "a wireless telegraph receiving system comprising an oscillation detector constructed and connected in accordance with the present invention," namely, his triode (U.S. Patent 841,387), shows an evacuated glass vessel (*D*) containing three conducting members: a metallic electrode (*F*) connected in series to a source of electricity that can heat it to incandescence; a conductor (*b*) made of platinum and, interposed between these two members, a grid-shaped platinum wire (*a*). Both drawings are available at http://www.uspto.gov.

that feeding part of the output from the triode back into its grid produces a self-regenerating oscillation in the circuit that can be used, when fed to an antenna, for broadcasting speech and music.

Consequently, the triode could be adapted for reception, amplification, and transmission. For more than half a century, all electronics—radios, televisions, and the first computers—depended on an increasing variety of vacuum tubes: tetrodes and pentodes, containing four and five electrodes in a high vacuum, and cathode ray tubes (for TV screen and display monitors) were eventually added to diodes and triodes. Domination of vacuum tubes ended only when transistors, superior amplifiers that operate on an entirely different principle, began taking over during the 1950s.

By 1913 the combination of better generators, antennas, amplifiers, and receivers provided a good foundation for commercial development of radio broadcasting, but large-scale diffusion of the new communication medium took place only after WWI (White 2003). Several reasons explain this delay. To begin with, only limited numbers of radio amateurs and military and ship operators had requisite transmitters and receivers. Public access to radio was delayed by restrictions imposed on the use of airwaves during WWI and by the necessity to wear headsets, a requirement that made the earliest reception a peculiar hobby rather than an easily accessible source of news and entertainment. That is why radio's expansive years came between the two world wars as transmission distances increased, numbers of radio stations multiplied, and every aspect of reception was greatly improved thanks to new tuning circuits, capacitors, microphones, oscillators, and loudspeakers. Radio's second great transformation came during the 1950s with the deployment of transistors that made it inexpensive and easily portable.

6

A New Civilization

> Every now and again something happens—no doubt it's
> ultimately traceable to changes in industrial technique,
> though the connection isn't always obvious—and the whole
> spirit and tempo of life changes, and people acquire a new
> outlook which reflects itself in their political behaviour, their
> manners, their architecture, their literature and everything
> else . . . And though, of course, those black lines across the
> page of history are an illusion, there are times when the
> transition is quite rapid, sometimes rapid enough for it to
> be possible to give it a fairly accurate date.
>
> George Orwell in a BBC broadcast on March 10, 1942

Orwell's observations make a perfect epigraph to fortify this book's arguments
about the unequaled pre-WWI technical saltation. Their discovery was ser-
endipitous: I began reading Orwell's wartime essays a few months before I was
to start this chapter. As soon as I came across the quoted passage, I felt as if
this book had been distilled, 60 years before it was written, into a paragraph.
Besides capturing the very intent and essence of this book, by asserting matter-
of-factly that great civilizational shifts are ultimately traceable to technical
changes, Orwell also believed that some of these momentous transformations

FRONTISPIECE 6. An image that embodies aspirations of a new era: "The Progress of
the Wheel: The Ousting of the Horse from London Thoroughfares," portrayed in *The
Illustrated London News*, June 27, 1903.

could be rather accurately dated. And although these might be seen as only inconsequential stylistic preferences, I was pleased that Orwell used three specific terms that are not commonly encountered in writings about epochal shifts. Rather than using the inappropriate, but now generally accepted, term technology, he wrote correctly about technique (a perceptive reader might have noticed that this is the only instance when technology appears in this book).

Orwell also talked about spirit and tempo of life, two elusive but critical concepts that I feel to be essential for real understanding of civilizations. And, serendipity squared, within days after reading Orwell's essay I came across a fitting frontispiece. Once I decided to precede every chapter of this book with an image that would capture both a key accomplishment and the spirit of innovation to be described in the subsequent pages I realized that it would not be easy to choose an appropriate illustration for these closing reflections. I rejected dozens of possibilities until I came across a 1903 engraving of a "continual procession of motors and cycles" that could be seen "any fine holiday afternoon and Sunday morning" along London's Kensington High Street.

This picture appears entirely unremarkable when seen from the vantage point of the early 21st century, yet it would have been quite unimaginable in the mid-1860s, and so it conveys perfectly the great technical saltation accomplished during the two pre-WWI generations. And the two portrayed conveyances and their users also tell us much about the attendant socioeconomic impacts. Phrases that come to mind capture the tempo and spirit of modern age: mobility (both in the physical and in the social sense), mass consumption, democratization, opportunity, equality, emancipation, rise of the middle class, recreation and leisure. And also aspiration and anticipation: so much had changed, but so many changes are ahead as those determined riders proceed on their way.

In structuring this chapter, I decided first to fill in some missing perspectives, and I do so in two different ways. But the principal subject of this chapter is to explain, in a terse and resonant manner, those social, economic, and behavioral waves that began to radiate across modern societies after new techniques had disturbed and transformed the traditional surface. Doing this in a fairly comprehensive way would require another book, and that is why my survey of new realities and enduring trends will concentrate only on two conjoined fundamental attributes and legacies of the era: the increase of energy consumption, and mechanized mass production and the delivery of new services.

Missing Perspectives

Writing this book was a constant exercise in restraint. Again and again I wanted to supply more technical details regarding particular inventions because those numbers and their improvement over time tell best the stories of unprecedented achievements and of continuous quest for technical perfection. At the same time, I would have preferred to prepare the ground for these details with generous explanations of conditions that prevailed before the great wave of the late 19th-century innovations, and to enliven the technical presentations by supplying much more personal information regarding the background, effort, and expectations of the era's leading inventors. And so, exercising restraint once again, I chose only two subjects for this section. Given the recent preoccupation with computers, I first describe the origins and early development of modern data processing. My second choice is to reflect a bit more systematically on personal triumphs, trials, and tragedies, as well as on notable idiosyncrasies, of some of the era's leading protagonists.

BC (Before Computers)

I did not set out to write a comprehensive history of technical advances brought by the Age of Synergy; instead, I concentrated on the four classes of fundamental innovations that have created and largely defined the 20th century. Even within those confines, there are many omissions and unavoidable simplifications. Undoubtedly, the most important set of innovations that I left out deals with what could be broadly described as modern information management.

Seen from the vantage point of the early 21st century, this may not seem to be an important omission. After all, the enormous post-1970 advances in computing—exemplified by Moore's law (see figure 1.1)—have greatly overshadowed even those accomplishments that were achieved during the first three decades of computer evolution before 1970. Those post-1970 developments—so promptly translated into widespread ownership of affordable personal machines and leading to the emergence and, after 1993, to very rapid adoption of the Internet—appear to relegate all of the precomputer data management to the category of inconsequentially primitive tinkering. This would be a wrong interpretation of historic reality.

Although semiconductors and microchips, the key components of modern computers, were invented only after WWII, we should not forget that for the first two-thirds of the 20th century, information management was directly beholden to innovations that were introduced well before 1914. The two pre-WWI generations were not as epoch-making for data management as they were for other advances described in this book, but they were much more important than the perspective skewed by recent accomplishments would lead us to believe. This conclusion is justified not only because of the fundamental fact that the age of electronics could not arise without reliable and affordable electricity generation. Two important pre-WWI advances included automated handling and evaluation of massive amounts of statistical information and the invention and widespread diffusion of mechanical and electromechanical calculating devices. Both of them retained their importance until the early 1970s, and I have vivid memories on both accounts.

The first new skill I had to learn after we arrived in the United States in 1969 was to program in FORTRAN. At that time, Penn State was a proud owner of a new IBM 360/67, and there was only one way to talk to the machine: to spend hours at a keypunch producing stacks of programming and data input cards, a technique introduced for the first time by Herman Hollerith (1860–1929) during the 1880s. And a few years later, when I needed to do some weekend calculations at home, I was still lugging from the university one of those portable yet not-so-light Burroughs machines. So here are at least brief reminders of the pre-WWI origins of two modern techniques that were instrumental in ushering the computing age.

Unfinished construction of Charles Babbage's analytical engine has been the best-studied advance of the early history of automated calculation: more than a dozen books were written about it and its creators (Babbage and Augusta Ada Byron, Countess of Lovelace); a partial prototype was completed and is now displayed in the Science Museum in London. Curiously, a much more successful effort by George (1785–1873) and Edvard (1821–1881) Scheutz, who succeeded in finishing and actually selling (with difficulty) two of their machines, which could not only do complex calculations but also print the results, remains generally unknown (Lindren 1990).

Both Babbage and the Scheutzes aimed too high, but many simple calculating machines were designed and offered for sale throughout the 19th century (Redin 2003; Chase 1980). Some of them were admirable examples of clever mechanical design, but none of the devices that were offered before 1886 was easy to use, and hence none of them was in demand by the growing office market. The necessity to enter numbers with levers and the absence of printers were the most obvious drawbacks. The first key-operated calculator was the Comptometer designed by Dorr Eugene Felt (1862–1930). Its prototype was housed in a wooden macaroni box and held together by staples and wire, and its production by Felt & Tarrant Manufacturing began in 1886. More than 6,000 units were sold during the subsequent 15 years (Boering 2003). By 1889 Felt added a printer to make the first Comptograph, but the machine never sold well.

A prototype of the first very successful adding and listing device was completed by 1884 by William Seward Burroughs (1857–1898), whose work as a bank clerk motivated him to ease the repetitiveness of tasks and to improve the accuracy of their calculations. In 1886 Burroughs and his three partners formed the American Arithmometer Company (renamed Burroughs Adding Machine Company in 1905), whose first adding machines had a printing mechanism activated by pulling a lever (figure 6.1). Initially modest sales rose rapidly after 1900 with an expanded product line, and their cumulative total reached 50,000 machines by 1907 (with more than 50 models); two decades later it surpassed 1 million units (Cortada 1993).

Although there was no shortage of competition, Burroughs machines were the world's dominant calculators throughout the first half of the 20th century. Class 1 models, introduced in 1905 and weighing nearly 30 kg, were famous because of their glass side walls showing the mechanism inside the machine. Subsequent important milestones included the first portable device (9 kg) in 1925, the first electric key-actuated machine three years later, and the first account machine with programmed control panel in 1950 (Hancock 1998). Two years after that a new era began as Burroughs built an electronic memory system for the ENIAC computer, but the company kept producing portable calculators throughout the 1960s (in 1986 it merged with Sperry to form Unisys). For more than eight decades, innumerable calculations, from astronomical ephemerids to artillery tables, were done by hundreds of different models of Burroughs machines and their competitors.

Hollerith's inventions were both derivative and original. His challenge was to find a practical automatic substitute for the highly labor-intensive, expensive, and increasingly protracted manual tabulation of U.S. Census data, which depended on marking rolls of paper in appropriate tiny squares and then adding them up. Hollerith began to work on a solution in 1882 (after he joined MIT), and in 1884 (after he moved to the U.S. Patent Office) he filed the first of his more than 30 patent applications, which detailed how the information recorded on punch cards would be converted into electrical impulses, which in turn would activate mechanical counters (figure 6.2). The storage medium chosen by Hollerith was first introduced in 1801 when Joseph-Marie Jacquard (1752–1834) programmed complex weaving patterns by the means of stiff pasteboard cards with punched holes. Hollerith's punched cards could store an individual's census data on a single card, and in his prototype he used a tram conductor's ticket punch to make the holes.

His great innovation was not just to have large numbers of these cards read by an automatic machine but to devise means of tabulating the results and to extract information on any particular characteristics or on their revealing combinations. Reading was done by a simple electromechanical device, as springmounted nails passed through the holes and made contacts. Punched cards were used to process a large variety of statistical information and to keep records that ranged from the U.S. censuses to particulars of prisoners in Hitler's Germany, where IBM's subsidiary Deutsche Hollerith Maschinen was taken over by the Nazis before WWII (Black 2001). The age of encoded paper ended

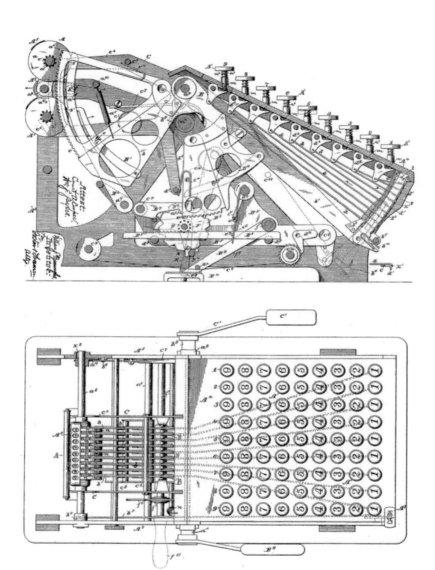

FIGURE 6.1. Longitudinal section and a plan view of a calculating machine invented by William S. Burroughs and granted U.S. Patent 388,116 in 1888. These images are available at http://www.uspto.gov.

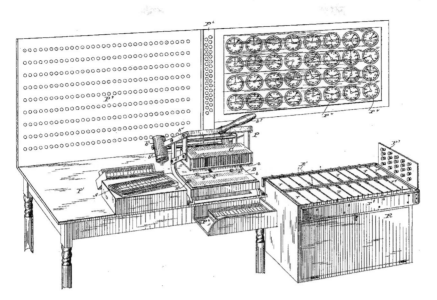

FIGURE 6.2. Holerith's 1889 machine for compiling statistics (U.S. Patent 395,781). Punched paper strips or cards placed on a stationary nonconducting bed-plate (B) with holes that contain embedded wires are read by lowering a movable platen of pins (C). The circuit wires from the bed-plate are connected to the switchboard (P3). Counters (P4, upper right hand) and sorting box (R) were other prominent features of the apparatus.

with the rise of magnetic memories, whose advantage includes not only enormous storage capacity per unit volume but also the ease of reuse by erasing the stored information, an obvious impossibility with punched cards.

Triumphs, Tragedies, Foibles

Although I did sprinkle the text of topical chapters with brief references to some interesting biographical facts, it is obvious that there is vastly more information that could be shared about fascinating or mundane private lives of great innovators and about their fate following their (often singular and brief) sojourns in the public eye. Not surprisingly, there were quite a few deserved triumphs as well as personal and family tragedies. And, as is rather common with creative individuals, many of them behaved in highly idiosyncratic ways, and while some of their beliefs, quirks, and foibles were quaint or amusing, others made them appear unbalanced and even psychotic.

One of the most admirable qualities of many creators of a new technical age, and one of a few readily quantifiable marks of their intellectual triumphs, was their prodigious inventiveness. Edison's unparalleled record of 1,093 U.S. and 1,239 foreign patents that were granted between 1868 and 1931 (additional 500–600 applications were unsuccessful or abandoned) is by far the best-known achievement in this illustrious category (TAEP 2002). Nearly 40% of his U.S. patents pertained to incandescent lights and to generation and transmission of electricity, with recorded sound, telegraphy and telephony, and batteries being the other major activities and dozens of patents were obtained for ore mining and milling and cement production.

Tesla's worldwide patent count surpassed 700. Frederick Lanchester, builder of the first British car and the inventor of disk brakes, held more than 400 patents. George Westinghouse, one of the creators of the electric era, amassed 361 patents, the most famous of which was the one for compressed air brake (U.S. Patent 88,929 in 1869) he designed after witnessing a head-on collision of trains between Schenectady and Troy. His other notable railroad-related inventions included automatic signaling and an automatic air and steam coupler; as already noted, he also had numerous patents in the field of new AC generation and transformation (Prout 1921). Among Nobel's 355 patents are substitutes for rubber and leather, perforated glass nozzles, production of artificial silk, and the world's first aluminum boat. Ernst Alexanderson had 344 patents, the final one granted in 1973 when he was 95 years old!

These examples show that triumphs of many great innovators of the Age of Synergy were not limited to a single class of devices or to a group of kindred advances within a particular field. Many multitalented thinkers and experimenters left behind entire catalogs of inventions and improvements strewn over their specialty, and frequently extending over several fields. The Siemens brothers contributed to such disparate advances as regenerative furnaces, dynamos, and intercontinental telegraphs. Werner will be always best known for his dynamo, but his inventions included also such unusual devices as electric range-finders, mine exploders, and a continuous alcoholmeter that was used by the Russian government to levy taxes on the production of vodka (Siemens 1893). Tesla's first U.S. patent, in 1888, was for electric motor (U.S. Patent 391,968; see figure 2.19); his last, in 1928 (U.S. Patent 1,655,114), was for what we would call today the vertical short takeoff and landing aircraft.

Before he designed the first practical electric starter for cars, Charles Kettering patented a driving mechanism for cash registers (U.S. Patent 924,616 in 1909) and later investigated the most suitable antiknocking additives for gasoline. Emile Berliner, the inventor of a loose-contact transmitter and gramophone, also patented a parquet carpet, porous cement tiles designed to improve acoustics of concert halls, lightweight internal combustion engine, and a tethered helicopter that actually lifted the weight of two adults in 1909 (LOC 2002). Besides amplitude modulation and heterodyne, Reginald Fessenden also patented electric gyroscope, sonar oscillator, and a depth finder.

Despite the keen perception and clever thinking that they displayed in technical matters, many inventors turned out be dismal businessmen. Edison spent virtually all of his fortune created by his numerous electrical inventions on his futile enterprise of mining and enriching iron ore in New Jersey. Lee De Forest was not the only inventor who spent a great deal of money on protracted law suits, but he was an exceptionally poor businessman: several of his companies failed, in 1903 he was accused of stealing one of Fessenden's inventions (and eventually found guilty of patent infringement), and a decade later he was tried with two of his business partners for misleading stock offerings (and found not guilty).

Other inventors were surprisingly susceptible to fraudulent claims of assorted spiritualist movements that were much in vogue during the late 19th century. Two of the era's leading British scientists—William Crookes and Oliver Joseph Lodge—became committed spiritualists. Lodge (figure 6.3), who shared his spiritualistic enthusiasm with his friend Arthur Conan Doyle, believed that the unseen universe is a great reality to which we really belong and shall one day return (Lodge 1928), and he hoped that after his death he will be able to send a message from "the other side" by using a medium. Carl Kellner, one of the inventors of sulfite papermaking process, was not only a devoted spiritualist but also a student of Asian mysticism (OTO 2003).

FIGURE 6.3. Oliver Joseph Lodge, one of the leading pioneers of radio age who promised to send the signals from "the other side" after his death.

Political and social judgments of many famous inventors and industrialists of the era were also questionable. Henry Ford's antisemitism was of such an intensity that in the early 1920s his framed photograph hung on the wall of Hitler's Munich office and copies of the German edition of *The International Jew*, a series of articles that appeared in Ford's newspaper *The Dearborn Independent* (1922), were displayed on the future Führer's table (Baldwin 2002). And in 1935, before the Fascist invasion of Ethiopia, Louis Lumière dedicated his photograph to *il Duce* "avec l'expression de ma profonde admiration," while in July 1941 the name of his brother Auguste appeared on the list of French ultracollaborators who created the Légion Volontaires Français to fight along with the Nazis (Cercle Marc Bloch 1999).

And while it is not surprising that a large and disparate group of pre-WWI innovators included individuals with dubious or even reprehensible beliefs, one of the last attributes one would associate with the creators of a new era would be their lack of imagination. Counterintuitively, this was not an uncommon failing. Recognized excellence in a particular field is no guarantee that every technical judgment would be acute and rewarding; it is perhaps the most fascinating idiosyncrasy of the creative process to see that many minds that were so inventive and so open to radical experimentation could be at the same time surprisingly resistant to other, and not even so shocking, ideas. One demonstration of this conservative behavior is the reluctance, or an outright refusal, to carry some inventions even a step further. There is no special term for this attitude, but Watt's syndrome may be a good description.

Watt's experiments and insights turned an extremely inefficient Newcomen's machine with a very limited range of applications into the first widely useful mechanical prime mover energized by fuel combustion—but his 1769 patent, and its subsequent 25-year extension, actually impeded the next step of steam-driven innovation. Because Watt was afraid to work with high-pressure steam, he not only made no attempts to develop steam-driven transportation but also actively discouraged William Murdock, the principal erector of his engines, from doing so (Robinson and Musson 1969). Watt's patent expired in 1800 and just a few years later Richard Trevithick and Oliver Evans had high-pressure boilers ready, and the diffusion of railways, steamships, and higher efficiency stationary engines was on the way. Some 80 years later, Karl Benz displayed a classic Watt syndrome by refusing to have anything to do either with very fast-running engines or vehicles other than motorized horse carriages.

Marconi's reluctance to broadcast anything else but Morse signals let others to take the lead in developing radio. And I have also noted, respectively, in chapters 2 and 1, Edison's two famous failures of imagination, his militant rejection of AC and his belief that electricity, rather than internal combustion engines, will be the dominant automotive prime mover. In the first instance, Edison reversed himself fairly fast, but the second conviction gripped him for most of the first decade of the 20th century. Instead of trying to invent, for example, a reliable, and lucrative, electric starter for internal combustion engines (the device that was eventually designed in 1911 by Kettering), he persevered for years in his futile quest for a superior battery.

Another display of failed imagination is a surprisingly common lack of confidence regarding the potential importance of one's own inventions. Combative confidence and exaggerated expectations might be expected as a part of aggressive inventive process—but not an almost inexplicable diffidence and technical timidity. Among the most famous examples are Elisha Gray's initial decision not to pursue his rights to patenting telephone because he believed that the device is just an interesting toy, Hertz's complaints to his students that the electromagnetic oscillations he just discovered cannot possibly have any practical use (think of today's universe of electronic devices!), and Louis Lumière's initial conviction that the cinema is an invention without any future as he expected that people will get bored watching his images of city scenes, trains, workers, and landscapes, most of which they could see just by walking around.

Besides the surprising failures of imagination, there were also some notable personal tragedies. While most of the era's inventors had fulfilling careers and led interesting lives, there were also suicides, premature deaths, and, more frequently, convictions of being unjustly treated. I have described Diesel's feelings of failure and his (almost certain) suicide (chapter 3). Ironically, by the time of Diesel's disappearance in 1913, his great invention was well set on its road to a commercial triumph, while an earlier sudden death may have changed the history of the world's most popular entertainment. In September 1890, Louis Le Prince boarded a train in Dijon, after a visit to his brother, and never made it to Paris. We will never know why, but we can ask: if Le Prince had lived for at least another five years, could he had perfected the camera with which he took in 1888 what are arguably the world's first moving pictures, and would he be known today as the inventor of cinema?

Alfred Nobel did not commit suicide and did not die a mysterious death, but one of his early experiments with nitroglycerine killed his youngest brother, Emil, and four other people, and his near-chronic depression, despondency, and searing self-appraisal would have made Freud shudder. This is how, at age 54, when refusing his brother's request to write an autobiography, Nobel saw himself (cited in Fant 1993:1):

Alfred Nobel—a pitiful creature, ought to have been suffocated by a humane physician when he made his howling entrance into his life. Greatest virtues: keeping his nails clean and never being a burden to anyone. Greatest weaknesses: having neither wife and kids nor sunny disposition nor hearty appetite. Greatest single request: to not be buried alive. Greatest sins: not worshiping Mammon. Important events in his life: none.

Two years later, in a letter to Sofie Hess, an Austrian flowers salesgirl whom he met as a 20-year-old in 1876 and who was his mistress for the next 18 years, he wrote (quoted in Fant 1993:318): "[W]hat a sad end I am going toward, with only an old servant who asks himself the whole time if he will inherit anything from me. He cannot know that I am not leaving a last will . . ." But, of course, he did, and probably the world's most famous one: it divided his diminished but still considerable riches among 19 relatives, coworkers, and acquaintances (20% of the total), various institutions (16%), and the Swedish Academy of Sciences in order to "constitute a fund, the interest on which shall be annually distributed in the form of prizes to those who, during the preceding year, shall have conferred the greatest benefit on mankind" (Nobel 1895).

While aging Nobel was depressed but lucid, the last two decades of Tesla's life were marked by a deepening psychosis. He told journalists of his great love for a white pigeon ("Yes, I loved her as man loves a woman . . ."), and, finally a few days before his death on January 7, 1943, he sent a messenger with a sealed envelope containing money and addressed to Samuel Clemens at 35 South Fifth Avenue. That was the address of Tesla's first New York laboratory of the late 1880s, and in 1943 Mark Twain, whom Tesla wanted to help ("He was in my room last night . . . He is having financial difficulties . . .") had been dead for 33 years (Cheney 1981).

Mergenthaler found he had tuberculosis even before he was 40; he moved from Baltimore to New Mexico, but there his house, and all of his papers, were destroyed by fire, and he died in Baltimore before his 46th birthday (Kahan 2000). Hertz died at the age of 36, after he became ill with an infection of the mouth and ears and underwent several unsuccessful surgeries. Wilbur Wright died of typhoid at age 45. Daimler's long and accomplished life had an ironic ending. He did not like to drive (and may have actually never driven at all), and his death came after he insisted, with his health failing, on being taken in poor weather in an open vehicle to inspect a possible site for a new factory. On the return trip he collapsed, fell out of the car, and died soon afterward.

But what beset his fellow inventor's family was infinitely more tragic. As Maybach's new powerful and elegant Mercedes 35 was breaking the world speed records in 1901 and 1902, his adolescent son Adolf (born in 1884) was succumbing to schizophrenia. His condition later deteriorated to such an extent that he could not be kept at home and had to be cared for in sanatoriums. Yet the worst was to come decades later: Adolf was murdered in 1940, 11 years after his father's death, as one of the thousands of victims of the Nazi *Euthanasieprogramm*. Who remembers this as the company is now advertising the rebirth of a great marquee whose vehicles now retail more than $350,000?

Not a few inventors discovered that even an extensive record of remarkable accomplishments did not automatically translate into widespread recognition and financial rewards. Fessenden felt slighted both in his native Canada and in the United States (Raby 1970). He got his first substantial reward only for his sonar patent, and received a large settlement for his pioneering radio inventions only in 1928, after years of litigation and just four years before his death. Much as today, willingness to engage in self-promotion and a dose of unorthodox behavior helped to capture press attention and to create an erroneous but greatly appealing image of a prototypical heroic inventor who appears to base his work on nothing but acute intuition and succeeds due to his uncommon perseverance.

Edison was the master of this game, while Tesla was too eccentric to play it successfully. This difference helps to explain why Edison remains, more than 70 years after his death, widely admired by the public that craves such suitably heroic figures and why Tesla's following—although highly, and some of it even fanatically, devoted to the memory of that master electrician—is much more cultlike, and why outside of his native Serbia (where he will always be a national hero) it is largely limited to scientists and to individuals who are intrigued by his research into ultrahigh-frequency discharges, large-scale energy transmissions without wires, and death rays. However, Tesla got one honor that eluded Edison: the unit of magnetic flux density carries his name, and so he joins the most select company of scientists and engineers—including Ampère, Coulomb, Faraday, Hertz, Ohm, and Volta—whose names were chosen for international scientific units.

But public admiration and official honors went not only to such skilled self-promoters as Edison or Marconi but also to inventors whose work is now known only to historians of science and engineering. In the words of his brother, William Siemens (see figure 4.5) "forced the public opinion of England to honour him in his lifetime, and in a still more striking manner after his death" (Siemens 1893:270). He was knighted, got honorary degrees from both Oxford and Cambridge, and on November 21, 1883, the London *Times* obituary spoke of his "singularly powerful and fertile mind." His funeral service took place in Westminster Abbey, and later a window in the cathedral was dedicated to his memory: England's highest honors for an immigrant German engineer.

I end these reflections on personalities by noting a few idiosyncrasies that have been a surprisingly frequent accompaniment of creative process. Edison could not just nap but sleep deeply just about anywhere—fully dressed in a much crumpled three-piece suit lying on wooden desks and benches (figure 6.4) and on bare floors, much like an ascetic Chinese sage with only a bent arm for his pillow. During the construction of the world's first electric network, his company kept a large stock of tubes in the cellar of the station at Pearl Street. Edison recalled that "as I was on all the time, I would take a nap of an hour or so in the daytime—and I used to sleep on those tubes in the cellar.

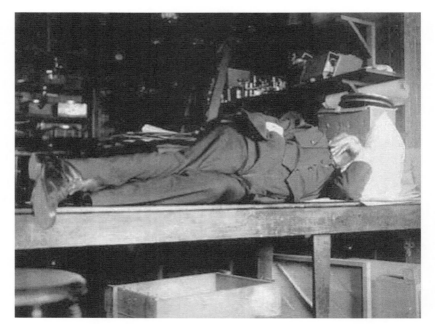

FIGURE 6.4. Thomas Edison snatching a nap on a wooden bench. This image is available from the National Park Service.

I had two Germans who were testing there, and both of them died of diphtheria, caught in the cellar, which was cold and damp. It never affected me" (quoted in Dyer and Martin 1929:400).

Cherish the thought of perhaps the most influential inventor of a new era reposing on iron pipes in a dingy cellar—and try to imagine one of today's CEOs getting involved in the same, pun unintended, down-to-earth fashion. In contrast, there were Tesla's obsessive neuroses about germs and cleanliness: he eventually required even his closest friends to stand at a distance lest they contaminate him—although he did not worry at all about close contact with thousands of pigeons, including many in his hotel rooms, that he cared for throughout his life (Cheney 1981). The Wright brothers were unsurpassed workaholics ("two of the workingest boys I ever saw," as one acquaintance put it) who had no friends outside the immediate family (Tobin 2003). In 1926, 14 years after Wilbur's death, their sister Katharine, 52 years old at that time, decided to marry Henry Haskell. Orville thought this a betrayal of the family, and he not only refused to attend the wedding but also cut all contacts with her until shortly before her death in 1929.

So Much Had Changed

Byrn (1896:82) noted in his essay on technical progress that

> [i]t is so easy to lose sight of the wonderful, whence familiar with it, that we usually fail to give the full measure of positive appreciation to the great things of this great age. They burst upon our vision at first like flashing meteors: we marvel at them for a little while, and then we accept them as facts, which soon become so commonplace and so fused into the common life as to be only noticed by their omission.

And what an omission that would be. The best way to appreciate the enormity of technical advances that took place between 1867 and 1914 is to try to construct a modern world devoid of just 10 major achievements introduced during that era.

No *electricity* and hence no nonpolluting, convenient, and inexpensive lights and no electric motors to power myriads of precisely controllable machines, appliances, and trains; no *internal combustion engines* and hence no fast and affordable motorized vehicles, no freedom of personal movement, and also no airplanes and no effortless long-distance travel; no *reproduction of sound* and hence no telephones and music recordings; no *photographic film* and hence no convenient cameras and no movies; no way to generate, broadcast, and receive *electromagnetic waves* and hence no radio or TV; no *steel alloys* and no *aluminum* and hence no skyscrapers, no affordable machines and appliances; no easy way to produce *inexpensive paper* or to *reproduce images* and hence no mass publication of books and periodicals; and no *nitrogen fertilizers* and hence very widespread malnutrition, shortened life spans, and a world that could not feed its population.

Byrn (1896:6) thought that this would leave "such an appalling void that we stop short, shrinking from the thought of what it would mean to modern civilization to eliminate from its life these potent factors of its existence." The void would be so profound that what would exist would be only a prelude to modern civilization that was already in place by the middle of the 19th century: a society relying on steam engines and draft horses, whose nights would be sparsely and dimly lit by kerosene and coal gas, whose dominant metal would be brittle cast iron, and whose best means of long-distance communication would be a telegraphic message.

And what was no less remarkable than this wealth of innovations was their lasting quality. Inevitably, not a few technical advances that were introduced during the Age of Synergy and that later came to dominate their respective markets were eventually replaced by new, superior 20th-century designs (e.g., internal combustion engines displaced by gas turbines in long-distance flights) or were relegated to small niche markets (the gramophone is still preferred by some audiophiles). But the opposite is true in a remarkably large number of cases where not only the basic design but even particular forms and modes of operation have endured with minimal amount of adjustments or even survived in virtually unchanged form.

Enduring Artifacts and Ubiquitous Innovations

Perhaps the most impressive way to illustrate enduring qualities of so many pre-WWI artifacts is a do-it-yourself exercise of comparing many objects and machines that we use today with the designs introduced, and rapidly improved, four to five generations ago. Most people will never have a chance to compare an 1880s Thomson dynamo or a large diesel engine built before 1910 with their modern counterparts. Consequently, such comparisons are done most conveniently with smaller objects of everyday use with which everybody is familiar. Light bulbs and spark plugs are excellent examples. Edison and Bosch could almost mistake one of today's standard incandescent lights or spark plugs for one of their own designs.

Modern light bulbs are machine-made rather than mouth-blown, their filaments are different, and Edison's lamps were not filled with inert gases. But when hidden by frosting (which was applied externally to some early lamps since the 1880s, internally by GE as early as 1903), the only outward feature that would give an old lamp immediately away would be the glass tip surmounting the globe. The shape, size, and proportions of GE's basic 100-W Soft White bulb made in 2000 are very similar to Edison's lamps made during the 1890s (figure 6.5). A gently angled neck widens into a spherical top whose diameter accounts for just more than half of the light bulb's overall length; a metal screw with four turns is another shared feature.

More important, the lamp's basic operational ratings are virtually identical to specifications set down by Edison and Upton in 1879: consuming 100 W of electricity at 120 V, a modern light draws the current of 0.83 A and has a resistance of 144.5 Ω. And if we move to 1913 and compare that year's tipless, internally frosted, gas-filled, coiled tungsten-filament lamp with today's 100-W Soft White, we have two almost identical items. There may be no better example of a relatively complex artifact that has remained basically unchanged during nine decades of the most rapid technical innovation in history, and that had been produced in billions of copies to become one of the most common possessions of the 20th century.

Spark plugs made a century apart share every key component: a terminal nut that connects to a spark plug wire, the metal-core center electrode that projects from the porcelain insulator nose, the ground electrode welded to the threaded part of the shell that forms the plug's reach, and the hexagonal section

FIGURE 6.5. Comparison of two light bulbs made a century apart: GE's standard frosted 100-W bulb made in the year 2000 (left) and Edison's low-voltage carbon-filament lamp from the 1890s (its flat buttonlike contact is a key dating feature).

in the upper part of the shell used to tighten the device by the spark plug wrench. One more example of an enduring design is decidedly low-tech (figure 6.6). In 1892 William Painter invented a clever way

> for the sealing of bottles by the use of compressible packing disks and metallic caps, which have flanges bent into reliable locking engagement with annular locking shoulders on the heads of bottles, while the packing-disk is in each case under heavy compression and in enveloping contact with the lip of the bottle (Painter 1892:1).

And crimped metal bottle caps are also an excellent example of innovations that are not examined in this book, which concentrates on fundamental, first-order advances and on their direct derivatives that often followed the primaries with an admirable speed. Electricity generation, internal combustion engines, inexpensive steels, aluminum, ammonia synthesized from its elements, and transmission of electromagnetic waves are the key primaries; electric motors, automobiles, airplanes, skyscrapers, recorded sound, and projected moving images are the ubiquitous derivatives. But there was much more to the period than introducing, rapidly improving, and commercializing those epoch-making inventions. There were many new inventions of simple objects of everyday use whose mass manufacturing was made possible, or commercially viable, thanks to the availability of new or better, and also cheaper, materials. Consequently, those inventions can be classed as secondary, or perhaps even tertiary, derivations.

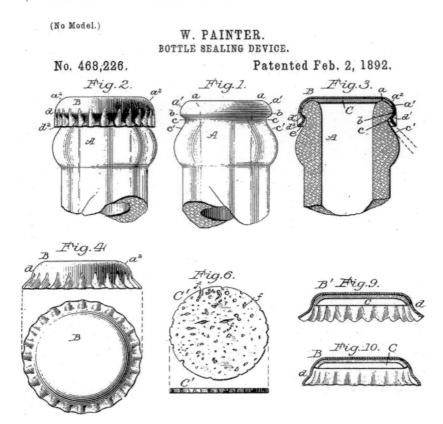

FIGURE 6.6. Drawings that accompanied William Painter's patent application for a bottle sealing device.

Their realization is based on one or more preceding fundamental innovations, but their commercial success owes no less to their ingenious designs. Many items belong to a large category of simple objects made from cheaper metals and better alloys. These include the following enduring classics, depicted in figure 6.7: the Gem paper clip (actually never patented) and other (patented) clip designs whose production required cheap steel wire and machines to bend it; scores (and eventually hundreds) of kinds of barbed wire (beginning with Joseph Glidden's pioneering 1874 twist) made of galvanized steel (which, unlike round or oval iron wire, was extremely durable because of its high homogeneity and tensile strength); William Hooker's 1894 spring mouse trap; and King Gillette's 1904 razor blades made from tempered steel a mere 0.15 mm thick. This metallic group also included staplers (Charles Henry Gould in 1868), sprinkler heads on showers (Harry Parmelee of New Haven

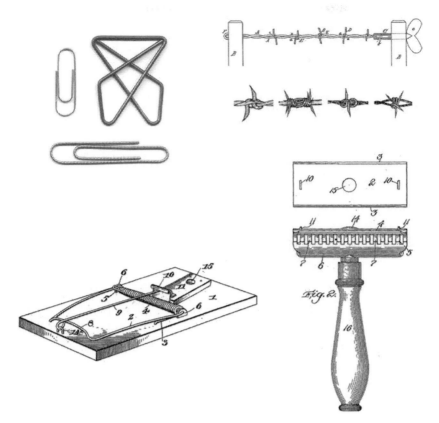

FIGURE 6.7. Artifacts made possible by cheap, good-quality steel: two sizes of Gem paperclips introduced during the 1890s and a large Ideal clip from 1902; Glidden's 1874 barbed wire and details of several other elaborate twists; Hooker's "animal trap" of 1894; and King Gillette's razor designed in 1901. Gems and an Ideal are scans of actual clips, other drawings are reproduced from their respective patent applications.

in 1874), Swiss Army knives (Victorinox company set up by Karl Elsener in 1891) and zippers (plastic teeth came decades later).

In addition, many other products of persistent tinkering had little or nothing to do with the epoch-shaping primary innovations that were taking place during those eventful decades. Genesis of these numerous items of everyday use introduced during the Age of Synergy should be seen as a further proof of the period's enormous wave of inventiveness. A reasonably complete list of these often very simple but enduring novelties would be tediously long. A few prominent examples include flat-bottom (instead of V-shaped) paper bags (1867), spring tape measures (1868), drinking straws (1888), and ice cream cones

(1904). The last two items also point to a large category of dietary and culinary innovations whose introduction created entirely new consumption habits.

Now world-famous branded foodstuffs that emerged in the United States during the two pre-WWI generations range from cheese to chocolate. Empire Cheese Company began selling Philadelphia Cream Cheese in 1880. In 1906 Will Keith Kellogg, brother of the notorious electric shock and enema lover John Harvey Kellogg, added sugar to his sibling's ascetic corn flakes (patented in 1896) and began marketing them as a breakfast cereal (Powell 1956). Campbell's soup trademark was registered in the same year; Milton Hershey began selling his Milk Chocolate Bar in 1894, and his rival Frank Mars opened for the business in 1911 (Brenner 1999). Any list of widely consumed new generic foodstuffs should be headed by two leading contributors to the late 20th-century increase of obesity: hamburgers, whose U.S. debut is variously dated between 1885 and 1904 (McDonald 1997), and American-style (sausage- and cheese-loaded) pizza (Gennaro Lombardi opened the first New York pizzeria in 1895).

The world of drink was enriched (if you share the taste for thirst-inducing, syrupy, excessively sweetened, and explosively carbonated liquids that leave peculiar aftertastes) or burdened (if your preferences run more into good tea, mineral water, or fruit juices) by Coca-Cola and Pepsi Cola. Coca-Cola was invented in 1886 by Atlanta physician John S. Pemberton. Pemberton sold rights to this "intellectual beverage and temperance drink" to a pharmacist, Asa Briggs Candler, in 1891, and by 1894 the first bottled cola was available in U.S. drugstores (Hoy 1986). Caleb Bradham's concoction of water, sugar, vanilla, essential oils, and cola nut extract was renamed Pepsi Cola in 1898. Yet another notable beverage brand that has endured is Canada Dry Ginger Ale, formulated in 1904 by a Toronto pharmacist John McLaughlin. That brand is now owned by Dr. Pepper/Seven Up, the first of these two excessively sweetened liquids being the oldest major U.S. soft drink, created in 1885 by Charles Alderton at Morrison's Old Corner Drug Store in Waco, Texas (Dr. Pepper/Seven Up 2003).

And there were also new pastimes ranging from a variety of board games (Hofer and Jackson 2003) made affordable by cheaper paper and printing to new sports. These included basketball (James Naismith's 1891 invention) and two of my favorites, cross-country skiing and tennis. Primitive skis, descendants of snowshoes, were around for millennia, but the sport of ski-running (or Nordic skiing) began in 1879 with the Huseby races near Christiania (today's Oslo), and 13 years later the first Holmenkollen festival attracted more than 10,000 spectators. Similarly, tennis has a venerable pedigree (medieval *jeu de paume*), but the modern version was played for the first time in December 1873 in north Wales and patented shortly afterward (G.B. Patent 685/1874) by Walter Clopton Wingfield. The only major modification came in 1877 on the occasion of the first All-England Lawn Tennis Championships at Wimbledon, when the original 26-m-long hourglass field became a 24 × 17 m rectangle; later the net height was lowered by 10 cm to 90 cm (Alexander 1974).

Given the multitude of innovations that flooded in during the Age of Synergy, it is not difficult to construct narratives of today's life whose minor material ingredients and mundane actions originated not just during that inventive period but, more restrictively, even just within a single eventful decade. Here is an example constrained to the 1880s and, in order to make the exercise more challenging, *excludes* any direct references to the decade's key primary innovations. A man wakes up late one day in one of America's large East Coast cities. First he makes a cup of Maxwell House coffee (a brand introduced by Joel Cheek, a Nashville hotelier in 1886). He hesitates between his favorite Aunt Jemima pancakes (his great-grandfather could have bought them for the first time in 1889) and Quaker Oats (the company began selling prefilled packages in 1884). His late breakfast is interrupted by a doorbell: an Avon lady is calling (they have been doing that since 1886), but his wife is out of town.

He finds that his only clean shirt needs a bit of ironing (Henry Seely patented that useful device in 1882), uses an antiperspirant (introduced in 1888), dresses up, and finds that he is out of brown paper bags (kraft process to make strong paper was first commercialized during the 1880s) to bring his customary lunch. He walks out to catch a streetcar (used to carry commuters since 1882), and once he gets downtown he enters a multistory steel-skeleton building (Jenney completed the first structure of this kind in Chicago in 1885) through a revolving door (Theophilus Van Kannel put the first one into a building lobby in Philadelphia in 1888). He stops in a small drugstore where a bored teenager is turning pages of *Cosmopolitan* (it appeared first in 1886), and buys a roll of film (celluloid replaced paper in 1889). He has to wait while the teenager fiddles with his cash register (introduced by James Ritty and John Birch in 1883).

Then he takes an elevator (Elisha Otis installed the first high-speed lift in New York in 1889) to his floor, but before getting to his office he stops at a vending machine (Percival Everitt introduced it in 1881) and buys a can of Coca Cola (formulated in Atlanta in 1886). The first thing he does in his office is make a long-distance phone call (possible since 1884), jots down a few notes with his ballpoint pen (John Loud patented the first one in 1888), pulls out a small Kodak camera (George Eastman's 1888 innovation) from a drawer, and inserts the film. As he reads a technical brief he comes across a strange word: he reaches for his abridged *Oxford English Dictionary* (publication of its first edition began in 1884) and contemplates the complexity of that amazingly hybridized language. We leave him at that point, the case clearly made: so many of our everyday experiences and actions were set, and so many artifacts that help us to cope with them, were introduced during the Age of Synergy that most people are not even remotely aware of the true scope of this quotidian debt.

Life Cycles of Innovations

Numerous as they are, those basically unchanged original designs constitute a minority of advances introduced during the two pre-WWI generations. Most of the innovations have undergone changes ranging from minor adjustments to fundamental transformations. As already stressed in chapter 1, this quest for improvement—motivated by factors that ranged from desire to capture larger markets to challenges of finding the most elegant engineering solutions—was one of the key marks of the era and one of its most important bequests to the 20th century. Its many forms took different approaches by reducing energy and material intensity and by providing longer durability, higher reliability, improved conversion efficiency, and greater versatility. As all of these aspects will be examined in detail in the companion volume, the only important matter I address here is the apparent inevitability and regularity of many key trends.

Perhaps the most remarkable aspect of the historical continuity of technical innovation is that so many subsequent advances appear to have the inexorability of water flowing downhill. Once the basic ideas were formulated by innovative thinking and tested by bold experiments, it was only a matter of time before they were perfected, diffused, and amplified to reach entirely new performance levels. Look at the photograph of a small Curtiss pusher biplane, piloted by Eugene Ely on the morning of January 18, 1911, as it is about to land on a temporary wooden platform that was laid over the deck and gun turret of the Pacific Fleet's armored cruiser *Pennsylvania* (figure 6.8; NHC 2003a). The ship is anchored off the San Francisco waterfront, with thousands of spectators ashore. An updraft lifts Ely's light plane just as it reaches the platform, but he compensates quickly, snags the arresting gear (just a series of ropes crossing the deck and weighted down by sand bags), and pulls to a smooth stop before reaching the safety barrier (NHC 2003b).

How that image evokes all those torpedo bombers coming back to their carriers to rearm and take off again on their missions to sink Japanese ships during WWII! How conceptually identical, despite all of the enormous intervening technical advances, is this scene compared with F-18s returning to America's nuclear-powered carriers stationed in the Persian Gulf (figure 6.8). These constants remain: a plane, a ship, a landing deck, an arresting gear, a skillful and alert pilot. Once imagined, and once shown to be practicable, the idea, in this instance not just figuratively, takes off and reaches an execution that was entirely unimaginable by its inventors but that is nevertheless unmistakably present in the original creation.

Another convincing comparative approach is to ask a simple question: would it have been possible to stop further developments of those newly launched techniques and freeze them at any arbitrary levels of performance and complexity? And the obvious answer is no. Hughes (1983) conceptualized

FIGURE 6.8. Development of an idea. Top: Eugene Ely's Curtiss biplane nears the landing platform on the USS *Pennsylvania*, anchored in San Francisco Bay, on January 18, 1911. Arresting lines were fastened to sand bags positioned along the deck. Photograph NH 77608 available at http://www.history.navy.mil/photos/images/h82000/h82737.jpg. Bottom: An F/A-18F Super Hornet nears landing aboard USS *Nimitz* in the Persian Gulf. U.S. Navy photo (030331-N-9228K-008.jpg) by Michael S. Kelly.

the process that follows an invention in four stages: transfer of new techniques to other places and societies, formative system growth followed by reaching a new momentum (that arises both from the accumulated mass of new machines and infrastructures and from the velocity of their diffusion), and finally, a qualitative change of mature techniques.

Once Edison mastered reliable electricity generation and once Tesla and Hertz opened the ways toward innovative electricity applications, it was only a matter of time before the entire electricity-driven universe of modern machines and devices, qualitatively so superior to the initial designs, was put in place. Timing of the key advances of this great transformation was undoubtedly contingent on many external factors, but their eventual attainment was a matter of very high probability. Similarly, once Karl Benz, Gottlieb Daimler, and Wilhelm Maybach mounted their high-speed gasoline motors on light carriagelike chassis, six-lane highways clogged by millions of increasingly sleeker steel machines were only a matter of three generations away.

Historical studies repeatedly demonstrate that the evolution of individual techniques or systems frequently follows an orderly progression much as living organisms do: initially slow growth accelerates and then slows down as it approaches a limit and eventually stops. Fittingly, it was during the Age of Synergy when Gabriel Tarde (1843–1904), a French sociologist, first described this S-shaped growth of innovations:

> A slow advance in the beginning, followed by a rapid and uniformly accelerated progress, followed again by progress that continues to slacken until it finally stops: these, then, are the three ages of those real social beings which I call inventions or discoveries (Tarde 1903:127).

This kind of progression also applies to life histories of all organisms, and the resulting patterns of development are very close to one of the growth, or logistic, curves. But this is not necessarily the case with life cycles of techniques and their performance parameters: technical advances do not have lives of their own, and their evolution is not governed by some internal calls that produce growth stages of relatively fixed proportions (Ayres 1969). This becomes particularly clear when attempts are made to fit one of the standard growth curves to a historic data set. Some innovations that originated during the Age of Synergy followed a fairly regular progression that conformed rather closely to an ideal curve, while others departed from it quite significantly.

Efficacy of incandescent lights between 1880 and 1930 is a good example of a fairly symmetrical S-curve, with the midpoint at around 1905, while the best fit for the highest conversion efficiency of steam turbines during the same period is obviously a straight line (figure 6.9). Many growth patterns of innovations introduced during the Age of Synergy, exemplified in figure 6.9 by the maximum U.S. steam turbogenerator ratings and by the highest U.S. trans-

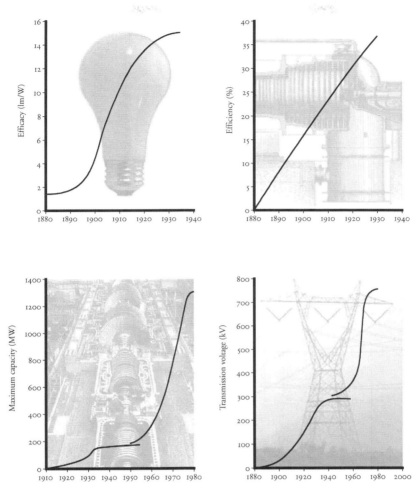

FIGURE 6.9. Efficacy of incandescent lights followed a fairly regular growth curve be-tween 1880 and 1930 (top left), while the efficiency of best steam turbines progressed in linear fashion (top right). Histories of the highest U.S. transmission voltages (bottom left) and the largest thermal turbogenerators (bottom right) show two successive growth curves. Based on data and figures in Smil (1994) and Termuehlen (2001).

mission voltages, form asymmetrical curves with the central inflection points shifted to the right as the early exponential growth rates were reduced by disruptions attributable to the interwar economic crisis. Indeed, both of these patterns could be better interpreted as two successive S-waves, with the first one reaching its plateau during the 1930s.

Commercial penetration rates of new techniques display orderly progressions that are dictated by the necessities of developing requisite manufacturing and distribution facilities and in many cases also putting in place the necessary infrastructures (roads, transmission lines). Not surprisingly, costly infrastructural needs or relatively high capital costs will tend to slow down the rate of adoption; for example, it was only after WWII when virtually all of America's rural households became electrified. In contrast, some industrial adoptions proceeded quite rapidly: electric motors captured half of the prime mover capacity in the U.S. manufacturing just three decades after Tesla's first devices were made by Westinghouse (see figure 2.21).

Once individual techniques, or their functional assemblages, reach the limits of their growth, they do not, like individual organisms, face inevitable decline and demise. Another biological analogy then becomes appropriate as they behave as mature, climax ecosystems that can maintain their performance for extended periods of time until their eventual displacement by a suite of superior techniques (a process that is analogical to a diffusion of new species in an ecosystem altered by climate). Some of these substitutions are gradual, even inexplicably tardy. Replacement of incandescent lights by fluorescents in households is a notable case of this slow pace: the former converters, despite their inferior efficiency, still dominate the market. Other substitutions, such as adoption of color TV or CDs, proceeded fairly rapidly.

Technical substitutions may be accelerated or hindered by economic and political factors, but, as with the growth of individual techniques, their progress is often very orderly as a new innovation enters the market, comes to dominate it, and then, shortly after its share peaks, begins to yield to a new technique. This wavelike progression has been also the case with the transitions to new sources of primary energies and new prime movers, the two processes that have been, together with ubiquitous mechanization and mass production, among the most notable markers of a new technical era.

Markers of a New Era

Qualitative appraisals of fundamental socioeconomic changes are done by systematically presenting all relevant trends, carefully choosing the descriptive adjectives, and thoughtfully selecting noteworthy examples to capture the wholes by resorting to the specifics. Quantifying epochal changes in macroeconomic terms is much more elusive, as aggregate measures and reductions to growth rates hide too much and subsume too many complexities in single figures. Moreover, many pre-WWI innovations had almost immediate, and far-reaching, economic and social impacts, while others needed many decades before they came to dominate their respective markets.

Consequently, no comparisons of economic and social indicators of the late 1860s with those of the immediate pre-WWI years should be taken as representative measures of advances attributable solely to the great technical saltation of the age. Inevitably, most of the economic gains of the 1870s had their investment and infrastructural roots in years preceding the beginning of this period, while most of the innovations introduced after 1900 made their full socioeconomic mark only after WWI. That is why in this section I proceed along both quantitative and qualitative lines.

This book's primary concern is to detail the extent and the lasting consequences of the unprecedented number of fundamental pre-WWI technical advances. There is no adequate means of quantifying that process, but the history of patenting in the country that came to dominate this modernization process offers a valuable proxy account of these accomplishments. The first numbered U.S. patent (for traction wheels) was issued in 1836, and the subsequent steady growth brought the total of successful applications granted to U.S. residents to almost 31,000 by the end of 1860. After a three-year dip in the early 1860s came a steep ascent, from 4,638 patents in 1864 to 12,301 patents in 1867, and then the annual grants reached a new plateau of high sustained inventiveness at 12,000–13,000 cases a year, a fact that provides additional support for my timing of the beginning of the Age of Synergy (USPTO 2002; figure 6.10).

Rapid growth of patenting recommenced by 1881, and soon afterward it

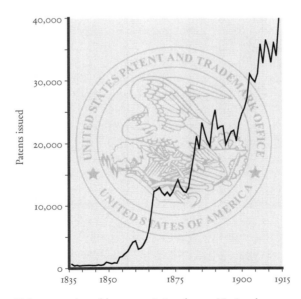

FIGURE 6.10. U.S. patents issued between 1836 and 1914. Notice the unprecedented rise in the annual rate of patenting that took place during the mid-1860s and the second wave of acceleration that began in 1881. Plotted from data in USPTO (2002).

formed a new, and somewhat more uneven, plateau (mostly with 19,000–23,000 grants per year) that lasted until the end of the 1890s. U.S. Patent 500,000, for a combined flush-tank and manhole, was issued in June 1893. The annual rate of 30,000 grants was surpassed for the first time in 1903; U.S. Patent 1,000,000 was issued in August 1911 (to Francis Holton for his vehicle tire), and before the beginning of WWI the total surpassed 1,100,000. Annual grants for the period 1900–1914 averaged more than 32,000 or more than three times the mean of the late 1860s, a convincing sign of intensifying inventive activity.

I do not hasten to undercut the conclusion I have just made, but I must reiterate (see chapter 1) that this simple quantitative focus may be somewhat misleading. The era's flood of unprecedented inventiveness also included a large number of not just trivial but also outright ridiculous patents that should not have ever been granted (Brown and Jeffcott 1932). Among the choicest examples of the latter category are mad combinations: of match-safe, pincushion, and trap (U.S. Patent 439,467 in 1890) and (yes, this is an exact citation) of grocer's package, grater, slicer, and a mouse and fly trap (U.S. Patent 586,025 in 1897). Chewing-gum locket, tapeworm trap, device for producing dimples, and electrical bedbug exterminator ("electricity will be sent through the bodies of the bugs, which will either kill them or startle them, so that they will leave the bedstead" according to the U.S. Patent 616,049 of February 7, 1898) are among my other favorites.

But there is also no doubt that many technical advances of the era had fairly prompt and profound economic effects that were reflected in some impressive growth and productivity gains. Aggregate national accounts are the favorite indicators of economic growth, but such figures have obvious limitations of coverage and comparability. Consequently, they should be seen merely as indicators of basic trends, not as accurate reflections of all economic activity. Standard series recalculated in constant monies show that between 1870 and 1913 the U.S. gross domestic product grew roughly 5.3 times and that the multiples were 3.3 for Germany, 2.2 for the United Kingdom, and 2.0 for France (Maddison 1995). This growth translated, respectively, to 3.9%, 2.8%, 1.9%, and 1.6% a year. Population growth—already fairly slow in Europe but rapid in the United States (mainly due to large immigration)—made per capita rates much more similar, with annual means of 1.8% for the United States, 1.6% for Germany, 1.5% for France, and just 1.0% for the United Kingdom.

Faster growth recorded by countries that started to industrialize aggressively only after 1850 is also shown by comparing the gains of gross domestic product per hour worked. Between 1870 and 1913, this indicator averaged 1.2% a year in the United Kingdom, while the German rate was 1.8% and the Japanese and U.S. averages reached 1.9% (Broadberry 1992). Rising productivity was accompanied by shorter work hours: their annual total in Western economies began to decline after 1860, from nearly 3000 to about 2500 by 1913, and by the 1970s they were below 1800 in all affluent countries (Maddison 1991). This trend was strongly influenced by declining employment in farming activities, but just before WWI agriculture still contributed about a third of gross domestic products in most Western countries and employed more people than did services sectors. And the term "service" called to mind the still very common household help rather than an array of activities that now account for the bulk of the Western economic product.

Average American wages rose rather slowly during the decades between the 1860s and WWI (BLS 1934), but their relatively modest progress must be seen against the falling cost of living. After a period of pronounced inflation, the American index of general price levels began to fall in 1867; then, three decades of deflation lowered it by about 45% by 1896, and renewed inflation drove it up by 40% by 1913 (NBER 2003). French industrial wages nearly doubled between 1865 and 1913, and German wages grew 2.6-fold, but the British earnings went up by only 40% (Mitchell 1998). Rising economic tides of the two pre-WWI generations lifted all the boats—but even in the relatively most affluent Western countries, average disposable incomes were still very low when measured by the standards of the late 20th century. Real incomes averaged no more than 10–15% of today's levels, and despite the falling cost of food, typical expenditures on feeding a family still claimed a large share (around half) of average disposable urban income.

But none of these indicators conveys adequately the epochal nature of the post-1860s developments. This is done best by focusing on the two trends that became both the key drivers and the most characteristic markers of the Age of Synergy. Most fundamentally, for the first time in human history the age was marked by the emergence of high-energy societies whose functioning, be it on mundane or sophisticated levels, became increasingly dependent on incessant supplies of fossil fuels and on rising need for electricity. Even more important, this process entailed a number of key qualitative shifts, and more than a century later it still has not run its full course even in the countries that were its pioneers. The same conclusion is true about the other fundamental trend that distinguishes the new era: mechanized mass production that has resulted in mass consumption and in growing global interdependence.

Mass production of industrial goods and energy-intensive agricultures that yield surpluses of food have brought unprecedented improvements to the overall quality of life, whether they are judged by such crass measures as personal possessions or by such basic existential indicators as morbidity and mortality. An increasing portion of humanity has been able to live in societies where a large share, even most, of the effort and time is allocated to providing a wealth of services and filling leisure hours rather than to producing food and goods. Both the rising energy needs and the mass production provided strong stimuli for the emergence of intensifying global interdependence, and this resulted in both positive feedbacks and negative socioeconomic and environmental effects.

High-Energy Societies

Thermodynamic imperatives and historical evidence are clear: rising levels of energy consumption do not guarantee better economic performance and higher quality of life—gross mismanagement of Russia's enormous energy wealth is perhaps the most obvious illustration of the fact—but they are the most fundamental precondition of such achievements. As long as human-controlled energy flows could only secure basic material necessities of life, there was no possibility of reliable food surpluses, larger-scale industrial production, mass consumption, prolonged education opportunities, high levels of personal mobility, and increased time for leisure. A high correlation between energy use and economic performance is the norm as individual countries go through successive stages of development, and the link applies to broader social achievements as well.

New sources of primary energy and new prime movers were essential to initiate this great transition. The shift began inauspiciously in the United Kingdom during the 17th century, and accelerated during the 18th century with the rising use of coal, invention of metallurgical coke for iron smelting, and James Watt's radical improvement of Newcomen's inefficient steam engine (Smil 1994). Even so, by 1860 coal production remained limited as the United Kingdom was the only major economy that was predominantly energized by that fossil fuel. Traditional biomass fuels continued to supply about 80% of the world's primary energy, and by 1865 the United States still derived more than 80% of its energy needs from wood and charcoal (Schurr and Netschert 1960).

Typical combustion efficiencies of household fireplaces and simple stoves were, respectively, below 5% and 15%. Steam engines, which were diffusing rapidly both in stationary industrial applications and in land and sea transportation, converted usually less than 5% of coal's chemical energy into reciprocating motion, and small water turbines were the only new mechanical prime movers that were fairly efficient. Although English per capita consumption of coal approached 3 t/year during the late 1860s (Humphrey and Stanislaw 1979), and the U.S. supply of wood and coal reached nearly 4 t of coal equivalent during the same time (Schurr and Netschert 1960), less than 10% of these relatively large flows were converted into space and cooking heat, light, and motion.

The Age of Synergy changed all of that, and the United States was the trendsetter. Expansion of the country's industrial production and the growth of cities demanded more coal, and in turn, industrial advances provided better means to extract more of it more productively. Internal combustion engines created a potentially huge market for liquid fuels, and newly introduced electricity generation was ready to use both coals and hydrocarbons (as well as water power) in order to satisfy a rapidly rising demand for the most convenient form of energy. Deviation-amplifying feedbacks of these developments resulted in an unprecedented increase of primary energy consumption, the category that includes all fossil fuels, hydroelectricity, and all biomass energies. Its U.S. total rose more than fivefold during the two pre-WWI generations, but the country's rapid population growth, from about 36 million people in 1865 to just more than 97 million in 1913, reduced this to less than a twofold (1.8 times) increase in per capita terms, a rise that prorates to annual growth of merely 1.2%.

This was a significant but hardly spectacular rise—and also a very misleading conclusion because this simple quantitative contrast hides great qualitative gains that characterized the energy use during the Age of Synergy. What matters most is not the total amount of available energy but the useful power that actually provides desired energy services. Substantial post-1870 improvements in this rate came from the combination of better performance of traditional conversions and introduction of new prime movers and new energy sources. A few sectoral comparisons reveal the magnitude of these gains that accompanied the epochal energy transition.

Higher efficiencies in household heating and cooking did not require any stunning inventions, merely better designs of stoves made with inexpensive steel, and large-scale replacement of wood by coal. Heat recirculation and tight structures of new coal, or multifuel, stoves of the early 20th century raised efficiencies commonly 40–50% above the designs of the 1860s. Typical efficiency of new large stationary steam engines rose from 6–10% during the 1860s to 12–15% after 1900, a 50% efficiency gain, and when small machines were replaced by electric motors, the overall efficiency gain was typically more than fourfold.

Because of transmission (shafting and belting) losses, only about 40% of power produced by a small steam engine (having 4% efficiency) would do useful work (Hunter and Bryant 1991), and another 10% of available power would be wasted due to accidental stoppages. Useful mechanical energy was thus only about 1.4% ($0.04 \times 0.4 \times 0.9$) of coal's energy content. Despite a relatively poor performance of early electricity generation (efficiencies of no more than 10% for a new plant built in 1913) and 10% transmission losses, a medium-sized motor (85% efficient) whose shafts were directly connected to drive a machine had overall energy efficiency of nearly 8% ($0.1 \times 0.9 \times 0.85$). Coal-generated electricity for a medium-size motor in the early 1910s thus supplied at least fives times as much useful energy as did the burning of the same amount of fuel to run a small steam engine of the 1860s.

Installing internal combustion engines in place of small steam engines would have produced at least two to three times as much useful energy from the same amount of fuel, and post-1910 efficiencies of steam turbines (the best ones surpassed 25% by 1913) were easily three times as high as those of steam engines of the 1860s. Higher energy efficiencies also made the enormous expansion of American ironmaking (nearly 30-fold between 1865 and 1913) possible. By 1900 coke was finally dominant, and its production conserved two-thirds of energy present in the charged coking coal (Porter 1924). The shift from charcoal of the 1860s to coke of 1913 brought 50% gain in energy efficiency, and better designs and heat management nearly halved the typical energy intensity of blast furnaces, from about 3 kg of coal equivalent per kilogram of pig iron in 1860s to about 1.6 kg by 1913 (Smil 1994).

This means that the overall energy costs of American pig iron production were reduced by about two-thirds. Finally, a key comparison illustrating the impressive efficiency gains in lighting: candles converted just 0.01% of paraffin's chemical energy into light, and illumination by coal gas (average yields of around 400 m³/t of coal, and typical luminosity of the gas at about 200 lm) turned no more than 0.05% of coal's energy into light. By 1913 tungsten filaments in inert gas converted no less than 2% of electricity into light, and with 10% generation efficiency and 10% transmission losses, the overall efficiency of new incandescent electric lighting reached 0.18% ($0.1 \times 0.9 \times 0.02$), still a dismally low rate but one nearly four times higher than for the gas lighting of 1860s!

Information available about pre-WWI sectoral energy consumption is not detailed enough to come up with accurate weighted average of the overall efficiency gain. But my very conservative calculations, using the best available disaggregations of final electricity use and the composition of prime movers, show that there was at least a twofold improvement of energy conversion in the U.S. economy between 1867 and 1913. America's effective supply of commercial energy thus rose by an order of magnitude (at least 11-fold) during the two pre-WI generations, and average per capita consumption of useful commercial energy had roughly quadrupled.

Such efficiency gains were unprecedented in history, and such rates of useful energy consumption provided the foundation for the country's incipient affluence and for its global economic dominance. In 1870 the United States consumed about 15% of the world's primary commercial energy and the country's output accounted for roughly 10% of the world's economic product; by 1913 the respective shares were about 45% and 20% (Maddison 1995; UNO 1956; Schurr and Netschert 1960). This means that the average energy intensity of the U.S. economic output rose during that period, an expected trend given the enormous investment in urban, industrial, and transportation infrastructures. Similar trends could be seen with energy intensities of Canadian or German economy.

No less important, this transition was coupled with qualitative improvements of energy supply. As noted in chapter 1, commercial energies began supplying more than half of the world's energy use sometime during the early 1890s; for the United States that milestone was reached during the early 1880s, and by 1914 less than 10% of the country's primary energy was from wood while about 15% of America's fossil energies came from crude oil and natural gas. The United States pioneered the transition from coal to hydrocarbons that was driven both by a rapid diffusion of a new prime mover (power installed in internal combustion engines surpassed that in all other prime movers before 1920) and by higher quality and greater flexibility of liquid fuels. Crude oil's energy density is nearly twice as high as that of good steam coal (42 vs. 22 GJ/t), and the fuel, and its refined products, is easily transported, stored, and used in any conceivable conversions, including flying.

At the beginning of the 20th century, oil resources of the pioneering fields of Pennsylvania (1859), California (1861), the Caspian Sea (Baku 1873; figure 6.11), and Sumatra (1885) were augmented by new major discoveries in Texas and in the Middle East (Perrodon 1981). On January 10, 1901, the Spindletop well southwest of Beaumont gave the first sign of oil production potential in Texas; by the end of the 20th century, the state still produces a fifth of America's oil (and more comes from its offshore fields in the Gulf of Mexico). The first giant oilfield in the Middle East was discovered on May 25, 1908, in Masjid-i-Suleiman in Iran; 90 years later the region was producing 30% of the world's crude oil and had 65% of all petroleum reserves (BP 2003).

The other key qualitative energy shift that was pioneered by the United States was the rising share of fossil fuel energy consumed indirectly as electricity. This share rose from, obviously, zero in 1881 to about 4% by 1913,

FIGURE 6.11. Wooden structures of oil wells in Baku, one of the principal centers of early crude oil production. Reproduced from *The Illustrated London News*, June 19, 1886. More than a century later, the still considerable untapped oil reserves of the Caspian Sea are, once again, a center of international attention.

surpassed 10% by 1950 and by the year 2000 was nearly 35% (Smil 2003). Finally, higher productivities in all energy industries translated into lower prices of fuels and electricity. Trends for crude oil and electricity were particularly impressive. When expressed in constant dollars, U.S. crude oil prices in 1910 were about 90% lower than during the early 1860s (see figure 1.9), and GE's household tariff for lighting fell by the same amount just during the two decades between 1892 and 1912.

Before leaving the subject of high-energy civilization, I must stress that all of the trends that began during the Age of Synergy and that I describe in this section either had continued for most of the 20th century or are still very much with us. Global shares of biomass energies fell to about 25% by 1950 and to no more than 10% by the year 2000 (Smil 2003). Per capita energy consumption still keeps rising even in the world's most affluent economies: during the 1990s it went up by nearly 5% in the United States and by almost 15% in Japan (BP 2003). Efficiencies of major energy conversions are still improving, although some techniques (gas-fired furnaces, aluminum smelting, the Haber-Bosch synthesis of ammonia) are now approaching their thermo-dynamic limits.

Transition from coal to hydrocarbons continues as the global share of crude oil and natural gas in the world's primary energy supply rose from less than 40% in 1950 to about 65% by the year 2000. Higher shares of fossil fuels are being converted to electricity: the global mean was 10% by 1950 and about 30% by the year 2000 (Smil 2003). And prices of crude oil, now the single most important fossil fuel worldwide, were falling until the early 1970s when the OPEC's actions suddenly reversed that secular trend—but their 1985 col-lapse returned them during the 1990s to levels that, when expressed in constant monies, were no higher than they were during the early 1920s or the late 1890s (see figure 1.9).

Mechanization and Mass Production

Higher energy flows used with increasing efficiencies by new prime movers were the key for the sweeping mechanization of tasks that ranged from crop harvesting to office work and from manufacturing to household chores. Even in the United Kingdom, this process got fully underway only during the Age of Synergy: at the time of the 1851 census the country's traditional craftsmen still greatly outnumbered machine-operating factory workers as there were more shoemakers than coal miners and more blacksmiths than ironworkers (Cameron 1985). Mechanization made mass production the norm in all mod-ernizing economies as it began to deliver food surpluses and affordable con-sumer goods while cutting the labor hours and enhancing the quality of life. Again, all of these trends are still very much with us, now in advanced stages

in affluent countries but still in midstream in such large modernizing economies as Brazil or China.

After 1870 the mechanization's reach grew ever more comprehensive, en-
compassing fundamental (steelmaking) as well as trivial (toy making) procedures. Enormous expansion of machine tool manufacturing brought the mechanization of tasks that ranged from wire drawing and twisting to metal milling
and shaping. Mechanization transformed activities that were both ubiquitously
visible (internal combustion and electric motors displacing horses in cities) and
largely hidden. An excellent example in the latter category is the speed with
which electricity-powered mechanical cutting diffused in the U.S. coal mining:
from nothing in the early 1880s to about 25% by 1900 and to half of all
produced coal by 1913 (Devine 1990b). The process was completed by the early
1950s when about 95% of all underground U.S. coal was cut from seams
mechanically.

But the first area where mechanization, and other new modes of production,
made the greatest difference to an average consumer was in the production of
food. The combination of better machines (inexpensive steel plows, efficient
harvesters and threshers), increased fertilizer use, improved storage, and cheaper
long-distance transportation brought steady increases in agricultural productivity. In the United Kingdom, Samuelson (1893) calculated that between 1861
and 1881 more than 110,000 farm workers were replaced by about 4,000 skilled
artisans that were making new field machines and another roughly 4,000 people who operated them. But, as with the energy transition, the United States
led the way, and pages could be filled with impressive comparisons of requirements for physical labor before the Age of Synergy and by its end, when an
increasingly mechanized activity was able to provide food surpluses by employing a shrinking share of the population (Schlebecker 1975; Smil 1994).

Between 1860 and 1914 the share of the U.S. farming population was halved
to just below 30% of the total, average time required to produce a ton of
wheat declined by about 45% to less than 40 hours, and the largest farms
could produce it with less than 10 hours of labor (Rogin 1931; McElroy, Hecht,
and Gavett 1964). New large grain mills (the first one with automatic steel
rollers was built by Cadwallader Washburn in 1879) produced superior flour,
and Andrew Shriver's introduction of the autoclave (U.S. Patent 149,256 in
1874) to use pressurized steam for the sterilization of food increased the capacity of canning 30-fold compared to the traditional method (Goldblith 1972).

New imports and mechanization made food cheaper also in Europe and in
the United Kingdom. By 1880s even working class English families were adding
not just jam, margarine, and eggs to their regular diet but also canned sardines,
coffee, and cocoa (Mitchell 1996). Bowley's (1937) detailed account shows that
compared to the 1860s, the average food basket of English families in 1913
contained three times as much meat and cheese, twice as much butter and
sugar, and four times as much tea. And there were also significant increases in

the consumption of fruits and vegetables, new imports of bananas and oranges, and the growing popularity of chocolate. And the paragon of stores that sell gourmet food opened in Paris at Place de la Madeleine in 1886, when Auguste Fauchon began to offer an unmatched selection of delicacies.

The second most important area where mechanization had the greatest impact on the quality of life for the largest number of people was in providing affordable infrastructures for better household and public hygiene. Modern plumbing, whose diffusion was supported by the latest scientific discoveries of waterborne pathogens, had enormous cumulative impacts on the reduction of morbidity and on the increase in longevity. Cheaper mass-produced metals, reinforced concrete and more powerful, more efficient electric pumps made it easier to build dams, conduits, and pipes and to bring treated drinking water (a process that became possible only with inexpensive large-scale production of chlorine) into dwellings and take the wastes out.

Continuous chlorination of drinking water began in the early years of the 20th century. Trenton, New Jersey, had the first large U.S. facility in 1908, and by 1910 the country had nearly 600 treatment sites serving 1.2 million people; by 1948 some 80 million Americans drank chlorinated water supplied by nearly 7,000 utilities (Thoman 1953). Chlorination brought rapid reduction in the incidence of waterborne infections, particularly of typhoid fever, whose pathogen, *Salmonella typhi*, was identified in 1880 by Karl Joseph Eberth. U.S. typhoid mortality fell from about 30/100,000 in 1900 to less than 3/100,000 by 1940, and the disease was virtually eliminated by 1950 (USBC 1975).

The first wave of building municipal sewage plants dates to the 1870s, and by 1913 most large Western European cities (North America lagged behind) had either sewage fields, settling tanks, grit removal, screens, or combinations of these techniques (Seeger 1999). There were other environmental health gains made possible by new techniques. Replacement of horses by engines and motors eliminated enormous volumes of excrement from cities, and this, together with paved streets, also greatly reduced the amount of airborne particulate matter. Although fuel-derived outdoor air pollution remained a serious urban problem for decades, the situation would have been much worse without more efficient stoves, boilers, and turbogenerators. Indoors, electric lights replaced air-polluting gas jets, and more affordable (and better quality) soap made more frequent washing of hands possible. As we now know, this simple chore remains the most cost-effective means of preventing the spread of many infectious diseases.

With adequate food supply and basic hygiene claiming a declining share of disposable income, rising shares of consumer expenditures began to shift first to purchases of more expensive foodstuffs (meat intake rose steadily), prepared meals, and drinks and then to acquisitions of an ever-widening array of personal and household goods whose quality was improving as mechanized mass manufacturing was turning out identical items at unprecedented speed. As

Landes (1969:289) put it, "[T]here was no activity that could not be mechanized and powered. This was the consummation of the Industrial Revolution."
Ingenious machines were designed to bend leather for shoes or steel wire for
paper clips, to blow glass to make light bulbs or to shape milk and beer bottles.

Impressive rise of retailing can be illustrated by higher revenues as well as
by the number and variety of stores. Berlanstein (1964) collated interesting
statistics about the growth of retail outlets for Ivry-sur-Seine, a Parisian
working-class suburb: between 1875 and 1911 the number of clothing stores per
1,000 inhabitants nearly quadrupled while the number of grocery stores quadrupled and that of stores selling drink and prepared food rose 4.5 times. And,
as already noted, consumers without access to richly stocked urban stores (see
figure 1.11) could rely on progressively greater choice available through mail-
order shopping, an innovation that was pioneered in 1872 by Aaron Montgomery Ward (1844–1913).

After 1890, mass consumption, particularly in the United States, began to
embrace many nonessential manufactures whose falling prices made them soon
broadly accessible. The Age of Synergy was thus the time of the democratization of possessions, habits, and tastes as machines, gadgets, and pastimes
spread with often dizzying pace. Unfortunately, as Veblen (1902:84) noted,
there is also a near universal tendency toward excessive consumption:

> The basis on which good repute in any highly organised industrial com-
> munity ultimately rests is pecuniary strength; and the means of showing
> pecuniary strength . . . are leisure and conspicuous consumption. Ac-
> cordingly, both of these methods are in vogue as far down the scale as
> it remains possible.

Premiere places of conspicuous consumption were new large department
stores where customers could get lost amidst the artful displays of goods. Zola
captured this milieu with unsurpassed perfection. As Octave Mouret, director
of *Au Bonheur des Dames*, surveyed the enormous crowd of females filling his
store on the occasion of the Great White Sale, he was aware that "his creation
was introducing a new religion, and while churches were gradually emptied
by the wavering of faith, they were replaced in souls that were now empty by
his emporium" (Zola 1883:416). More than a century later, the only notable
difference is that huge department stores have been displaced by much larger
shopping centers sheltering scores or hundreds of smaller stores under one
roof. And the modern habit of walking around in visibly branded clothes or
with other prominently labeled merchandise also goes back more than a century: in 1899 Louis Vuitton began to put his elegantly lettered initials on his
hand-crafted products, and this practice is now imitated across all price ranges.

But in retrospect, it is clear that the most important item of conspicuous
consumption and expanding accumulation was the ownership of cars. No other

mass-produced item turned out to have such wide-ranging effects on the structure and performance of affluent economies, on the spatial organization of society, and on so many social and cultural habits. Only a few hundred rich eccentrics would buy the first motor cars in the early 1890s; only thousands of well-off doctors, businessmen, and engineers were eager to get one of Olds's early models a decade later—but nearly half a million individuals and families made that purchase every year just before WWI began. Extraordinarily large productivity gains were behind the car's rapid mass penetration. Data from Ford cost books show that 151 hours was needed to make a car in 1906, 39 in 1914, and 37 in 1924 (Ling 1990)—but productivity was rising in every sector of the economy.

In 1909 Thomas B. Jeffrey, a Wisconsin manufacturer of the Rambler car, noted that the "mortgage has gone from the Middle Western farm, and to take its place there is the telephone, the heating system, the water supply, improved farm machinery, and the automobile" (cited in Ling 1990:169). By the mid-1920s car making became America's leading industry in terms of product value, and it has retained this primacy throughout the 20th century. During the late 1990s, U.S. car sales were more than 20% of all wholesale business and more than 25% of all retail, and automakers were the largest purchasers of steel, rubber, glass, machine tools, and robots. Adding crude oil extraction and refining, highway construction and maintenance, roadside lodging and eating, and car-dependent recreation activities leaves no doubt that automobiles have been the key factor of the Western economic growth. No other machine has done so much for the still spreading *embourgeoisement* and the rise of the middle class.

By the time American car sales reached hundreds of thousands a year, many newly mass-produced goods were quite affordable—but the range of accumulated personal possession was still fairly limited. As we have seen, all of the products that had eventually become such inalienable ingredients of modernity—electric lights, telephones, cars—were yet to diffuse to most of the population, and there were large differences even among neighboring countries. For example, by 1913 there were 26 people per telephone in Germany, but in France the ratio was still almost 135 (Mitchell 1998). The same was true about basic living conveniences: in 1913, fewer than 20% of Americans had flush toilets.

Regardless of the differences in the specific national rates of technical progress, cities everywhere were the greatest beneficiaries. Progress in sanitation, communication, and transportation solved or eased the three sets of key obstacles to their further growth. In turn, cities had the leading role in stimulating innovation and facilitating its diffusion (Bairoch 1991). The Age of Synergy was thus overwhelmingly an urban phenomenon, and this generalization is true not only about the outburst of technical inventiveness but also about the period's incredible artistic creativity. Just try to extend the experiment played with the absent inventions to music, literature, and painting created during the two pre-WWI generations.

Imagine that we would not have any, or most, compositions by Brahms, Bruckner, Debussy, Dvořák, Gounod, Mahler, Puccini, Rachmaninov, Ravel, and Tchaikovsky. Imagine that the novels, stories, and poems of Chekhov, Kipling, Maupassant, Rilke, Tolstoy, Twain, Verlaine, Verne, Wilde, and Zola would not exist. Remove from the world of images the paintings of French impressionists and the canvases that immediately preceded and followed that glorious era: gone are all or most of the works of Braque, Cézanne, Gauguin, Manet, Matisse, Monet, Renoir, Rousseau, Seurat, Signac, and van Gogh.

This admirable outpouring of artistic creativity (particularly its fin de siècle phase) helped to widen the opportunities for visiting art exhibitions and for attending musical performances. But every pastime became more popular as leisure became a widely shared social phenomenon only during the two pre-WWI generations, when its many forms became a notable part of the process of modernization (Marrus 1974). Some of its manifestations were captured in such unforgettable impressionistic masterpieces as Pierre-August Renoir's *Déjeuner des canotiers* (1881 oil painting of a boating party gathered around a canopy-covered and wine- and fruit-laden table) or Georges Seurat's pointillistic gem *Un dimanche après-midi à l'Ille de la Grande Jatte* that was painted in 1884–1885 and depicts Parisians enjoying a summer afternoon along the banks of the Seine.

Costs of these leisure activities varied greatly. Even some simple pastimes were rather expensive. To ride the first great wheel built by George W. Ferris (1859–1896) for the 1893 Columbian Exposition in Chicago (figure 6.12) cost 50 cents at a time when a Chicago laborer was paid 15 cents per hour (BLS 1934). But this was a popular safe thrill, and during the 20th century it was replicated by hundreds of eponymous structures, most recently by the world's tallest Ferris, London's Millennium Wheel. Other pastimes were cheap: Frank Brownell's 1900 Brownie camera cost a dollar, a perfect gift for children to take snapshot of their friends and pets.

And the Age of Synergy saw the birth of pastimes that targeted the largest possible passive participation as well as of elite pursuits that required not just a great deal of money but also uncommon skills. Animated cartoon films (introduced in 1906) are an excellent and enduring example of the first category, America's Cup of the other. America's unbroken string of pre-WWI Cup victories generated a great deal of public attention in its defense, and in 1914 experts were delighted when the tradition of secrecy was lifted and technical details of competing yachts became available before the race (Anonymous 1914), which the war's outbreak postponed until 1920. And, of course, the Olympic Games were reborn in 1904.

Extensive contemporary writings dealt with many negative consequences of mechanization and mass production. People who reacted with condemnation

FIGURE 6.12. George Washington Ferris supported his 75-m-diameter wheel by two 42-m steel towers, and the nearly 14-m-long axle was the single largest steel forging at that time. Two reversible 750-kW engines powered the rotation of 36 wooden cars, each with 60 seats. This original Ferris wheel was reassembled at the St. Louis Exposition in 1904, and it was scrapped two years later. Reproduced from *Scientific American* of July 1, 1893.

and outrage to their often appalling surroundings included not only social critics (from Thomas Carlyle to Karl Marx and from John Ruskin to Matthew Arnold) and perceptive novelists but also dedicated photographers who documented urban squalor and misery (Riis 1890). As Kranzberg (1982) rightly noted, this bleak interpretation of industrialization was still prevalent among many scholars a century later. Some modern critics have argued that industrialization deepened, rather than relieved, human poverty and suffering; others regretted the "irretrievable destruction of much of the beauty of the countryside" and the "loss of the peacefulness of mind which the gentle and unaltering rhythm of country life can bring" (Fleck 1958:840).

These are indefensible views as the reality was much more nuanced. No historical study can convey it better than did Zola's monumental Rougon-Macquart cycle, perhaps the unsurpassed witness of the post-1860 era, with its astonishing sweep of desperation and hope, poverty and riches, suffering and triumphs, in milieus both rural and urban and in settings ranging from an almost unimaginably brutal world of vengeful peasantry to the intrigues of Parisian *nouveaux riches*. The supposed rural harmonies hid a great deal of poverty, misery, and autocratic abuse, and many pre-industrial landscapes were neither untouched nor well cared for. And for every demonstration of undeniably reprehensible urban reality—unemployment, exploitation, low wages, unsanitary living conditions, noise, air and water pollution, illiteracy, lack of security—there are countervailing examples (Hopkins 2000).

Most significantly, millions of new jobs that paid far better than any work that could be obtained in the countryside were created by risk-taking entrepreneurs. Some of them were industrialists with exemplary social conscience. Robert Bosch instituted an eight-hour working day in 1906, four years later made Saturday afternoon free, and transferred most of the company's profits to charities (Bosch GmbH 2003). This charitable pattern endures: the company is 92% owned by Robert Bosch Foundation, which funds many social, artistic, and scientific activities. And while the overall social progress was not as rapid as many would have liked, it was undeniable.

Besides gradual introduction of shorter work hours and higher purchasing power, the two pre-WWI generations also saw diffusion of compulsory grade school attendance and the beginning of pension and health insurance schemes in many countries (first in Bismarck's Germany in the early 1880s, followed soon by Austria). For all of these reasons, Ginzberg (1982:69) rightly observed that Karl Marx "was better as a critic than as a prophet" as he did not anticipate the substantial gains in the quality of life that would eventually result from mechanization and increased productivity.

Statistical bottom lines speak for themselves: in no previous period of Western history did so many people become adequately fed (e.g., British intake of high-quality animal protein roughly doubled), basically educated (as grade school attendance became compulsory: in the United Kingdom up to age 10 in 1880, up to 12 by 1899), and able to enjoy the modicum of material affluence (more than one set of clothes and bedding, more pieces of furniture) than they did during the two pre-WWI generations. But, above all, they benefited from unprecedented declines in infant mortality, which was the main reason for rising life expectancy. Between the 1860s and 1914, Western infant mortality was roughly halved to just more than 100/1,000 newborns while life expectancy at birth increased by 20–25% to around 50 years, with the British means going from 40 years during the late 1860s to 53 years by 1913 (Steckel and Floud 1997).

Recent laments about negative consequences of globalization often sound as if the process of economic interdependence was the invention of the closing decades of the 20th century. In reality, parts of Europe participated in relatively intensive long-range trade already during the early modern era that preceded industrialization. This process widened and intensified during the latter half of the 19th century, and the trend toward economic globalization was firmly in place by the beginning of the 20th century, only to be derailed and slowed down until the 1950s, when it resumed at a much accelerated pace.

Both the epochal transition from biomass to fossil fuels and the diffusion of mechanization and mass production led inevitably to greater dependence on nonlocal resources. Coal deposits can be found in scores of countries around the world, but many coal basins contain seams of rather inferior quality, and so excellent steam and metallurgical coals became an early item of international trade. Major oil and gas fields are much less equitably distributed, and hence the universal transition from coals to hydrocarbons had to be accompanied by rising shipments of oil and, once technical progress made that possible, also by large-scale trade in natural gas. Ores, needed in increasing quantities by rapidly expanding iron and steel industry and nonferrous metallurgy, are even more unevenly distributed; for example, only three countries (Australia, Guinea, and Brazil) produce 60% of the world's bauxite (Plunkert 2002).

Virtually every country, aside from a few city states, could produce all (nearly all) food to assure adequate nutrition for its population, but this autarky would be very expensive, and in most cases it would result in a limited choice of foodstuffs. More affluent societies with larger disposable incomes thus naturally create markets for a greater variety of food, and this demand is best satisfied by specialized producers who have comparative advantage thanks to the climate, long experience in particular cultivation, or low labor costs. Consequently, international trade in foodstuffs began expanding beyond the shipments of low-volume and luxury items during the 1870s as soon as long-distance transportation of bulky commodities (grains, refrigerated meat from North America, Australia, and Argentina) could rely on large-capacity iron-hull ships.

Between 1870 and 1913, the share of exports in the total economic product rose by 50% (to about 12%) in 10 of the most industrialized countries, and by

1913 more than 25% of the United Kingdom's, Australia's, and Canada's gross national products originated from exports, and the shares were about 20% for Germany and Italy (Maddison 1995). This means that some countries were relatively nearly as dependent on exports in 1900 as they were in 2000. The period also saw the rise of a new kind of enterprise, multinational corporations that made their products in a number of countries rather than just exporting from one location. Robert Bosch was one of the first producers with a clear global aspiration: by 1913 his operations extended to 20 countries, and, still uncharacteristic for that time, his company derived more than 80% of its revenues from sales outside Germany. Several companies that were established before WWI and that eventually expanded into the world's leading multinationals have been mentioned in this book: General Electric, Brown Boveri, General Motors, Ford, Siemens, Marconi—and also the two top cola makers, Coca and Pepsi.

Remarkably, only two of today's 10 largest multinationals (when ranked by their revenues in the year 2000) were not set up before 1914: Walmart Stores (no. 2) and Toyota Motor Corporation (no. 10). First-ranked Exxon-Mobil comes from John D. Rockefeller's Standard Oil Trust that was organized in 1882 and dissolved by the Congress in 1911. Origins of the third, fourth, and fifth largest companies, all of them automakers (GM, Ford, and Daimler Chrysler), have already been described. Royal Dutch/Shell (no. 6) resulted from a 1907 merger of two young companies (Shell 1892, Royal Dutch 1903), while BP (no. 7) was set up in 1901 by William Knox D'Arcy in order to explore oil concession in Persia. GE, whose origins go back to Edison's first electric company of 1878, ranked eighth worldwide in 2000, and Mitsubishi, set up as Tsukomo Shokai by Yataro Iwasaki in 1870 (Mitsubishi 2003), was ninth.

More than a century later, all of these companies continue to be at the forefront of globalization, and the interdependence generated by this now so pervasive business pattern creates many positive feedbacks as well as many regrettable trends. Principal manifestations of the process will be examined in some detail in the companion volume. There is only one more matter I want to address here—in chapter 7 I return, briefly, to those remarkable pre-WWI years in order to convey some of the perceptions, feelings, hopes, and fears of those who lived in those momentous times and reflected on the accomplishments, promise, and perils of technical advances that were creating a new civilization.

7

Contemporary Perceptions

> It has been a gigantic tidal wave of human ingenuity and
> resource so stupendous in its magnitude, so complex in its
> diversity, so profound in its thought, so fruitful in its wealth,
> so beneficent in its results, that the mind is strained and em-
> barrassed in its effort to expand to a full appreciation of it.
>
> Edward W. Byrn, *Scientific American* (1896)

This book pays homage to the astonishing concatenation of epochal innovations that were introduced and improved during the two pre-WWI generations
and whose universal adoption created the civilization of the 20th century. I
have marshaled a great deal of evidence to justify this—in retrospect so inevitable—judgment. But I do not want to leave this fascinating subject without
asking, and attempting to answer, one last obvious question: to what extent
were the people who lived at that time aware that they were present at the
creation of a new era? This question must have a complex answer that spans
the entire range of responses from "not at all" through "in some ways" to "very
much so." Reasons for this are obvious.

Short as it was in relative terms, the span of two generations is too long a
period to be dominated by any consistent perception. Those two generations

FRONTISPIECE 7. By 1914 this kind of an early-morning commuting scene—female office workers, shop assistants and dressmakers are shown here arriving at a Paris termi-
nus—was common in all large Western cities. Reproduced from *The Illustrated London
News*, April 11, 1914.

saw both the pronounced ups and downs of economic fortunes and some notable social and political upheavals (many quite unrelated to technical advances) that greatly influenced the public mood. And even in the countries at the forefront of technical advances, the chores and challenges of everyday existence claimed inevitably much more attention than matters that, at least initially, appeared to have little bearing on everyday life. The earliest stages of building a new civilization were inevitably shaped more by many attributes of the preceding era that coexisted for many decades with some thoroughly modern practices.

Coexistence of old and new, of improvements eagerly anticipated and welcomed and widely distrusted or ridiculed, was thus entirely natural. The enormity of the post-1860s saltation was such that people alive in 1913 were further away from the world of their great-grandparents who lived in 1813 than those were from their ancestors in 1513. At the same time, everyday experiences of the two pre-WWI generations had inevitably much in common with the preindustrial norms: horses were still the most important prime movers in Western agriculture and in urban traffic, cars looked like, and were initially called, horseless carriages, and as they rushed out to see their first airplane in a summer sky many village children still ran barefoot, saving their shoes for fall and winter. Many epochal inventions appeared to be just fascinating curiosities, legerdemains of scientists and engineers with little practical importance for poor families.

Some innovations had rapidly conquered many old markets or created entirely new ones. Others had to undergo relatively prolonged maturation periods as their convenience, cost, and flexibility were not immediately obvious compared to long-established devices and practices. Among the many examples in this category are Otto's early gas engine (during the early 1870s nobody foresaw that their gasoline-fueled successors will eventually energize millions of vehicles) and the first automobiles of the late 1880s (there was nothing to indicate that those expensive and unreliable toys bought by a few rich adventurers will mutate in just two decades into the paragon of mass ownership). And, finally, the novelty of some developments was so much outside the normal frames of references that those ideas were not perceived as epochal but rather simply as mad.

This was the case, most prominently, with the first airplanes. When in 1906 Alfred Harmsworth (Lord Northcliffe, the publisher of *Daily Mail*) offered £1,000 for the first pilot to cross the English Channel, *Punch* immediately ridiculed the challenge:

> Deeply impressed as always with the conviction that the progress of invention has been delayed by lack of encouragement, *Mr. Punch* has decided to offer . . . £10,000 to . . . the first aeronaut who succeeds in flying to Mars and back within a week. (Anonymous 1906:380)

But the Channel prize was won by Blériot just four years later (figure 7.1) and in 1913 Lord Northcliffe offered £10,000 for the first crossing of the Atlantic— this time with no parodies from *Punch*. The new challenge was taken up in early 1914 by Rodman Wanamaker, who commissioned Glenn Curtiss to build a flying boat powered by a 150-kW engine that would be capable of making

FIGURE 7.1. Blériot's 1909 monoplane in flight and a head-on view of its propeller and three-cylinder engine. Library of Congress images (LC-USZ62-94564 and LC-USZ62-107356).

that trip in a single flight of 12–15 hours, but the plan had to be shelved because of WWI.

If the public was confused and unsure, so were many innovators. As we have also already seen, some of them were surprisingly diffident and dismissive regarding the worth and eventual impact of their efforts. But other creators of the Age of Synergy were mindful of their place in creating a new world. As they looked back, they were proud of their accomplishments, and many of them had no doubt that even greater changes were ahead, that the worth of their inventions was more in the promise of what they would do rather than in what they had already done. Still, I suspect that most of them would be surprised to see how the great technical transformations of the 20th century left so many of their great innovations fundamentally intact.

And these feelings of confident expectations were not limited to creative and decision-making elites. The frontispiece of this closing chapter exudes the confident demeanor of young chic Parisian ladies as they spill from a railway terminal to the city streets in the spring of 1914 on their way to jobs as typists, telephonists, and shop assistants. The image captures what was for them already an everyday routine but what would have been an unthinkable display of independence and opportunity in 1864. Although they were still only modestly paid, they were breaking a key feature of the traditional social order. They must have known that they are living in an era that was unlike anything else in the past, and they expected further great changes.

This shift in female work was one of the signal achievements of the age. Several trends combined to make it possible: modern plumbing, heating, and electric appliances eased the household chores and were doing away with domestic servants, while at the same time telephones, typewriters, calculators, and keypunches were opening up new opportunities for female employment, and electric trains and subways were making it possible to commute to downtowns where the new office jobs were concentrated. During the 1860s, most of the working women were employed as household servants, and in the United Kingdom that share was still as high as 85% by 1891 (Samuelson 1896). By 1914 household servants were a rapidly shrinking minority of female labor force everywhere in the Western world as a new class of increasingly independent wage-earning females was expanding and creating, very visibly, a new social and economic reality.

Although in an expected minority, many astute observers were aware of the epoch-making nature of new inventions as they correctly foresaw their long-term impact. They were able to see the advantages of new ways quite clearly even decades before their general acceptance. For example, Rose (1913:115), writing about the rapid diffusion of agricultural tractors (a process that was not completed even in the United States until the early 1960s), felt that there is no question but we have entered upon a new era in agriculture. The farmer desires the comforts and advantages of the city dweller, and these he gets easily and cheaply with the small gasoline engine . . . It multiplies his capacity and gives him either more leisure or enables him to farm a larger area and increases his income. Power farming has just begun, and the start is encouraging.

Some innovations were so indisputably revolutionary that even uninitiated public was made almost instantly aware of their importance years before it had any chance to buy or to use these gadgets or machines or to benefit from the new processes. Although rather primitive even in comparison with designs that followed just a few years later, Bell's first telephones were greeted almost immediately with general enthusiasm. Edison's work on electric light received plenty of public attention mainly because of detailed, and technically competent, coverage by New York newspapers. And once electricity made industrial motors the leading prime movers, there was no doubt about the importance of that change: "Electricity came in between the big wheel at the prime mover and the little wheels at the other end, and the face of material civilization was changed almost in a day" (Collins 1914:419).

Parsons's turbine had a key role in the swift creation of this new reality. As already described, he found a very unorthodox way to demonstrate that it was also a superior prime mover in naval propulsion by a daring public show, as his small *Turbinia* outpaced every warship at Queen Victoria's Diamond Jubilee Naval Review in 1897. This technical stunt was guaranteed to generate admiring headlines. And just a few years later *The Illustrated London News*— a weekly that was normally much given to portraits of royalty, accounts of state visits, and engravings of angelic children—devoted an entire page of its large format to the explanation of this new system of naval propulsion (Fisher 1903). More than that, the article also included a detailed drawing of the arrangement of stationary and moving blades in Parsons machine, a kind of illustration that is hard to imagine in today's *People* magazine.

Perhaps the best contemporary appraisal of the era's technical progress was offered by Edward Byrn in his 1896 essay written for the *Scientific American's* semicentennial contest (Byrn 1896). Although Byrn's assessment covered the years 1846–1896, most of the remarkable achievements noted in the essay originated after 1866, and his overall characterization of the period as "an epoch of invention and progress unique in the history of the world" would have been only strengthened if he were to write a similar retrospective in 1913. Byrn (1896: 82) argued that the progress of invention during the second half of the 19th century was "something more than a merely normal growth or natural development" and (see this chapter's epigraph) found it difficult to convey the true magnitude of the period's overwhelming contributions.

In a complete agreement with my arguments (made entirely independently more than a century later), Byrn singled out the decade of 1866–1876 as "the beginning of the most remarkable period of activity and development in the history of the world" and headed the list of its great inventions with the perfection of the dynamo. Going through his long list of most notable advances made during the subsequent three decades would be to retrace most of the ground that is covered in this book. What is more interesting is to list some items on Byrn's list that have not been mentioned here: compressed-air rock drills, pressed glassware, machines for making tin cans, hydraulic dredges, enameled sheet ironware for cooking, Pullman railway cars, and artificial silk from pyroxyline.

Scientific American ran another essay contest in 1913. In his winning entry William I. Wyman ranked the 10 greatest inventions of the preceding 25 years by looking for the advances that were "most revolutionary in character in the broadest fields, which affected most our mode of living, or which opened up the largest new sources of wealth" (Wyman 1913:337), and dated them according to their successful commercial introduction. Wyman's list—electric furnace (1889), steam turbine (1894), gasoline automobile (1890), moving pictures (1893), wireless telegraphy (1900), aeroplane (1906), cyanide process (1890), linotype machines (1890), induction motor (1890), and electric welding (1889)—contains only one item that did not make it into my selection of key pre-WWI inventions, the cyanide process of gold extraction.

The process was patented in 1888 by three Glaswegians, chemist John S. MacArthur and physician brothers Robert and William Forrest, and it revolutionized the art of color metallurgy: silver and copper could be also produced by it (Wilson 1902). Its rapid adoption brought the trebling of global gold output by 1908 and made South Africa the world's largest producer. This had major socioeconomic effects on societies whose monetary policies were based on the gold standard. More than a century after its introduction, the cyanide process—whereby ground ores are mixed with diluted cyanide [$Ca(CN)_2$, KCN, or $NaCN$] and the metal is then precipitated from the soluble $Au(CN)_2$ by addition of powdered zinc—remains the leading technique of gold extraction (Cornejo 1984), but the metal's role in the world's economy has been marginal ever since Richard Nixon took the United States off the gold standard in 1971.

Unlike Wyman and others, I refused to do any rankings in this book because interactions of many major and minor components that drive complex dynamic systems make such orderings irrelevant. Still, a closer inspection will show that even Wyman's most questionable choices are defensible. He justified his lead ranking of electric arc furnace because of its multiple impacts: radical transformation of steel industry and indispensability for producing aluminum, and at the time of his writing, electric arc was the only promising means of fixing atmospheric nitrogen. The last role was displaced, after 1913, by the

Haber-Bosch process, but the other two key contributions have become far more prominent as today's electric arc furnaces produce more than a third of the world's steel and all of its aluminum.

Similarly, Wyman's inclusion of electric welding was well justified by the ability of the process to join what were previously considered unweldable metals (brass, bronze, cast iron) and to produce shapes that could be made previously only by laborious riveting. In 1964 the technique was transformed by the introduction of plasma welding that uses gas heated by electric arc to extremely high temperatures for accurate and high-quality applications ranging from work on precision instruments to repairs of gas turbines. As a result, Wyman's identification of the most revolutionary technical advances stood the test of time quite admirably. His excellent essay, as well as Byrn's retrospective, show that many well-informed observers were not only aware of the era's unique contributions but also could accurately identify its most far-reaching innovations without the benefit of longer historical perspectives that would have made it much easier to point out the cases of successful, and lasting, impacts.

In contrast, one thing that could not go unnoticed by anybody who lived during those eventful decades of rapid mechanization was the widespread demise of artisanal work. This change elicited a spectrum of feelings from enthusiasm to grudging acceptance to obvious and deeply felt regrets. A difference of opinion between the two protagonists of Zola's grand and tragic *L'Assommoir*—Gervaise, a washerwoman, and Goujet, a metalworker—captures vividly some of these emotions. Goujet, after watching silently as a bolt-making machine was churning out perfect copies, turned in a resigned way to Gervaise and said how that sight makes him feel small. His only solace was that the machines might eventually help to make everybody richer (as they indeed did). But Gervaise scoffed at this as she found the mechanical bolts poorly made: "You understand," she exclaimed with passion, "they are too well made . . . I like yours better. In them one can at least feel the hand of an artist" (Zola 1877:733).

And although we have now been living for generations in the society whose prosperity rests on mass production of perfect copies, so many of us still repeatedly feel something of Gervaise's regret. Indeed, we are willing to pay the premium for the greatly diminished range of artisanal products, cherishing the traces left by the hands of their creators (or naively trusting fraudulent claims of some manufacturers that that indeed is the case). But we are, inescapably, ready consumers of countless low-priced mass-produced items, and we do not even seriously consider that things should be done otherwise or that there should be opposition, violent or Gandhi-like, to reverse this pattern.

This almost unquestioned acceptance became one of the surprising norms of the entire Age of Synergy. In 1811, when young Ned Ludd smashed a knitting machine and launched a rebellion against the supremacy of mechanization

(Bailey 1998), the entire process of industrialization was still in its beginnings as machines dominated only certain segments of a few industries in just a handful of countries. Two and three generations later they were ubiquitous, providing countless more reasons to oppose the diffusion of dehumanizing ways of mass industrial production—but the Luddite sentiment was not overt among the men and women whose labor created the new tools of production and served them in order to flood the markets with new goods: its most evident champions were social critics and writers, such as John Ruskin with his bleak view of the future (Fox 2002).

This reality reminds us how different was the entire worldview and how difficult it is to capture the prevailing attitudes even when the historical distance is just a matter of four generations. Above all—and in a sharp contrast with increasing frequency of immature behavior in modern society that displays so many infantile traits—most people were not burdened by unrealistic expectations. As young adults my grandparents had electric lights but no telephone, and I do not think they actually made a single phone call even late in their lives. Their diet was adequate but simple, their necessary daily walks long, and their schooling, much as their possessions, basic. My parents shared a great deal of these frugal realities, and I still experienced some of them as a child in post-WWII Europe: rationed food, a new book as an expensive and much prized possession, long daily walks or ski trips to school, delights of just a few weeks of ripe summer fruit and freshly picked wild strawberries.

At a time when the ratio of phones to people (babies included) is approaching one in the world's richest countries, when there are just two people per car (*vehicle* is a more accurate term in the world of monstrous SUVs), when food costs are barely more than a tenth of the average disposable family income (even though so many items are carted out of the season halfway around the globe) and when books are being replaced by Web browsing, no amount of quiet reflection will help people growing up with these realities to grasp the modalities of pre-telephone, pre-internal combustion engine, pre-supermarket, pre-PC life. We face the same problem when trying to insert ourselves into the minds of our pre-WWI ancestors, who were surrounded by the nascent manifestations of a new civilization but benefited from them only to a limited extent.

At least I am sure that my grandparents did not feel deprived because they lived and died (decades after Bell's and Benz's inventions) without telephones or automobiles. But the absence of a telephone or an automobile would be the least of it. Again, I think of my grandfathers, men who worked very hard with little reward to create the modern world. My paternal grandfather helped to energize it—first as a coal miner in German deep mines (figure 7.2), then as a *Steiger* (foreman) in Bohemian hard coal pits. My maternal grandfather helped to provide its key material foundation as he built and repaired Martin-Siemens steel furnaces in Škoda Works, at that time one of Europe's largest industrial enterprises. Naturally, by today's standards both were poor but proud of their taxing and dangerous work, both were keen readers and bright, independent thinkers, neither saw himself as a victim, neither looked for a salvation in simplistic Marxist slogans of class struggle or in new social utopias.

As always, intellectuals were much more willing to worship these radical solutions: their convictions led soon to the three generations of the Soviet Communist empire and later to its Maoist replica, two grand "new society" experiments that were paid for with the lives of more than 50 million people. Others were convinced that progress, so obvious in the staggering sum of new technical advances, can lead only to better and better outcomes, with technocracy and democracy triumphant (Clarke 1985). This appeared to be self-evident to such prominent creators of America's industrial success as Thomas Edison and Andrew Carnegie, who wrote that in America "the drudgery is ever being delegated to dumb machines while the

FIGURE 7.2. My grandfather, Václav Smil, at the beginning of the 20th century in Homberg am Rhein. After generations of black coal mining east of the Rhine, in the Ruhr region, extraction began also on the river's western bank, across from Duisburg. Coal was discovered at Homberg in 1854 at the depth of 174 m, and the first shaft was opened in 1876. Photograph by Otto Meltzer.

brain and muscle of men are directed into higher channels" (Carnegie 1886:215). But that was not, obviously, the only, not even the dominant, perception: with gains came losses and worries, and too many everyday realities were hardly uplifting.

New machines and rising combustion of fossil fuels meant that even rich city dwellers could not avoid ubiquitous noise and worsening air pollution. New industries and expanding cities were also polluting water and claiming fertile agricultural land on unprecedented rates. Successful innovations that created new economic opportunities were also eliminating entire classes of labor force. And high profits, including Carnegie's immense fortune, often rested not only on the deployment of new techniques but also on treating labor in ways that seem today quite intolerable. After the violent Homestead strike of 1892, Carnegie cut the steelworkers' wages by 25–40% without changing the burden of 12-hour shifts seven days a week (Krause 1992). So much for the "higher channels" he invoked just six years earlier.

Nearly a generation later, working conditions of new Homestead immigrants remained hard: "Their labor is the heaviest and roughest in the mill, handling steel billets and bars, loading trains, working in cinder pits; labor that demands mostly strength but demands that in large measure . . . Accidents are frequent, promotions rare" (Byington 1910:133). But their wages had increased, placing their households in the upper third of the U.S. income scale, and Carnegie, retired and dispensing some $350 million (about $6 billion in 2000 US$) to charities, created a benefit fund for the employees of the company. These contrasts provide an excellent example of complexities that work against simplistic conclusions.

On an abstract level, rapid technical advances highlighted the widening gap between the impressive designs and capabilities of machines and engineering and scientific solutions on one hand and the prevailing social and economic arrangements on the other. Wells (1905:102) believed that

> were our political and social and moral devices only as well contrived to their ends as a linotype machine, an antiseptic operating plant, or an electric tramcar, there need now at present moment be no appreciable toil in the world, and only the smallest fraction of the pain, the fear, and the anxiety that now makes human life so doubtful in its value.

But just three years before that, in his address to the Royal Institution on the discovery of the future, he offered another appraisal of the past and the future of a new civilization. This one left no doubt that one of the era's most provocative thinkers was acutely, and accurately, aware of its special place in human history, of its evolutionary-revolutionary nature, of its immense promise (Wells 1902b:59–60):

> We are in the beginning of the greatest change that humanity has ever undergone. There is no shock, no epoch-making incident—but then there is no shock at a cloudy daybreak. At no point can we say, "Here it commences, now; last minute was night and this is morning." But insensibly we are in the day . . . And what we can see and imagine gives us a measure and gives us faith for what surpasses the imagination.

References

Abbott, Charles G. 1934. *Great Inventions.* New York: Smithsonian Institution.

Abeles, Paul W. 1949. *The Principles and Practice of Prestressed Concrete.* London: Charles Lockwood.

Adriaanse, Albert, et al. 1997. *Resource Flows: The Material Basis of Industrial Economies.* Washington, DC: World Resources Institute.

AeroVironment. 2004. High-altitude solar-electric airplanes. Monrovia, CA: Aero-Vironment. *http://www.aerovironment.com/area-aircraft/unmanned.html*

Aitken, Hugh G. J. 1976. *Syntony and Spark—The Origins of Radio.* New York: John Wiley.

Alcoa (Aluminum Company of America). 2003. *Alcoa.* Pittsburgh, PA: Alcoa http://www.alcoa.com

Alexander, George E. 1974. *Lawn Tennis: Its Founders and Its Early Years.* Lynn, MA: H. O. Zimman.

Allen, Robert C. 1981. Entrepreneurship and technical progress in the northeast coast pig iron industry: 1850–1913. *Research in Economic History* 6:35–71.

Almond, J. K. 1981. A century of basic steel: Cleveland's place in successful removal of phosphorus from liquid iron in 1879, and development of basic converting in ensuing 100 years. *Ironmaking and Steelmaking* 8:1–10.

Anderson, Edwin P., and Rex Miller. 1983. *Electric Motors.* Indianapolis, IN: T. Audel.

Andreas, John C. 1992. *Energy-efficient Electric Motors: Selection and Application.* New York: M. Dekker.

Anonymous. 1877. The telephone. *Scientific American* 36:1.

Anonymous. 1883. We have some very interesting figures. *Nature* 28:281.

Anonymous. 1889a. Twenty years. *Nature* 41:1–5.

Anonymous. 1889b. The electric lighting of London. *Illustrated London News* 95(2636): 526–527.

Anonymous. 1901. The telephone auto-commutator. *Scientific American* 84:85.

Anonymous. 1902. The New York Edison Power Station. *Scientific American* 87:152.

Anonymous. 1906. Mr. Punch's great offers. *Punch* 131:380.

Anonymous. 1913a. The first Diesel locomotive. *Scientific American* 109:225–226.

Anonymous. 1913b. The motor-driven commercial vehicle. *Scientific American* 109: 168,170.

Anonymous. 1913c. The electric production of steel. *Scientific American* 109:88–89.

Anonymous. 1914. The defense of the "America's" Cup. *Scientific American* **110**:865–866.

Arrillaga, Jos. 1983. *High Voltage Direct Current Transmission*. London: Institution of Electrical Engineers.

ASHRAE (American Society of Heating, Refrigerating and Airconditioning Engineers). 2001. *Handbook of Fundamentals*. Atlanta, GA: ASHRAE.

Ashton, Thomas S. 1948. *The Industrial Revolution, 1760–1830*. Oxford: Oxford University Press.

ASME (American Society of Mechanical Engineers). 2003. Edison "Jumbo" Engine-driver Dynamo (1882). New York: ASME. http://www.asme.org/history/roster/H048.html

Ayres, Robert U. 1969. *Technological Forecasting and Long-Range Planning*. New York: McGraw-Hill.

Bailey, B. F. 1911. *The Induction Motor*. New York: McGraw-Hill.

Bailey, Brian J. 1998. *The Luddite Rebellion*. New York: New York University Press.

Bairoch, Paul. 1991. The city and technological innovation. In: Paul Higonnet, ed., *Favorites of Fortune,* Cambridge, MA: Harvard University Press, pp. 159–176.

Baldwin, Neil. 2002. *Henry Ford and the Jews: The Mass Production of Hate*. New York: Public Affairs.

Bancroft, Hubert H. 1893. *The Book of the Fair*. Chicago: Bancroft Co. http://columbus.gl.iit.edu/bookfair/bftoc.html

Banks, Eric. 2001. *E-Finance: The Electronic Revolution*. New York: John Wiley.

Bannard, Walter. 1914. Modern electric engine starters and electric illumination systems. *Scientific American* **110**:18–19.

Bannister, R. L., and G. J. Silvestri. 1989. The evolution of central station steam turbines. *Mechanical Engineering* **111**(2):70–78.

Basalla, George. 1988. *The Evolution of Technology*. Cambridge: Cambridge University Press.

BASF. 1911. *Patentschrift Nr 235421 Verfahren zur synthetischen Darstellung von Ammoniak aus den Elementen*. Berlin: Kaiserliches Patentamt.

Batchelor, Ray. 1994. *Henry Ford, Mass Production, Modernism and Design*. Manchester: Manchester University Press.

Beales, Hugh L. 1928. *The Industrial Revolution, 1750–1850: An Introductory Essay*. London: Longmans, Green & Co.

Beauchamp, Kenneth G. 1997. *Exhibiting Electricity*. London: Institution of Electrical Engineers.

Beaumont, W. Worby. 1902 and 1906. *Motor Vehicles and Motors: Their Design Construction and Working by Steam Oil and Electricity*. 2 volumes. Westminster: Archibald Constable & Co.

Beck, Theodore R. 2001. Electrolytic production of ammonia. In: Zoltan Nagy, ed., *Electrochemistry Encyclopedia*. Cleveland, OH: Case Western Reserve University. http://electrochem.cwru.edu/ed/encycl/

Behrend, Bernard A. 1901. *The Induction Motor: A Short Treatise on Its Theory and Design, with Numerous Experimental Data and Diagrams*. New York: Electrical World and Engineer.

Bell, Alexander G. 1876a. Letter from Alexander Graham Bell to Alexander Melville Bell, March 10, 1876. Family Papers, Folder: Alexander Melville Bell, Family Correspondence, Alexander Graham Bell, 1876. http://memory.loc.gov/cgi-bin/ampage?collId=magbell&fileName=005/00500211/bellpage.db&recNum=0

Bell, Alexander Graham. 1876b. *Improvement in Telegraphy. Specification Forming Part of Letters Patent No. 174,465, Dated March 7, 1876.* Washington, DC: U.S. Patent Office. http://www.uspto.gov

Bell, I. Lothian. 1884. *Principles of the Manufacture of Iron and Steel.* London: George Routledge & Sons.

Belrose, John S. 1995. Fessenden and Marconi: their differing technologies and transatlantic experiments during the first decade of this century. In: Institute of Electrical Engineers, *100 Years of Radio,* London: Institute of Electrical Engineers, pp. 32–43.

Belrose, John S. 2001. A radioscientist's reaction to Marconi's first transatlantic wireless experiment—revisited. In: IEEE Antennas & Propagation Society, *2001 Digest Volume One,* Ann Arbor, MI: IEEE, pp. 22–25.

Berlanstein, Lenard R. 1964. *The Working People of Paris, 1871–1914.* Baltimore, MD: Johns Hopkins University Press.

Berliner, Emile. 1888. *The Gramophone.* Montreal: Berliner Gramophone Co. (paper read before the Franklin Institute, May 16, 1888, reprinted in 1909).

Bessemer, Henry. 1905. *Autobiography.* London: Offices of *Engineering.*

BIA (Büro-und Industrie-Anrüstungen Vertriebs). 2001. *Wiring Devices.* Weiterstadt: BIA. http://www.biagmbh.com/

Biermann, Christopher J. 1996. *Handbook of Pulping and Papermaking.* San Diego: Academic Press.

Bijker, Wiebe E. 1995. *Of Bicycles, Bakelites, and Bulbs: Toward a Theory of Sociotechnical Change.* Cambridge, MA: MIT Press.

Billington, David P. 1989. *Robert Maillart's Bridges: The Art of Engineering.* Princeton, NJ: Princeton University Press.

Binczewski, George J. 1995. The point of a monument: A history of the aluminum cap of the Washington Monument. *Journal of Metals* 47(11):20–25.

Birch, Alan. 1967. *The Economic History of the British Iron and Steel Industry 1784–1879.* London: Cass.

Black, Edwin. 2001. *IBM and the Holocaust.* New York: Crown Publishers.

Black, Joseph. 1803. *Lectures on the Elements of Chemistry Delivered in the University of Edinburgh by the Late Joseph Black.* London: Longman & Rees.

BLS (Bureau of Labor Statistics). 1934. *History of Wages in the United States from Colonial Times to 1928.* Washington, DC: USGPO.

Boddy, William. 1977. *The History of Motor Racing.* London: Orbis.

Boering, Brooke W. 2003. *Comptometer: Biography of a Machine.* http://www2.cruzio.com/~vagabond/ComptHome.html

Boggess, Trent. 2000. *All Model T's Were Black.* Centerville, IN: Model T Ford Club of America. http://www.mtfca.com/encyclo/P-R.htm#paint4

Bohr, Niels. 1913. On the constitution of atoms and molecules. *Philosophical Magazine* **26**(series 6):1–25.

Bolton, William. 1989. *Engineering Materials Pocket Book.* Boca Raton, FL: CRC Press.

Bond, A. Russell. 1914. Going through the shops—I. *Scientific American* **110**:8–10.

Bondyopadhyay, Probir K. 1993. Investigations on the correct wavelength of transmission of Marconi's December 1901 transatlantic wireless signal. In: *IEEE Antennas and Propagation Society International Symposium,* Vol. 1, Ann Arbor, MI: IEEE, pp. 72–75.

Borchers, Wilhelm. 1904. *Electric Smelting and Refining: The Extraction and Treatment of Metals by Means of the Electric Current.* London: Charles Griffin & Co.

Borlaug, Norman. 1970. *The Green Revolution: Peace and humanity.* A speech on the occasion of the awarding of the 1970 Nobel Peace Prize in Oslo, Norway, December 11, 1970. Stockholm: Nobel e-Museum. http://www.nobel.se/peace/laureates/1970/borlaug-lecture.html

Bosch GmbH. 2002. 100 years of Bosch spark plugs: the evolution of a sparkling idea. http://www.boschusa.com/News/

Bosch GmbH. 2003. Ein Blick in die Geschichte. http://archive.bosch.com/de/archive/

Boselli, Primo. 1999. *Una Vita per la Radio: Guglielmo Marconi, Cronologia Storica, 1874–1937.* Firenze: Medicea.

Bouwsma, William J. 1979. The Renaissance and the drama of Western history. *American Historical Review* **84**:1–15.

Bowers, Brian. 1998. *Lengthening the Day: A History of Lighting Technology.* Oxford: Oxford University Press.

Bowley, Arthur L. 1937. *Wages and Income in the United Kingdom since 1860.* Cambridge: Cambridge University Press.

Bowring, Samuel A., et al. 1993. Calibrating rates of early Cambrian evolution. *Science* **261**:1293–1298.

Boylston, H. M. 1936. *An Introduction to the Metallurgy of Iron and Steel.* New York: John Wiley.

BP. 2003. *BP Statistical Review of World Energy 2003.* London: BP. http://www.bp.com/worldenergy

Brachner, Alto. 1995. *Röntgenstrahlen: Entdeckung, Wirkung, Anwendung.* München: Deutsches Museum.

Brandt, Leo, and Carlo Schmid. 1956. *The Second Industrial Revolution.* Bonn: Social Democratic Party of Germany.

Braudel, Fernand. 1950. Pour une économie historique. *Revue Économique* **1**:37–44.

Brayer, Elizabeth. 1996. *George Eastman: A Biography.* Baltimore, MD: Johns Hopkins University Press.

Brennan, Mairin. 2000. What's that stuff? *Chemical & Engineering News* **78**:42.

Brenner, Joel G. 1999. *The Emperors of Chocolate: Inside the Secret World of Hershey and Mars.* New York: Random House.

Bridge, James. 1903. *Inside History of Carnegie Steel Company.* New York: Aldine.

Bright, Arthur A., Jr. 1949. *The Electric Lamp Industry: Technological Change and Economic Development from 1800 to 1947.* New York: Macmillan.

Brinkley, Douglas. 2003. *Wheels for the World: Henry Ford, His Company, and a Century of Progress.* New York: Viking.

Broadberry, Steven N. 1992. *The Productivity Race: British Manufacturing in International Perspective, 1850–1990.* Cambridge: Cambridge University Press.

Brock, William H. 1992. *The Norton History of Chemistry.* New York: W. W. Norton.

Brown, Alford E., and H. A. Jeffcott. 1932. *Beware of Imitations!* New York: Viking Press.

Bruce, Robert V. 1973. *Bell: Alexander Graham Bell and the Conquest of Solitude.* Boston: Little, Brown.

Burwell, Calvin C. 1990. Transportation: electricity's changing importance over time. In: Sam H. Schurr et al., eds. *Electricity in the American Economy.* New York: Greenwood Press, pp. 209–231.

Byington, Margaret F. 1910. *Homestead: The Households of a Mill Town.* New York: Charities Publication Committee.

Byrn, Edward W. 1896. The progress of invention during the past fifty years. *Scientific American* **75**:82–83.

Byrn, Edward W. 1900. *The Progress of Invention in the Nineteenth Century.* New York: Munn & Co.

Calvert, James B. 2001. *Inside Transformers.* Denver, CO: J. B. Calvert. http://www .du.edu/~jcalvert/tech/transfor.htm

Cameron, Rondo. 1982. The Industrial Revolution: A misnomer. *History Teacher* **15:** 377–384.

Cameron, Rondo. 1985. A new view of European industrialization. *Economic History Review* **38:**1–23.

Campbell, Harry R. 1907. *The Manufacture and Properties of Iron and Steel.* New York: McGraw-Hill Publishing.

Carnegie, Andrew. 1886. *Triumphant Democracy.* New York: Charles Scribner's Sons.

Carnot, Sadi. 1824. *Réflexions sur la Puissance Motrice du Feu et Sur les Machines Propres a Developper Cette Puissance.* Paris: Librairie Bachelier.

Carr, Charles C. 1952. *ALCOA: An American Enterprise.* New York: Rinehart & Co.

Casson, Herbert N. 1910. *The History of the Telephone.* Chicago: A. C. McClurg & Co.

Caterpillar. 2003. *Caterpillar© 797.* Peoria, IL: Caterpillar. http://www.caterpillar.com /about_cat/news/03_products_n_services/_attachment/797_update.html

Cercle Marc Bloch. 1999. *Affaire Lumière: Contre Marc Bloch.* Lyon: Cecrle Marc Bloch. http://www.angelfire.com/biz2/rlf69/CR/lumiere.html

CETS (Commission on Engineering and Technical Systems). 1993. *Materials Research Agenda for the Automobile and Aircraft Industries.* Washington, DC: National Academy of Science.

Chanute, Octave. 1894. *Progress in Flying Machines.* New York: American Engineer and Railroad Journal.

Chase, George C. 1980. History of mechanical computing machinery. *Annals of the History of Computing* **2:**198–226.

Chauvois, L. 1967. *Histoire merveilleuse de Zénobe Gramme.* Paris: Albert Blanchard.

Chen, Yu et al. 2003. Electronic paper: flexible active-matrix electronic ink display. *Nature* **423:**136.

Cheney, Margaret. 1981. *Tesla: Man out of Time.* New York: Dorsett Press.

CineFrance. 2003. The early years (1895–1909). Paris: CineFrance. http://home.att .net/~cinefance/hist1895_1909.htm

Cipolla, Carlo M. 1966. *Guns, Sails and Empires.* New York: Pantheon Books.

Clark, George B. 1981. Basic properties of ammonium nitrate fuel oil explosives (ANFO). *Colorado School of Mines Quarterly* **76:**1–32.

Clark, John, Christopher Freeman, and Luc Soete. 1981. Long waves and technological developments in the 20th century. In: D. Petzina and G. van Roon, eds., *Konjunktur, Krise, Gesselschaft: Wirtschaftliche Wechsellagen und sozialer Entwicklung im 19 und 20 Jahrhundert,* Stuttgart: Klett-Cotta, pp. 132–169.

Clarke, I. F. 1985. American anticipations. *Futures* **17:**390–402.

Clausius, Rudolf. 1867. *Abhandlungen über die mechanische Wärmetheorie.* Braunschweig: F. Vieweg.

Clerk, Dugald. 1909. *The Gas, Petrol, and Oil Engine.* London: Longmans, Green.

Clerk, Dugald. 1911. Oil engine. In: *Encyclopaedia Britannica,* 11th ed., Vol. 20. Cambridge: Cambridge University Press, pp. 25–43.

Coe, Brian. 1977. *The Birth of Photography.* New York: Taplinger Publishing.

Coe, Brian. 1988. *Kodak Cameras: The First Hundred Years.* Hove: Hove Foto Books.

Collins, James H. 1914. The electric industry and the young man. *Scientific American* **110:**419.

Coltman, John W. 1988. The transformer. *Scientific American* **258**(1):86–95.

Conreur, Gérard. 1995. *Les annees Lumière: De l'âge de pierre à l'âge d'or du cinéma en France.* Paris: Editions France-Empire.

Constable, A. R. 1995. The birth pains of radio. In: Institute of Electrical Engineers, *100 Years of Radio,* London: Institute of Electrical Engineers, pp. 14–19.

Coraglia, Giorgio. 2003. Linotype & Linotipisti: l'Arte di Fondere I Pensieri in Piombo. http://digilander.libero.it/linotype/index.htm

Cornejo, Leonardo M. 1984. *Fundamental Aspects of the Gold Cyanidation Process: A Review.* Golden, CO: Colorado School of Mines.

Cortada, James W. 1993. *Before the Computer: IBM, NCR, Burroughs and Remington Rand and the Industry They Created.* Princeton, NJ: Princeton University Press.

Cotter, Arundel. 1916. *The Story of Bethlehem Steel.* New York: Moody.

Covington, Edward J. 2002. *Early Incandescent Lamps.* Millfield, OH: frognet.net. http://www.frognet.net/~ejcov/index.html

Cox, James A. 1979. *A Century of Light.* New York: Benjamin.

Crafts, N. F. R., and C. K. Harley. 1992. Output growth and the British Industrial Revolution. *Economic History Review* **45:**703–730.

Craig, Norman C. 1986. Charles Martin Hall—the young man, his mentor, and his metal. *Journal of Chemical Education* **63:**557–559.

Crookes, William. 1892. Some possibilities of electricity. *Fortnightly Review* **102:**173–181.

Crookes, William. 1899. *The Wheat Problem.* London: Chemical News Office.

Culick, F. E. C. 1979. The origins of the first powered, man-carrying airplane. *Scientific American* **241**(1):86–100.

Cummins, C. Lyle. 1989. *Internal Fire.* Warrendale, PA: Society of Automotive Engineers.

DaimlerChrysler. 1999. *150th Birthday of Bertha Benz.* Stuttgart: DaimlerChrysler. http://www.daimlerchrysler.com/news/top/1999/t90503b_e.htm

DaimlerChrysler. 2003. *Wilhelm Maybach—The King of Design.* Stuttgart: DaimlerChrysler. http://www.daimlerchrysler.com/specials/maybach/maybach5_e.htm

Dalby, William E. 1920. *Steam Power.* London: Edward Arnold.

Daniel, Eric D., C. Denis Mee, and Mark H. Clark, eds. 1999. *Magnetic Recording: The First 100 Years.* Ann Arbor, MI: IEEE Press.

David, P. A. 1991. The hero and the herd in technological history: Reflections on Thomas Edison and the Battle of the Systems. In: P. Higonett, D. S. Landes, and H. Rosovsky, eds., *Favorites of Fortune: Technology, Growth and Economic Development since the Industrial Revolution,* Cambridge, MA: Harvard University Press, pp. 72–119.

Davis, A. B. 1981. *Medicine and Technology: An Introduction to the History of Medical Instrumentation.* Westport, CT: Greenwood Press.

Davy, Humphry. 1840. *The Collected Works of Sir Humphry Davy.* London: Smith, Elder & Co. Cornhill.

Dearborn Independent. 1922. The international Jew, the world's foremost problem [Series]. Dearborn, MI: Dearborn Publishing.

De Vries, Jan. 1994. The Industrial Revolution and the industrious revolution. *Journal of Economic History* **54**:249–270.

De Vries, Tjitte. 2002. The 'Cinématographe Lumière' a Myth? http://www.xs4all.nl/~whichm/myth.html

Delhumeau, Gwenael. 1999. *L'invention du béton armeé—Hennebique 1890–1914.* Paris: Norma.

Deutz. 2003. *We Move Your World.* Cologne: Deutz. http://www.deutz.de

Devine, Warren D. 1990a. Early developments in electroprocessing: new products, new industries. In: Sam H. Schurr, et al., *Electricity in the American Economy: Agent of Technological Progress,* New York: Greenwood Press, pp. 77–98.

Devine, Warren D. 1990b. Coal mining: underground and surface mechanization. In: Sam H. Schurr et al., eds., *Electricity in the American Economy,* New York: Greenwood Press, pp. 181–208.

Dickinson, H. W. 1939. *A Short History of the Steam Engine.* Cambridge: Cambridge University Press.

Diderot, Denis, and Jean Le Rond D'Alembert. 1751–1777. *L'Encyclopedie ou Dictionnaire Raisonne des Sciences des Arts et des Metiers.* Paris: Ave Approbation and Privilege du Roy.

Dieffenbach, E. M., and R. B. Gray. 1960. The development of the tractor. In: U.S.

Diesel, Eugen. 1937. *Diesel: Der Mensch, das Werk, das Schicksal.* Hamburg: Hanseatische Verlagsanstalt.

Diesel, Rudolf. 1893. *Theorie und Konstruktion eines rationellen Wärmemotors zum Ersatz der Dampfmaschinen und der heute bekannten Verbrennungsmotoren (Theory for the Construction of a Rational Thermal Engine to Replace the Steam Engine and the Currently Known Internal Combustion Engines).* Berlin: Springer Verlag.

Diesel, Rudolf. 1913. *Die Entstehung des Dieselmotors (The Origin of the Diesel Engine).* Berlin: Verlag von Julius Springer.

Donkin, Bryan. 1896. *A Text-book on the Gas, Oil, and Air Engines.* London: C. Griffin & Co.

Donovan, John J. 1997. *The Second Industrial Revolution: Business Strategy and Internet Technology.* Upper Saddle River, NJ: Prentice Hall.

Dr. Pepper/Seven Up. 2003. *Brands.* Plano, TX: Dr Pepper/Seven Up. http://www.dpsu.com

Dunsheath, Percy. 1962. *A History of Electrical Industry.* London: Faber & Faber.

Dvorak, August, et al. 1936. *Typewriting Behavior.* New York: American Book Co.

Dvorak, Josef. 1998. Carl Kellner. http://homepage.sunrise.ch/homepage/prkoenig/kellner.htm

Dyer, Frank L., and Thomas C. Martin. 1929. *Edison: His Life and Inventions.* New York: Harper & Brothers.

Dyson, James. 1998. *Against the Odds: An Autobiography.* London: Orion Books.

Edge, Stephen R. F. 1949. Papermaking. In: Jocelyn Thope and M. A. Whiteley, eds., *Thorpes's Dictionary of Applied Chemistry,* Vol. 9, London: Longmans, Green & Co., pp. 215–225.

Edison, Thomas A. 1876. *Improvement in Autographic Printing. Specification Forming Part of Letters Patent No. 180,857, dated August 8, 1876.* Washington, DC: USPTO. http://www.uspto.gov

Edison, Thomas A. 1880a. *Electric Lamp. Specification Forming Part of Letters Patent No. 223,898, Dated January 27, 1880.* Washington, DC: U.S. Patent Office. http://www.uspto.gov

Edison, Thomas A. 1880b *Electric Light. Specification Forming Part of Letters Patent No. 227,229, Dated May 4, 1880.* Washington, DC: U.S. Patent Office. http://www.uspto.gov

Edison, Thomas A. 1881. *System of Electric Lighting. Specification Forming Part of Letters Patent No. 239,147, Dated March 22, 1881.* Washington, DC: U.S. Patent Office. http://www.uspto.gov

Edison, Thomas A. 1883. *System of Electrical Distribution. Specification Forming Part of Letters Patent No. 274,290, Dated March 20, 1883.* Washington, DC: U.S. Patent Office. http://www.uspto.gov

Edison, Thomas A. 1884. *Electrical Indicator. Specification Forming Part of Letters Patent No. 307,031, Dated October 21, 1884.* Washington, DC: U.S. Patent Office. http://www.uspto.gov

Edison, Thomas A. 1889. The dangers of electric lighting. *North American Review* **149:** 625–634.

Edwards, Junius D. 1955. *The Immortal Woodshed.* New York: Dodd, Mead, & Co.

EIA (Energy Information Administration). 1996. *Residential Lighting Use and Potential Savings.* Washington, DC: EIA. http://www.eia.doe.gov/emeu/lighting/contents.html

EIA. 2000. *Energy Use.* Washington, DC: EIA. http://www.eia.doe.gov/emeumecs/iab/

EIA. 2002. *International Energy Annual 2000.* Washington, DC: EIA. http://www.eia.doe.gov/emeu/iea/elec.html

Einstein, Albert. 1905. Zur Elektrodynamik bewegter Körper. *Annalen der Physik* **17:** 891–921.

Einstein, Albert. 1907. Relativitätsprinzip und die aus demselben gezogenen Folgerungen. *Jahrbuch der Radioaktivität* **4:**411–462.

Eldredge, Niles, and Stephen J. Gould. 1972. Punctuated equilibria: An alternative to phyletic gradualism. In: T. J. M. Schopf, ed., *Models in Paleobiology.* San Francisco: Freeman, Cooper, pp. 82–115.

Electricity Council. 1973. *Electricity Supply in Great Britain: A Chronology—From the Beginnings of the Industry to 31 December 1972.* London: Electricity Council.

Ellis, L. W. 1912. The International Motor Contest at Winnipeg. *Scientific American* **107:**113, 124.

e-Manufacturing Networks. 2001. *John T. Parsons, "Father of the Second Industrial Revolution," Joins e-Manufacturing Networks Inc. Board of Advisors.* http://www.e-manufacturing.com/2001/pr/2001-18.htm

Escales, Richard. 1908. *Nitroglycerin und Dynamit.* Leipzig: Veit & Co.

Ewing, James A. 1911. Steam engine. In: *Encyclopaedia Britannica,* 11th Ed., Vol. 25, Cambridge: Cambridge University Press, pp. 818–850.

Fahie, J. J. 1899. *A History of Wireless Telegraphy.* London: Blackwood.

Fant, Kenne. 1993. *Alfred Nobel: A Biography* (trans. M. Ruuth). New York: Arcade Publishing.

FAO (Food and Agriculture Organization). 2003. *Forestry.* Rome: FAO. http://apps.fao.org

Faraday, Michael. 1839. *Experimental Researches in Electricity.* London: Richard & John Edward Taylor.

Feldenkirchen, Wilfried. 1994. *Werner von Siemens*. Columbus: Ohio State University Press.

Fessenden, Helen M. 1940. *Fessenden, Builder of Tomorrows*. New York: Coward-McCann.

Ffrench, Yvonne. 1934. *News from the Past: The Autobiography of the Nineteenth Century*. New York: Viking Press.

Figuier, Louis. 1888. *Les Nouvelles Conquêtes de la Science: L'Électricité*. Paris: Manpir Flammarion.

Fisher, A. Hugh. 1903. Steam turbines for Channel vessels: The system explained. *Illustrated London News* **122:**898.

Fleck, Alexander. 1958. Technology and its social consequences. In: Charles Singer, ed., *A History of Technology,* Vol. 5, Oxford: Clarendon Press, pp. 814–841.

Fleming, John A. 1901. *The Alternate Current Transformer*. London: *The Electrician* Printing & Publishing.

Fleming, John A. 1911. Electricity. In: *Encyclopaedia Britannica,* 11th ed., Vol. 9. Cambridge: Cambridge University Press, pp. 179–193.

Fleming, John A. 1934. *Memories of a Scientific Life: An Autobiography*. London: Marshall, Morgan & Scott.

Flink, James J. 1975. *The Car Culture*. Cambridge, MA: MIT Press.

Flink, James J. 1988. *The Automobile Age*. Cambridge, MA: MIT Press.

Flower, Raymond, and Michael W. Jones. 1981. *100 Years of Motoring: An RAC Social History of Car*. Maidenhead: McGraw-Hill.

FMC (Ford Motor Co.). 1908. *Ford Motor Cars*. Detroit, MI: FMC. http://www.mtfca.com

FMC. 1909. *Ford Motor Cars*. Detroit, MI: FMC. http://www.mtfca.com

Ford, Henry. 1922. *My Life and Work*. New York: Doubleday.

Fores, Michael. 1981. The myth of a British Industrial Revolution. *History* **66:**181–198.

Fox, Nicols. 2002. *Against the Machine: The Hidden Luddite Tradition in Literature, Art, and Individual Lives*. Washington, DC: Island Press.

Frazier, Ian. 1997. Typewriter man. *Atlantic Monthly* **280**(11):81–92.

Freeman, Christopher, ed. 1983. *Long Waves in the World Economy*. London: Frances Pinter.

Friedel, Robert, and Paul Israel. 1986. *Edison's Electric Light*. New Brunswick, NJ: Rutgers University Press.

Frizot, Michel. 1998. *The New History of Photography*. Köln: Künemann.

Garcke, Emil. 1911a. Electric lighting. In: *Encyclopaedia Britannica,* 11th ed., Vol. 9. Cambridge: Cambridge University Press, pp. 651–673.

Garcke, Emil. 1911b. Telephone. In: *Encyclopaedia Britannica,* 11th ed., Vol. 26. Cambridge: Cambridge University Press, pp. 547–557.

Garratt, Gerald R. M. 1994. *The Early History of Radio from Faraday to Marconi*. London: Institution of Electrical Engineers.

Gay, Albert, and C. H. Yeaman. 1906. *Central Station Electricity Supply*. London: Whittaker & Co.

GEH (George Eastman House). 2001. Technology time line. Rochester, NY: Kodak. http://www.geh.org/fm/timeline-cameras/

Giedion, Siegfried. 1948. *Mechanization Takes Command*. New York: Oxford University Press.

Ginzberg, Eli. 1982. The mechanization of work. *Scientific American* **247**(3):67–75.

Goldblith, Samuel A. 1972. Controversy over the autoclave. *Food Technology* **26**(12):62–65.

Gomery, Douglas. 1996. The rise of Hollywood. In: Geoffrey Nowell-Smith, ed., *The Oxford History of World Cinema*. Oxford: Oxford University Press, pp. 43–53.

Gordon, Robert J. 2000. Does the "New Economy" measure up to the great inventions of the past? *Journal of Economic Perspectives* **14**:49–74.

Gould, Stephen J. 1989. *Wonderful Life: The Burgess Shale and the Nature of History*. New York: Norton.

Grahame, Kenneth. 1908. *The Wind in the Willows*. New York: Charles Scribner's Sons.

Gray, Thomas. 1890. Telephony. In: *Encyclopaedia Britannica*, 9th ed., Vol. 23. Chicago: R. S. Peale & Co., pp. 127–135.

Grotte, Jupp, and Bernard Marrey. 2000. *Freyssinet: La Précontrainte et l'Europe, 1930–1945*. Paris: Éditions du Linteau.

Gunston, Bill. 1986. *World Encyclopaedia of Aeroengines*. Wellingborough: Patrick Stephens.

Gunston, Bill. 1999. *The Development of Piston Aero Engines*. Sparkford: Patrick Stephens.

Gunston, Bill. 2002. *Aviation: The First 100 Years*. Hauppauge, NY: Barron's Educational Series.

Haber, Fritz. 1909. *Letter to the Directors of the BASF, 3 July 1909*. Fritz Haber, Allgemeine Correspondenz II, no. 92. Ludwigshafen: BASF Unternehmensarchiv.

Haber, Fritz. 1911. *Patentschrift Nr 238450 Verfahren zur Darstellung von Ammoniak aus den Elementen durch Katalyses unter Druck beerhöhter Temperatur*. Berlin: Kaiserliche Patentamt.

Haber, Fritz. 1920. *The Synthesis of Ammonia from Its Elements*. Nobel Lecture, June 3, 1920. Stockholm: Nobel e-Museum. http://www.nobel.se/chemistry/laureates/1918/haber-lecture.html

Hammer, William J. 1913. The William J. Hammer Historical Collection of Incandescent Electric Lamps. Reprinted from *The Transactions of the New York Electrical Society*, No. 4. http://www.bulbcollector.com/William_Hammer.html

Hammond, Rolt. 1964. *The Forth Bridge and Its Builders*. London: Eyre & Spottiswoode.

Hancock, Michael. 1998. Burroughs Adding Machine Company: History 1857–1993. http://www.dotpoint.com/xnumber/hancock7.htm

Hands, R. 1996. Ammonium nitrate on trial. *Nitrogen* **219**:15–18.

Hawkins, Laurence A. 1951. *William Stanley (1858–1916)—His Life and Work*. New York: Newcomen Society of America.

Haynes, William. 1954. *American Chemical Industry: Background and Beginnings*. New York: D. Van Nostrand.

Hendricks, Gordon. 2001. *Eadweard Muybridge: The Father of the Motion Picture*. Mineola, NY: Dover.

Henley, William E. 1919. *Poems by William Ernest Henley*. New York: Charles Scribner's Sons.

Hertz, Heinrich. 1887. Über sehr schnell elektrischen Schwingungen. *Annalen der Physik* **21**:421–448.

Hertz, Heinrich. 1893. *Electric Waves; Being Researches on the Propagation of Electric Action with Finite Velocity through Space* (English trans. by D. E. Jones). London: Macmillan.

Hertz, Mathilde, and Charles Süsskind, eds. 1977. *Heinrich Hertz: Erinnerungen, Briefe, Tagebücher.* Weinheim: Physik-Verlag.

Heywood, John B. 1988. *Internal Combustion Engine Fundamentals.* New York: McGraw-Hill.

Hiltpold, Gustav F. 1934. *Erzeugung und Verwendung motorischer Kraft.* Zürich: Girsberger.

Hofer, Margaret K., and Kenneth T. Jackson. 2003. *The Games We Played.* Princeton, NJ: Princeton Architectural Press.

Hoff, Nicholas J. 1946. A short history of the development of airplane structures. *American Scientist* **1946**:212–225, 370–388.

Hogan, William T. 1971. *Economic History of the Iron and Steel Industry in the United States.* Lexington, MA: Lexington Books.

Holdermann, Karl. 1954. *Im Banne der Chemie: Carl Bosch Leben und Werk.* Düsseldorf: Econ-Verlag.

Holmes, Harry N. 1914. Powdered coal for fuel. *Scientific American* **110**:330–331.

Holzmann, Gerard J., and Björn Pehrson. 1994. The first data networks. *Scientific American* **270**(1):124–129.

Honda. 2002. *Motorcycles: Overview.* http://world.honda.com/motorcycle/overview.htmlb

Honda Engines. 2003. GX Series. http://www.honda-engines.com/gxgx240spe.htm

Hopkins, Eric. 2000. *Industrialisation and Society: A Social History, 1830–1951.* London: Routledge.

Hoshide, R. K. 1994. Electric motor do's and don'ts. *Energy Engineering* **91**:6–24.

Hossli, Walter. 1969. Steam turbines. *Scientific American* **220**(4):100–110.

Hounshell, David A. 1981. Two paths to the telephone. *Scientific American* **244**(1):157–163.

Houston, R. E. 1927. *Model T Ford Production.* Centerville, IN: Model T Ford Club of America. http://www.mtfca.com

Howell, J. W., and H. Schroeder. 1927. *The History of the Incandescent Lamp.* Schenectady, NY: Maqua Co.

Hoy, Anne H. 1986. *Coca-Cola: The First Hundred Years.* Atlanta, GA: Coca-Cola Co.

Hughes, David E. 1899. Researches of Professor D. E. Hughes, F. R. S., In electric waves and their application to wireless telegraphy, 1879–1886. Appendix D of Fahie, J. J., *A History of Wireless Telegraphy,* London: Blackwood, pp. 305–316.

Hughes, Thomas P. 1983. *Networks of Power.* Baltimore, MD: Johns Hopkins University Press.

Humphrey, William S., and Joe Stanislaw. 1979. Economic growth and energy consumption in the UK, 1700–1975. *Energy Policy* **7**:29–42.

Hunter, Louis C. 1979. *A History of Industrial Power in the United States, 1780–1930.* Vol. 1: *Waterpower in the Century of the Steam Engine.* Charlottesville, VA: University Press of Virginia.

Hunter, Louis C., and Lynwood Bryant. 1991. *A History of Industrial Power in the United States, 1780–1930.* Vol. 3: *The Transmission of Power.* Cambridge, MA: MIT Press.

Hyundai. 2003. *History.* Seoul: Hyundai. http://www.hhi.co.kr/english/aboutus/history.asp

IAI (International Aluminium Institute). 2000. London: IAI. http://www.world-aluminum.org/history/

IAI. 2003. Electrical power used in primary aluminium production. London: IAI. *http://www.world-aluminium.org/iai/stats/formServer.asp?form=7*

ICOLD (International Commission on Large Dams). 1998. *World Register of Dams.* Paris: ICOLD.

IISI (International Institute of Steel Industry). 2003. *World Steel in Figures.* Brussels: IISI. http://www.worldsteel.org/media/wsif/wsif2003.pdf

Iles, George. 1906. *Inventors at Work.* New York: Doubleday, Page & Co.

Ingels, M. 1952. *Willis Haviland Carrier, Father of Air Conditioning.* New York: Arno Press.

Intel. 2003. Moore's law. Santa Clara, CA: Intel. http://www.intel.com/research/silicon/mooreslaw.htm

Israel, Paul. 1998. *Edison: A Life of Invention.* New York: John Wiley.

ITA (International Trade Administration). 2002. *Electric Current Abroad.* Washington, DC: U.S. Department of Commerce. http://www.ita.doc.gov

Ives, Frederic E. 1928. *The Autobiography of an Amateur Inventor.* Philadelphia: Private print.

Jacot, Bernard L., and D. M. B. Collier. 1935. *Marconi—Master of Space: An Authorized Biography of the Marchese Marconi.* London: Hutchinson.

Jehl, Francis. 1937. *Menlo Park Reminiscences.* Dearborn, MI: Edison Institute.

Jewett, Frank B. 1944. *100 Years of Electrical Communication in the United States.* New York: American Telephone & Telegraph.

Johansson, Carl H., and P. A. Persson. 1970. *Detonics of High Explosives.* London: Academic Press.

Johnson, Elmer D. 1973. *Communication: An Introduction to the History of Writing, Printing, Books and Libraries.* Metuchen, NJ: Scarecrow Press.

Jolly, W. P. 1974. *Sir Oliver Lodge.* Rutherford, NJ: Fairleigh Dickinson University Press.

Jones, Howard M. 1971. *The Age of Energy: Varieties of American Experience 1865–1915.* New York: Viking Press.

Josephson, Matthew. 1959. *Edison: A Biography.* New York: McGraw Hill.

Joule, James P. 1850. *On Mechanical Equivalent of Heat.* London: R. & J. E. Taylor.

Kahan, Basil. 2000. *Ottmar Mergenthaler: The Man and the Machine.* New Castle, DE: Oak Knoll.

Kanemitsu, Nishio. 2001. *The Spark Plug.* Lausanne: FontisMedia.

Keenan, Thomas J. 1913. How trees are converted into paper. *Scientific American* **109:** 256–258.

Kelly, Thomas D., and Michael D. Fenton. 2003. *Iron and Steel Statistics.* Washington, DC: USGS. http://minerals.usgs.gov/minerals/pubs/of01–006/ironandsteel.html

Kennedy, Edward D. 1941. *The Automobile Industry: The Coming of Age of Capitalism's Favorite Child.* New York: Reynal & Hitchcock.

King, Clarence D. 1948. *Seventy-five Years of Progress in Iron and Steel.* New York: American Institute of Mining and Metallurgical Engineers.

King, W. I. 1930. The effects of the new industrial revolution upon our economic welfare. *Annals of the American Academy of Political and Social Science* **149**(1):165–172.

Kino International. 2003. *The Lumière Brothers First Films.* New York: Kino on Video.

Kiple, Kenneth F., and Kriemhild C. Ornelas, eds. 2000. *The Cambridge World History of Food.* Cambridge: Cambridge University Press.

Kirsch, David A. 2000. *The Electric Vehicle and the Burden of History*. New Brunswick, NJ: Rutgers University Press.

Kline, Ronald R. 1992. *Steinmetz: Engineer and Socialist*. Baltimore, MD: Johns Hopkins University Press.

Kodak. 2001. History of Kodak. Rochester, NY: Kodak. http://www.kodak.com/US/en/corp/aboutKodak/kodakHistory

Kondratiev, Nikolai D. 1935. The long waves in economic life. *Review of Economic Statistics* **17**:105–115.

Kramer, Deborah A. 1997. *Explosives*. Washington, DC: USGS. http://minerals.usgs.gov/minerals/pubs/commodity/explosives/

Kranzberg, Melvin. 1982. The industrialization of Western society, 1860–1914. In: Bernhard, Carl G., et al., eds., *Science and Technology and Society in the Time of Alfred Nobel*, Oxford: Pergamon Press, pp. 209–230.

Krause, Paul. 1992. *The Battle for Homestead, 1880–1892: Politics, Culture, and Steel*. Pittsburgh, PA: University of Pittsburgh Press.

Kuhn, Thomas. 1962. *The Structure of Scientific Revolutions*. Chicago: Chicago University Press.

Kurzweil, Ray. 1990. *The Age of Intelligent Machines*. Cambridge, MA: MIT Press.

Kurzweil, Ray. 1999. *The Age of Spiritual Machines: When Computers Exceed Human Intelligence*. New York: Viking.

Lamson, Alexander. 1890. Press the button. *Beacon* 2(19):154.

Landau, Sarah B., and Carl W. Condit. 1996. *Rise of the New York Skyscraper, 1865–1913*. New Haven, CT: Yale University Press.

Landes, David. 1969. *The Unbound Prometheus: Technological Change and Industrial Development in Western Europe from 1750 to the Present*. Cambridge: Cambridge University Press.

Langen, Arnold. 1919. *Nicolaus August Otto, der Schöpfer des Verbrennungsmotors*. Stuttgart: Franck.

Law, Edward F. 1914. *Alloys and Their Industrial Applications*. London: Charles Griffin & Co.

Lebrun, Maurice. 1961. *La Soudure, le Brasage et l'Oxycoupage des Métaux; 3500 ans d'Histoire*. Paris: Académie de la Marine.

Legros, Lucien A., and John C. Grant. 1916. *Typographical Printing-Surfaces: The Technology and Mechanisation of Their Production*. New York: Longmans, Green & Co.

Lemelson-MIT Program. 1996. *1996 Invention Index*. Cambridge, MA: MIT. http://web.mit.edu/invent/n-pressreleases/n-press-96index.html

LeMo (Lebendiges virtuelles Museum Online). 2003. *Wilhelm Maybach*. http://www.dhm.de/lemo/html/biografien/MaybachWilhelm/

Lewchuk, Wayne. 1989. Fordism and the moving assembly line: The British and American experience, 1895–1930. In: Nelson Lichtenstein and Stephen Meyer, eds., *On the Line: Essays in the History of Auto Work*, Urbana, IL: University of Illinois Press, pp. 17–41.

Lewis, David L. 1986. The automobile in America: The industry. *Wilson Quarterly*, Winter, 47–63.

Liebig, Justus von. 1840. *Die chemie in ihrer Anwendung auf Agricultur und Physiologie*. Braunschweig: Verlag von F. Vieweg und Sohn.

Lilley, Samuel. 1966. *Men, Machines and History*. New York: International Publishers.

Lindren, Michael. 1990. *Glory and Failure: The Difference Engines of Johann Müller, Charles Babbage and Georg and Edvard Scheutz.* Cambridge, MA: MIT Press.

Ling, Peter J. 1990. *America and the Automobile: Technology, Reform and Social Change.* Manchester: Manchester University Press.

Litan, Robert E., and Alice M. Rivlin, eds. 2001. *The Economic Payoff from the Internet Revolution.* Washington, DC: Brookings Institution Press.

LOC (Library of Congress). 1999. *Inventing Entertainment: The Motion Pictures and Sound Recordings of the Edison Companies.* Washington, DC: LOC. http://memory.loc.gov/ammem/edhtml/

LOC. 2002. *Emile Berliner and the Birth of Recording Industry.* Washington, DC: LOC. http://memory.loc.gov/ammem/berlhtml/

Lodge, Oliver. 1894. The work of Hertz and some of his successors. *Electrician* **27**:221–222, 448–449.

Lodge, Oliver. 1908. *Signalling through Space without Wires.* London: *Electrician* Printing & Publishing.

Lodge, Oliver. 1928. *Why I Believe in Personal Immortality.* London: Cassell.

Lucchini, Flaminio. 1966. *Pantheon—Monumenti dell' Architettura.* Roma: Nuova Italia Scientifica.

Lüders, Johannes. 1913. *Dieselmythus: Quellenmässige Geschichte der Entstehung des heutingen Ölmotors.* Berlin: M. Krayn.

Luff, Peter P. 1978. The electronic telephone. *Scientific American* **238**(3):58–64.

MacKeand, J. C. B., and M. A. Cross. 1995. Wide-band high frequency signals from Poldhu? In: Institute of Electrical Engineers, *100 Years of Radio,* London: Institute of Electrical Engineers, pp. 26–31.

MacLaren, Malcolm. 1943. *The Rise of the Electrical Industry During the Nineteenth Century.* Princeton, NJ: Princeton University Press.

MacLeod, M. 2001. *The Electric Chair.* New York: Courtroom TV Network. http://www.crimelibrary.com/classics4/electric

Maddison, Angus. 1991. *Dynamic Forces in Capitalist Development.* Oxford: Oxford University Press.

Maddison, Angus. 1995. *Monitoring the World Economy 1820–1992.* Paris: Organization for Economic Cooperation and Development.

MAN (Maschinenfabrik Augsburg Nürnberg). 2003. *The MAN Group: An Overview.* Nürnberg: MAN. http://www.man.de/ueberblick/ueberblick_e.html

Mannesmannröhren-Werke AG. 2003. *A Brief History of Mannesmann.* Mülheim ad Ruhr: Mannessmann.

Marchetti, Cesare. 1986. Fifty-year pulsation in human affairs. *Futures* **18**:376–388.

Marconi, Guglielmo. 1909. *Wireless Telegraphic Communication.* Nobel Lecture December 11, 1909. Stockholm: Nobel e-Museum. http://www.nobel.se/physics/laureates/1909/marconi-lecture.html

Marconi Corporation. 2003. *Our History.* London: Marconi Corporation. http://www.marconi.com

Marrus, M. R. 1974. *The Emergence of Leisure.* New York: Harper & Row.

Marshall, Arthur. 1917. *Explosives.* London: J. & A. Churchill.

Martin, Thomas C. 1894. *The Inventions Researches and Writings of Nikola Tesla.* New York: Electrical Engineer.

Martin, Thomas C. 1922. *Forty Years of Edison Service, 1882–1922: Outlining the Growth and Development of the Edison System in New York City.* New York: New York Edison Co.

Maxwell, James C. 1865. A dynamical theory of the electromagnetic field. *Philosophical Transactions of the Royal Society London* **155**:495–512.

Maxwell, James C. 1873. *A Treatise on Electricity and Magnetism.* Oxford: Clarendon Press.

May, George S. 1975. *A Most Unique Machine: The Michigan Origins of the American Automobile Industry.* Grand Rapids, MI: William B. Eerdmans Publishing.

May, George S. 1977. *R. E. Olds: Auto Industry Pioneer.* Grand Rapids, MI: William B. Eerdmans Publishing.

Mayer, Julius R. 1851. *Bemerkungen über das mechanische Aequivalent der Wärme.* Heilbronn: J. V. Landherr.

McClure, Henry H. 1902. Messages to mid-ocean: Marconi's own story of his latest triump. *McClure's Magazine,* April, 525–527.

McDonald, Ronald. 1997. *The Complete Hamburger—The History of America's Favorite Sandwich.* New York: Birch Lane Press.

McElroy, Robert C., Reuben W. Hecht, and Earle E. Gavett. 1964. *Labor Used to Produce Field Crops.* Washington, DC: USDA.

McMenamin MAS and McMenamin DLS. 1990. *The Emergence of Animals: The Cambrian Breakthrough.* New York: Columbia University Press.

McNeill, William H. 1989. *The Age of Gunpowder Empires, 1450–1800.* Washington, DC: American Historical Association.

Melville, George W. 1901. The engineer and the problem of aerial navigation. *North American Review,* December, 825.

Mendeleev, Dimitrii I. 1891. *The Principles of Chemistry.* London: Longmans, Green.

Mengel, Willi. 1954. *Ottmar Mergenthaler and the Printing Revolution.* Brooklyn, NY: Mergenthaler Linotype Co.

Mensch, Gerhard. 1979. *Stalemate in Technology.* Cambridge, MA: Ballinger.

Menzel, Peter. 1994. *Material World: A Global Family Portrait.* New York: Sierra Club Books.

Mercedes-Benz. 2003. *Innovations.* Berlin: Daimler-Chrysler. http://www.mercedes -benz.com

MES (Mitsui Engineering and Shipbuilding). 2001. *MES Completes World's Largest Class New Type Diesel.* Tokyo: Mitsui & Co. http://www.jsmea.or.jp/e-news/win2002/news_05.pdf

Meyer, Robert. 1964. *The First Airplane Diesel Engine: Packard Model DR-980 of 1928.* Washington, DC: Smithsonian Institution.

Michelin. 2003. *History.* Clermont Ferrand: Michelin. http://www.michelin.com/corporate/en/histoire/

Mitchell, Brian R. 1998. *International Historical Statistics Europe 1750–1993.* London: Macmillan.

Mitchell, Sally. 1996. *Daily Life in Victorian England.* Westport, CT: Greenwood Press.

Mitry, Jean. 1967. *Histoire du Cinéma I (1895–1914).* Paris: Éditions universitaires.

Mitsubishi. 2003. *The Man Who Started It All.* Tokyo: Mitsubishi. http://www.mitsubishi.or.jp/e/h/feature1/intr_sph.html

Mittasch, Alvin. 1951. *Geschichte der Ammoniaksynthese.* Weinheim: Verlag Chemie.

Mokyr, Joel. 1990. *The Lever of Riches: Technological Creativity and Economic Progress.* New York: Oxford University Press.

Mokyr, Joel. 1999. The Second Industrial Revolution, 1870–1914. In: V. Castronovo, ed., *Storia dell'Economia Mondiale*, Rome: Latreza, pp. 219–245. http://www.faculty .ecn.northwestern.edu/faculty/mokyr/castronovo.pdf

Mokyr, Joel. 2002. *The Gifts of Athena: Historical Origins of Knowledge Economy*. Princeton, NJ: Princeton University Press.

Montgomery Ward. 1895. *Montgomery Ward & Co. Catalogue and Buyer's Guide No. 57, Spring and Summer 1895*. Chicago: Montgomery Ward & Co.

Moore, G. 1965. Cramming more components onto integrated circuits. *Electronics* **38**(8): 114–117.

Moravec, Hans P. 1999. *Robot: Mere Machine to Transcend Mind*. New York: Oxford University Press.

Mott, Frank L. 1957. *A History of American Magazines, 1885–1905*. Cambridge, MA: Harvard University Press.

MTFCA (Model T Ford Club of America). 2003. *Model T Encyclopedia*. Centerville, IN: MTFCA. http://www.mtfca.com

MTS (Midwest Tungsten Service). 2002. *Tungsten Wire History*. Willowbrook, IL: MTS. http://www.tungsten.com.tunghist.html

Musser, Charles. 1990. *The Emergence of Cinema: The American Screen to 1907*. New York: Charles Scribner's Sons.

Musson, Albert E. 1978. *The Growth of British Industry*. New York: Holmes & Meier.

Muybridge, Eadweard. 1887. *Animal Locomotion: An Electro-photographic Investigation of Consecutive Phases of Animal Movements*. Philadelphia, PA: University of Pennsylvania.

NAE (National Academy of Engineering). 2000. *Greatest Engineering Achievements of the Twentieth Century*. Washington, DC: NAE. http://www.greatachievements.org/

Nagengast, Bernard. 2000. It's a cool story. *Mechanical Engineering* **122**(5): 56–63.

NASM (National Air and Space Museum). 2000. *Langley Aerodrome A*. Washington, DC: NASM. http://www.nasm.si.edu/nasm/aero/aircraft/langleyA.htm

NBER (National Bureau of Economic Research). 2003. *Index of the General Price Level*. http://www.nber.org/databases/macrohistory/rectdata/04/docs/m04051.txt

Needham, Joseph, et al. 1965. *Science and Civilisation in China*, Vol. 4, Pt. II: *Physics and Physical Technology*. Cambridge: Cambridge University Press.

Needham, Joseph, et al. 1971. *Science and Civilisation in China*, Vol. 4, Pt. III: *Civil Engineering and Nautics*. Cambridge: Cambridge University Press.

Needham, Joseph, et al. 1986. *Science and Civilisation in China*, Vol. 5, Pt. VII: *Military Technology: The Gunpowder Epic*. Cambridge: Cambridge University Press.

Németh, J. 1996. *Landmarks in the History of Hungarian Engineering*. Budapest: Technical University of Budapest.

Newby, Frank, ed. 2001. *Early Reinforced Concrete*. Burlington, VT: Ashgate.

NHC (Naval Historical Center). 2003a. *Eugene Ely's Flight to USS* Pennsylvania, *18 January 1911: Landing on the Ship*. Washington, DC: NHC. http://www.history .navy.mil/photos/images/h82000/h82737.jpg

NHC. 2003b. *Eugene Ely's Flight to USS* Pennsylvania, *18 January 1911: Narrative and Special Image Section*. Washington, DC: NHC. http://www.history.navy.mil/ photos/events.ev-1910s/ev1911/ely-pa.htm

Nilsson, Bengt V. 2000. *Ernst Frederik Werner Alexanderson*. Stockholm: Telemuseum. http://www.telemuseum.se/historia/alex/

NMMA (National Marine Manufacturers Association). 1999. *U.S. Recreational Boating.* Chicago: NNMA. http://www.nnma.org

Nobel, Alfred, 1868. *Improved Explosive Compound.* U.S. Patent 78,317. Washington, DC: USPTO. http://www.uspto.gov

Nobel, Alfred. 1895. *Alfred Nobel's Will.* Stockholm: Nobel Foundation. http://www.nobel.se/nobel/alfred-nobel/biographical/will/will-full.html

Nowell-Smith, Geoffrey, ed. 1996. *The Oxford History of World Cinema.* Oxford: Oxford University Press.

Nye, David E. 1990. *Electrifying America: Social Meaning of a New Technology.* Cambridge, MA: MIT Press.

O'Brien, Geoffrey. 1995. First takes. *New Republic* **212:**33–37.

O'Connor, J. J. 1913. The genealogy of the motorcycle. *Scientific American* **109:**12–13.

Odhner, C. T. 1901. *The Nobel Prize in Physics 1901.* Stockholm: Nobel e-Museum. http://www.nobel.se/physics/laureates/1901/press.html

Ohm, Georg S. 1827. *Die galvanische Kette, mathematisch bearbeitet.* Berlin: T. H. Rieman.

Olmstead, Alan L., and Paul Rhode. 1988. An overview of California agricultural mechanization, 1870–1930. *Agricultural History* **62:**86–112.

Ordóñez, José A. F. 1979. *Eugène Freyssinet.* Barcelona: Coop Industrial Trabajo.

Osborn, Fred M. 1952. *The Story of the Mushets.* London: T. Nelson.

Osler, A. 1981. *Turbinia.* Newcastle-upon-Tyne: Tyne and Wear County Council Museums. http://www.asme.org/history/brochures/h073.pdf

OTO (Ordo Templi Orientis). 2003. *Carl Kellner.* http://www.oto.de/hist_Kellner.html

Otto, Nicolaus A. 1877. *Improvement in Gas-Motor Engines. Specification Forming Part of Letters Patent No. 194,047, Dated August 14, 1877.* Washington, DC: USPTO. http://www.uspto.gov

Ovington, Earle L. 1912. The Gnome rotary engine. *Scientific American* **107:**218–219,230.

Painter, William. 1892. *Bottle-Sealing Device. Specification Forming Part of Letters Patent No. 468,226, Dated February 2, 1892.* Washington, DC: U.S. Patent Office. http://www.uspto.gov

Parsons, Charles A. 1911. *The Steam Turbine.* Cambridge: Cambridge University Press.

Parsons, Robert H. 1936. *The Development of Parsons Steam Turbine.* London: Constable & Co.

Passer, Harold C. 1953. *The Electrical Manufacturerrs, 1875–1900: A Study in Competition, Entrepreneurship, Technical Change and Economic Growth.* Cambridge, MA: Harvard University Press.

Paton, J. 1890. Gas and gas-lighting. In: *Encyclopaedia Britannica,* 9th ed., Vol. 10. Chicago: R. S. Peale & Co., pp. 87–102.

Payen, Jacques. 1993. *Beau de Rochas: Sa Vie, son Oeuvre.* Digne-les-Bains: Editions de Haute-Provence.

Perez, Carlota. 2002. *Technological Revolutions and Financial Capital.* Cheltenham: Edward Elgar.

Perkins, Frank C. 1902. High-speed German railway at Zossen. *Scientific American* **86:** 91–92.

Perrodon, Alain. 1981. *Histoire des Grandes Découvertes Pètrolieres.* Paris: Elf Aquitaine.

Perry, H. W. 1913. Teams and motor trucks compared. *Scientific American* **108:**66.

Pessaroff, Nicky. 2002. An electric idea. . . . Edison's electric pen. *Pen World International* **15**(5):1–4. http://www.penworld.com/Issues2002/aprmay02/edison.html

Peters, Tom F. 1996. *Building the Nineteenth Century.* Cambridge, MA: MIT Press.

Petroski, Henry. 1993. On dating inventions. *American Scientist* **81**:314–318.

Phillips, David C. 1996. *Art for Industry's Sake: Halftone Technology, Mass Photography and the Social Transformation of American Print Culture, 1880–1920.* New Haven, CT: Yale University. http://dphillips.web.wesleyan.edu/halftone/

Platt, Harold L. 1991. *The Electric City: Energy and the Growth of the Chicago Area, 1880–1930.* Chicago: University of Chicago Press.

Plunkert, Patricia. 2002. Bauxite and alumina. In: *Mineral Commodity Summaries,* Washington, DC: U.S. Geological Survey, pp. 30–31.

Pocock, Rowland F. 1995. Improved communications at sea: A need and a new technology. In: Institute of Electrical Engineers, *100 Years of Radio,* London: Institute of Electrical Engineers, pp. 57–61.

Pohl, E., and R. Müller, eds. 1984. *150 Jahre Elektromotor, 1834–1984.* Würzburg: Vogel-Verlag.

Pope, Franklin L. 1894. *Evolution of the Electric Incandescent Lamp.* New York: Boschen & Wefer.

Popović, Vojin, R. Horvat, and N. Nikić, eds. 1956. *Nikola Tesla: Lectures, Patents, Articles.* Beograd: Nikola Tesla Museum.

Porter, Horace C. 1924. *Coal Carbonization.* New York: Chemical Catalog Co.

Powell, Horace B. 1956. *The Original Has This Signature—W. K. Kellogg.* Englewood Cliffs, NJ: Prentice-Hall.

Press, Frank, and Raymond Siever. 1986. *Earth.* New York: W. H. Freeman.

Prout, Henry G. 1921. *A Life of George Westinghouse.* New York: American Society of Mechanical Engineers.

Raby, Ormond. 1970. *Radio's First Wave: The Story of Reginald Fessenden.* Toronto: Macmillan.

Radovsky, M. 1957. *Alexander Popov—Inventor of Radio.* Moscow: Foreign Languages Publishing House.

Rae, John B. 1971. *The Road and the Car in American Life.* Cambridge, MA: MIT Press.

Ratzlaff, John T., and Leland I. Anderson. 1979. *Dr. Nikola Tesla Bibliography.* San Carlos, CA: Ragusan Press.

Redin, James. 2003. *A Brief History of Mechanical Calculators.* http://www.dotpoint.com/xnumber/mechanical1.htm

Rhodes, Frederick L. 1929. *Beginnings of Telephony.* New York: Harper.

Richardson, Kenneth. 1977. *The British Motor Industry 1896–1939.* London: Macmillan.

Riedel, Gerhard. 1994. *Der Siemens-Martin-Ofen: Rückblick auf eine Stahlepoche.* Düsseldorf: Stahleisen.

Riis, Jacob. 1890. *How the Other Half Lives.* New York: Charles Scribner's Sons.

Rinderknecht, Peter, ed. 1966. *75 Years Brown Boveri, 1891–1966.* Baden: Brown, Boveri & Co.

Ristori, Emanuel J. 1911. Aluminium. In: *Encyclopaedia Britannica,* 11th ed., Vol. 1. Cambridge: Cambridge University Press, pp. 767–772.

Rittaud-Hutinet, Jacques. 1995. *Les frères Lumière: l'invention du cinéma.* Paris: Flammarion.

Roberts, Peter. 1976. *Any Color So Long as It's Black . . . The First Fifty Years of Automobile Advertising.* London: David & Charles Newton Abbott.

Robinson, Eric, and Albert E. Musson. 1969. *James Watt and the Steam Revolution.* New York: Augustus M. Kelley.

Robson, Graham. 1983. *Magnificent Mercedes: The Complete History of the Marque.* New York: Bonanza Books.

Rogge, Michael. 2003. More Than One Hundred Years of Film Sizes. http://www.xs4all.nl/~whichm/filmsize.html

Rogin, L. 1931. *The Introduction of Farm Machinery.* Berkeley, CA: University of California Press.

Rolls, Charles S. 1911. Motor vehicles. In: *Encyclopaedia Britannica,* 11th ed., Vol. 18. Cambridge: Cambridge University Press, pp. 914–921.

Rose, Philip S. 1913. Economics of the farm tractor. *Scientific American* **109:**114–115.

Rostow, Walt W. 1971. *The Stages of Economic Growth: A Non-Communist Manifesto.* Cambridge: Cambridge University Press.

Rott, N. 1990. Note on the history of the Reynolds number. *Annual Review of Fluid Mechanics* **22:**1–11.

Rushmore, David B. 1912. The electrification of the Panama Canal. *Scientific American* **107:**397, 405–406.

Samuelson, James. 1893. *Labour Saving Machinery.* London: K. Paul, Trench, Trüber & Co.

Samuelson, James, ed. 1896. *The Civilisation of Our Day.* London: Sampson, Low, Marston & Co.

Sauvage, Léo. 1985. *L'Affaire Lumière Du Mythe à l'Histoire: Enquête sur les Origines du Cinéma.* Paris: L'Herminier.

Schaaf, Larry J. 1992. *Out of the Shadows: Herschel, Talbot, and the Invention of Photography.* New Haven, CT: Yale University Press.

Schlebecker, John T. 1975. *Whereby We Thrive: A History of American Farming, 1607–1971.* Ames, IO: The Iowa State University Press.

Schumpeter, Joseph A. 1939. *Business Cycles: A Theoretical, Historical, and Statistical Analysis of the Capitalist Process.* New York: McGraw-Hill.

Schurr, Sam H., and Bruce C. Netschert. 1960. *Energy in the American Economy 1850–1975.* Baltimore, MD: Johns Hopkins University Press.

Schurr, Sam H., et al. 1990. *Electricity in the American Economy: Agent of Technological Progress.* New York: Greenwood Press.

Scott, R. H. 1951. *Mechanism and Operation of Modern Linotypes.* London: Linotype and Machinery Ltd.

Seeger, Hendrik. 1999. The history of German waste water treatment. *European Water Management* **2:**51–56.

Seifer, Marc J. 1996. *Wizard: The Life and Times of Nikola Tesla—Biography of a Genius.* New York: Birch Lane.

Selden, George B. 1895. *Road-Engine. Specification Forming Part of Letters Patent No. 549,160, Dated November 5, 1895.* Washington, DC: U.S. Patent Office. http://www.uspto.gov

Sellen, Abigail, and Richard Harper. 2002. *The Myth of the Paperless Office.* Cambridge, MA: MIT Press.

Shaeffer, R. E. 1992. *Reinforced Concrete: Preliminary Design for Architects and Builders.* New York: McGraw-Hill.

Shankster, H. 1940. Dynamite. In: Jocelyn Thope and M. A. Whiteley, eds., *Thorpes's Dictionary of Applied Chemistry*, London: Longmans, Green & Co., Vol. 4, pp. 239–245.

Sharlin, Harold I. 1967. Applications of electricity. In: Melvin Kranzberg and Carroll W. Pursell, Jr., eds., *Technology in Western Civilization*, Vol. 1. New York: Oxford University Press, pp. 563–578.

Siemens, C. William. 1882. Electric lighting, the transmission of force by electricity. *Nature* **27**:67–71.

Siemens, Werner von. 1893. *Personal Recollections of Werner von Siemens* (trans. W. C. Coupland). New York: D. Appleton & Co.

Simpson, George G. 1983. *Fossils and the History of Life*. New York: Scientific American Library.

Sinnott, Terri, and John Bowditch. 1980. *Edison "Jumbo" Engine-Driven Dynamo and Marine-Type Triple Expansion Engine-Driven Dynamo*. Detroit, MI: ASME. http://www.asme.org/history/brochures/h048-h049.pdf.

Sittauer, Hans L. 1972. *Gebändigte Explosionen*. Berlin: Transpress Verlag für Verkehrswesen.

SMA (Société de Motorisations Aéronautiques). 2003. *Jet A for General Aviation*. Lognes-Marhe-La-Vallée: SMA. http://www.smaengines.com/en/faq.shtml

Smil, Vaclav. 1991. *General Energetics*. New York: John Wiley.

Smil, Vaclav. 1994. *Energy in World History*. Boulder, CO: Westview Press.

Smil, Vaclav. 1999a. Nitrogen in crop production: An account of global flows. *Global Biogeochemical Cycles* **13**:647–662.

Smil, Vaclav. 1999b. China's great famine: 40 years later. *British Medical Journal* **7225**:1619–1621.

Smil, Vaclav. 2000a. Perils of long-range energy forecasting: Reflections on looking far ahead. *Technological Forecasting and Social Change* **65**:251–264.

Smil, Vaclav. 2000b. *Feeding the World: A Challenge for the Twenty-First Century*. Cambridge, MA: MIT Press.

Smil, Vaclav. 2001. *Enriching the Earth: Fritz Haber, Carl Bosch and the Transformation of World Food Production*. Cambridge, MA: MIT Press.

Smil, Vaclav. 2002. *The Earth's Biosphere: Evolution, Dynamics and Change*. Cambridge, MA: MIT Press.

Smil, Vaclav. 2003. *Energy at the Crossroads: Global Perspectives and Uncertainties*. Cambridge, MA: MIT Press.

Smith, Cyril S. 1967. Metallurgy: Science and practice before 1900. In: Melvin Kranzberg and Carroll W. Pursell, Jr., eds., *Technology in Western Civilization*, Vol. 1. New York: Oxford University Press, pp. 592–636.

Smith, Edward S. 1911. Heavy commercial vehicles. In: *Encyclopaedia Britannica*, 11th Ed., Vol. 18, Cambridge: Cambridge University Press, pp. 921–930.

Smith, G. W. 1954. *Dr. Carl Gustaf Patrick de Laval*. New York: Newcomen Society in North America.

Smith, N. 1980. The origins of the water turbine. *Scientific American* **242**(1):138–148.

Smith, William W. 1966. *Midway, Turning Point of the Pacific*. New York: Crowell.

Smithsonian Institution. 2003. *Promoting Edison's Lamp*. Washington, DC: Smithsonian Institution. http://americanhistory.si.edu/lighting/promo19.htm

Song, Y. 1966/1673. *Tiangong kaiwu* (*The Creations of Nature and Man*) (E. Sun and S. Sun, trans.). State College, PA: Pennsylvania State University Press.

Southward, John. 1990. Typography. In: *Encyclopaedia Britannica,* 9th ed., Vol. 23. Chicago: R. S. Peale & Co., pp. 697–710.

Stanley, William. 1912. Alternating-current development in America. *Journal of the Franklin Institute* **173**:561–580.

Steckel, Richard H., and Roderick Floud, eds. 1997. *Health and Welfare during Industrialization.* Chicago: University of Chicago Press.

Steenberg, B. 1995. Carl David Ekman—Pioneer. *Nordisk Pappershistorisk Tidskrift* **23**(4):9–16.

Sterne, Harold. 2001. *A Catalogue of Nineteenth Century Printing Presses.* New London: British Library.

Stoltzenberg, Dietrich. 1994. *Fritz Haber: Chemiker, Nobelpreisträger, Deutscher, Jude.* Weinheim: VCH.

Stone, Richard. 1993. *Introduction to Internal Combustion Engines.* Warrendale, PA: Society of Automotive Engineers.

Straub, Hans. 1996. *Die Geschichte der Bauingenieurkunst.* Basel: Birkhäuser Verlag.

Strunk, Peter. 1999. *Die AEG: Aufstieg und Niedergang einer Industrielegende.* Berlin: Nicolai.

Sung, Hsun-chang. 1945. *Thermal Cracking of Petroleum.* Ann Arbor, MI: University of Michigan.

Suplee, Henry H. 1913. Rudolf Diesel: An appreciation. *Scientific American* **109**:306.

TAEP (Thomas A. Edison Papers). 2002. *Edison's U.S. Patents, 1880–1882.* Piscataway, NJ: Rutgers University. http://edison.rutgers.edu/patente2.htm

Tarde, Gabriel. 1903. *The Laws of Imitation.* New York: Henry, Holt & Co.

Taylor, Charles F. 1984. *The Internal-Combustion Engine in Theory and Practice.* Cambridge, MA: MIT Press.

Taylor, Michael J. H., ed. 1989. *Jane's Encyclopedia of Aviation.* New York: Portland House.

Telefunken. 2003. *Telefunken: 100 Jahre, 1903–2003.* Berlin: Telefunken. http://www.telefunken.de

Temin, Peter. 1964. *Iron and Steel in Nineteenth-Century America.* Cambridge, MA: MIT Press.

Temin, Peter. 1997. Two views of the British Industrial Revolution. *Journal of Economic History* **57**:63–82.

Temple, Robert. 1986. *The Genius of China: 3,000 years of Science, Discovery and Invention.* New York: Simon and Schuster.

Termuehlen, Heinz. 2001. *100 Years of Power Plant Development.* New York: ASME Press.

Tesla, Nikola. 1888. *Electro-magnetic Motor. Specification Forming Part of Letters Patent No. 391,968, Dated May 1, 1888.* Washington, DC: U.S. Patent Office. http://www.uspto.gov

Tesla, Nikola. 1894. *Electro-magnetic Motor. Specification Forming Part of Letters Patent No. 524,426, Dated August 14, 1894.* Washington, DC: U.S. Patent Office. http://www/uspto.gov

Tesla, Nikola. 1900. *Apparatus for Transmission of Electrical Energy. Specification Forming Part of Letters Patent No. 649,621, Dated May 1900.* Washington, DC: U.S. Patent Office. http://www/uspto.gov

TGVweb. 2000. *Under the Hood of a TGV.* Pisa: TGVweb. http://mercurio.iet.unipi.it/tgv/motrice.html

Thoman, J. R. 1953. *Statistical Survey of Water Supply and Treatment Practices.* Washington, DC: U.S. Public Health Service.

Thomson, William. 1852. On a universal tendency in nature to the dissipation of mechanical energy. *Proceedings of the Royal Society of Edinburg* **3**:139.

Thurston, Robert H. 1878. *A History of the Growth of the Steam-Engine.* New York: D. Appleton & Co.

Tilghman, Benjamin C. 1867. *Improved Mode of Treating Vegetable Substances for Making Paper-Pulp. Letters Patent No. 70,485, Dated November 5, 1867; Antedated October 26, 1867.* Washington, DC: U.S. Patent Office. http://www.uspto.gov

Tobin, James. 2003. *To Conquer the Air: The Wright Brothers and the Great Race for Flight.* New York: Free Press.

Toynbee, Arnold. 1884. *Lectures on the Industrial Revolution.* London: Rivingtons.

Turney, Peter. 1999. Increasing evolvability considered as a large-scale trend in evolution. In: A. S. Wu, ed., *Proceedings Workshop on Evolvability at the 1999 Genetic and Evolutionary Computation Conference,* Orlando, FL: GECCO, pp. 43–96.

Twain, Mark. 1889. *A Connecticut Yankee in King Arthur's Court.* New York: Charles L. Webster & Co.

Typewriter Topics. 1924. *The Typewriter: History & Encyclopedia.* New York: Typewriter Topics.

UNDP (United Nations Development Programme). 2003. *Human Development Report 2003.* New York: UNDP. http://www.undp.org

UNO (United Nations Organization). 1956. World energy requirements in 1975 and 2000. In: *Proceedings of the International Conference on the Peaceful Uses of Atomic Energy,* Vol. 1, New York: UNO, pp. 3–33.

UPS (United Parcel Service). 2003. *Company History.* Atlanta, GA: UPS. http://www.ups.com

USAF Museum. 2003. *History Gallery.* Wright-Patterson Air Force Base, OH: USAF. http://www.wpafb.af.mil/museum/history

USBC (U.S. Bureau of the Census). 1975. *Historical Statistics of the United States: Colonial Times to 1970.* Washington, DC: U.S. Department of Commerce.

USBC. 1954. *U.S. Census of Manufacturers: 1954.* Washington, DC: USGPO.

USCFC (U.S. Centennial Flight Commission). 2003. The first powered flight-1903. http://www.1903to2003.gov

USDC (US Department of Commerce). 2002. *Electric Current Abroad.* Washington, DC: USDC. http://www.ita.doc.gov/media/Publications/pdf/current2002FINAL.pdf

USPTO (U.S. Patent and Trademark Office). 2002. *U.S. Patent Activity: Calendar Years 1790–2001.* Washington, DC: USPTO. http://www.uspto.gov/web/offices/ac/ido/oeip/taf/h_counts.htm

Van Duijn, Jacob J. 1983. *The Long Wave in Economic Life.* London: George Allen & Unwin.

Vasey, Ruth. 1996. The world-wide spread of cinema. In: Geoffrey Nowell-Smith, ed., *The Oxford History of World Cinema,* Oxford: Oxford University Press, pp. 53–61.

Vasko, Tibor, Robert Ayres, and L. Fontvieille, eds. 1990. *Life Cycles and Long Waves.* Berlin: Springer-Verlag.

Veblen, Thorstein. 1902. *The Theory of the Leisure Class.* London: Macmillan.

Verhoeven, John D., A. H. Pendray, and W. E. Dauksch. 1998. The key role of impurities in ancient Damascus steel blades. *Journal of Metals* **50**(9):58–64.

Vlasic, Bill, and Bradley A. Stertz. 2000. *Taken for a Ride: How Daimler-Benz Drove off with Chrysler.* New York: W. Morrow.

Vyvyan, Richard N. 1933. *Wireless Over Thirty Years.* London: G. Routledge.

Walz, Werner, and Harry Niemann. 1997. *Daimler-Benz: Wo das Auto Anfing.* Konstanz: Verlag Stadler.

Ward's Communications. 2000. *2000 Motor Vehicle Facts & Figures.* Southfield, MI: Ward's Communications.

Wells, Herbert George. 1902a. *Anticipations of the Reaction of Mechanical and Scientific Progress upon Human Life and Thought.* New York: Harper.

Wells, Herbert George. 1902b. *The Discovery of the Future: A Discourse Delivered to the Royal Institution on January 24, 1902.* London: T. Fisher Unwin.

Wells, Herbert George. 1905. *A Modern Utopia.* New York: Charles Scribner's Sons.

Welsbach, Carl A. von. 1902. History of the invention of incandescent gas-lighting. *Chemical News* **85**:254–256.

White, Thomas H. 2003. *United States Early Radio History.* http://earlyradiohistory.us

Whitt, Frank R. and David G. Wilson. 1982. *Bicycling Science.* Cambridge, MA: MIT Press.

Wildi, Theodore. 1981. *Electrical Power Technology.* New York: John Wiley.

Williams, D. S. D., ed. 1972. *The Modern Diesel: Development and Design.* London: Newnes-Butterworths.

Williams, Michael. 1982. *Great Tractors.* Poole: Blandford Press.

Wilson, Eugene B. 1902. *Cyanide Processes.* New York: J. Wiley & Sons.

Womack, James P., Daniel T. Jones, and Daniel Roos. 1990. *The Machine That Changed the World: The Story of Lean Production.* New York: Rawson Associates.

Wordingham, Charles H. 1901. *Central Electrical Stations.* London: C. Griffin & Co.

Wright, Orville. 1953. *How We Invented the Airplane.* New York: David McKay.

Wyatt, J. W. 1911. Paper manufacture. In: *Encyclopaedia Britannica,* 11th ed., Vol. 20. Cambridge: Cambridge University Press, pp. 727–736.

Wyman, William I. 1913. What are the ten greatest inventions of our time? *Scientific American* **109**:337–339.

Zapffe, Carl A. 1948. *A Brief History of the Alloy Steel.* Cleveland, OH: American Society for Metals.

Zellers, John Adam. 1948. *The Typewriter: A Short History, on Its 75th Anniversary, 1873–1948.* New York: Newcomen Society of England, American Branch.

Zola, Émile. 1877. *L'Assommoir.* In: Henri Mitterand, ed., *Émile Zola: Oeuvres Complétes,* Paris: Cercle du Livre Précieux (1966).

Zola, Émile. 1883. *Au Bonheur des Dames.* Paris: G. Charpentier [English translation by R. Buss. 2001. *The Ladies Delight.* London: Penguin Books].

Name Index

Subject Index